INFRARED DETECTION TECHNIQUES
FOR SPACE RESEARCH

ASTROPHYSICS AND
SPACE SCIENCE LIBRARY

A SERIES OF BOOKS ON THE RECENT DEVELOPMENTS

OF SPACE SCIENCE AND OF GENERAL GEOPHYSICS AND ASTROPHYSICS

PUBLISHED IN CONNECTION WITH THE JOURNAL

SPACE SCIENCE REVIEWS

VOLUME 30

INFRARED
DETECTION TECHNIQUES
FOR SPACE RESEARCH

PROCEEDINGS OF THE FIFTH ESLAB/ESRIN SYMPOSIUM
HELD IN NOORDWIJK, THE NETHERLANDS,
JUNE 8–11, 1971

Edited by

V. MANNO

Space Science Department (ESLAB),
European Space Research and Technology Centre, Noordwijk, The Netherlands

and

J. RING

Imperial College of Science and Technology, 10 Prince's Gardens, London S.W.7, England

D. REIDEL PUBLISHING COMPANY

DORDRECHT-HOLLAND

Library of Congress Catalog Card Number 70–179894

ISBN-13: 978-94-010-2887-5   e-ISBN-13: 978-94-010-2885-1
DOI: 10.1007/978-94-010-2885-1

FOREWORD

Infrared Astronomy is a relatively new subject but it has already radically altered our ideas about astronomical sources. Recent progress in this subject is the result of improved detection techniques, particularly the use of detectors at liquid helium temperatures. Unfortunately, the terrestrial atmosphere greatly restricts Infrared astronomers by allowing them to detect radiation only in narrow transmission windows and by presenting a foreground emission which limits the faintness of observable sources.

It is only from aircraft or balloon altitudes that we can begin to observe faint sources over the complete range of wavelengths between the visible and the radio regions. Few such observations have yet been made and none from satellites, although the latter vehicle will offer complete freedom from atmospheric effects.

New developments and intermediate steps will be required before the ultimate aim of flying in space can be achieved. It is not surprising therefore that the Fifth Eslab/Esrin Symposium should deal with this problem. This book contains the proceedings of the Symposium and faithfully records all discussions. The Symposium covered the present situation and future perspectives of IR techniques. International leaders in the field reviewed the results to date and the possible developments in telescope systems, detectors, cryogenics, filters, and interferometers. Individual contributions were made by European and U.S. scientists in each of these fields. The last half day of the Symposium was devoted to a General Discussion among the participants on the needs for future developments and on the present and prospective capabilities of aircraft, balloons, rockets and spacecraft.

The editors are pleased to note that the Symposium appears to have achieved its aim of prompting frank and extensive discussions and they wish to thank all who contributed to this success.

J. RING

V. MANNO

TABLE OF CONTENTS

4. CRYOGENICS

5. FILTERS

6. INTERFEROMETERS

7. GENERAL DISCUSSION (directed by Prof. F. Kneubühl)

LIST OF PARTICIPANTS

Ade, P. A. R., Queen Mary College, Mile End Road, London, E.1, England

Aiguabella, R., Space Science Department, ESTEC, Domeinweg, Noordwijk, Holland

Anderegg, M., Space Science Department, ESTEC, Domeinweg, Noordwijk, Holland

Andresen, R. D., Space Science Department, ESTEC, Domeinweg, Noordwijk, Holland

Arens, I., Space Science Department, ESTEC, Domeinweg, Noordwijk, Holland

Auriemma, G., Consiglio Nazionale delle Ricerche, Via E. Fermi 9–11/C.P. 67, 00044 Frascati, Italy

Bader, M., Ames Research Center, Moffett Field, Calif. 94035, U.S.A.

Baltes, H. P., Laboratorium für Festkorperphysik, Hönggerberg, CH-8049 Zürich, Switzerland

Bangert, W., Meteorologisches Institut, Barbarastrasse 16, 8 München 13, Germany

Becklake, E. J. S., Marconi Space and Defence Systems, Camberley, Surrey, England

Beckman, J. E., Queen Mary College, Mile End Road, London, E.1, England

Biraud, Y., Groupe IR Spatiale, Observatoire de Meudon, 92 Meudon, France

Boeckel, J. J. van, Space Science Department, ESTEC, Domeinweg, Noordwijk, Holland

Boggess, N. W., Code SG, NASA Headquarters, Washington D.C. 20546, U.S.A.

Bolle, H.-J., Meteorologisches Institut, Barbarastrasse 16, 8 München 13, Germany

Breckenridge, R. W., Arthur D. Little, Acorn Park, Cambridge, Mass. 02140, U.S.A.

Breton, J., Centre Nationale des Etudes Spatiales, BP No. 4, 91 Bretigny-sur-Orge, France

Chanin, G., Service d'Aeronomie, CNRS, BP No. 3, 91 Verrières-le-Buisson, France

Clegg, P. E., Queen Mary College, Mile End Road, London, E.1, England

Cohen, M., The Observatories, Madingley Road, Cambridge, CB3 OHA, England

Coron, N., Groupe IR Spatiale, Observatoire de Meudon, 92 Meudon, France

Dall'Oglio, G., Laboratorio per lo studio delle Radiazioni—Universitá dí Firenze, Via Carlo Bini 44, Florence, Italy

Domingo, V., Space Science Department, ESTEC, Domeinweg, Noordwijk, Holland

Douma, M., Imperial College of Science and Technology, 10 Prince's Gardens, London, S.W.7, England

Duinen, R. J. van, University of Groningen, PO Box 800, Groningen, Holland

Durney, A. C., Space Science Department, ESTEC, Domeinweg, Noordwijk, Holland

Epchtein, N., Groupe IR Spatiale, Observatoire de Meudon, 92 Meudon, France

Feuerbacher, B. P., Space Science Department, ESTEC, Domeinweg, Noordwijk, Holland

Fitton, B., Space Science Department, ESTEC, Domeinweg, Noordwijk, Holland

Friedlander, M. W., Washington University, St. Louis, Mo 63130, U.S.A.

Gaide, A., ESRO, 114 Avenue Charles de Gaulle, 92 Neuilly/Seine, France

Gay, J., Groupe IR Spatiale, Observatoire de Meudon, 92 Meudon, France

Gauffre, G., ONERA, 29 Avenue de la Division Leclerc, 92 Chatillon, France

Girard, A., ONERA, 29 Avenue de la Division Leclerc, 92 Chatillon, France

Glass, I. S., Royal Greenwich Observatory, Hailsham, Sussex, England

Gonfalone, A., Space Science Department, ESTEC, Domeinweg, Noordwijk, Holland

Graauw, Th. de, Astrophysical Laboratory, Zonnenburgh 2, Utrecht, Holland

Grard, R., Space Science Department, ESTEC, Domeinweg, Noordwijk, Holland

Grewing, M., Astronomische Institut der Universität, Popplesdorfer Allee 49, 53 Bonn, Germany

Hanel, R. A., NASA, Goddard Space Flight Center, Code 620, Greenbelt, Md. 20771, U.S.A.

Harang, O., The Auroral Observatory, PO Box 387, 9001 Tromsø, Norway

Heidemann, M., Meteorologisches Institut, Barbarastrasse 16, 8 München 13, Germany

Hoekstra, R., Sterrekundig Instituut te Utrecht, Beneluxlaan 21, Utrecht, Holland

Hoffmann, W., Goddard Institute for Space Studies, 2880 Broadway, New York, N.Y. 10025, U.S.A.

Hofmann, W., Max Planck Institut für Astronomie, 6900 Heidelberg – Königstuhl, Germany

Huizinga, J. S., Queen Mary College, Mile End Road, London, E.1, England

Irslinger, C., Physikalisches Institut, Hermann Herder Strasse 3, 7800 Freiburg 1. Br., Germany

Jaeschke, R., Space Science Department, ESTEC, Domeinweg, Noordwijk, Holland

Jameson, R. F., University of Leicester, Leicester LE1 7RH, England

Jennings, R. E., University College, Gower Street, London, W.C.1, England

Jones, D., Space Science Department, ESTEC, Domeinweg, Noordwijk, Holland

Joseph, R. D., Washington University, St. Louis, Mo. 63130, U.S.A.

Kneubühl, F., Laboratorium für Festkorperphysik, Hönggerberg, CH-8049 Zürich, Switzerland

Knott, K., Space Science Department, ESTEC, Domeinweg, Noordwijk, Holland

Köhn, D., Space Science Department, ESTEC, Domeinweg, Noordwijk, Holland

Kopp, E., Space Science Department, ESTEC, Domeinweg, Noordwijk, Holland

Laroche, D., Engins Matra, 93 Avenue Victor Hugo, 92 Rueil-Malmaison, France

Laude, L., Space Science Department, ESTEC, Domeinweg, Noordwijk, Holland

Lemke, D., Max Planck Institut für Astronomie, 6900 Heidelberg – Königstuhl, Germany

Léna, P., Groupe IR Spatiale, Observatoire de Meudon, 92 Meudon, France

Lucas, A., Space Science Department, ESTEC, Domeinweg, Noordwijk, Holland

Manno, V., Space Science Department, ESTEC, Domeinweg, Noordwijk, Holland

Marsden, P. L., University of Leeds, Leeds LS2 9JT, England

Marsh, J. C. D., Hatfield Polytechnic, Bayfordbury, Hertfordshire, England

Martin, D. H., Queen Mary College, Mile End Road, London, E.1, England

Meiner, R. C., Space Science Department, ESTEC, Domeinweg, Noordwijk, Holland

Melchiorri, F., Laboratorio per lo studio delle Radiazioni – Universitá di Firenze, Via Carlo Bini 44, Florence, Italy

Mercer, J. B., University of Leeds, Leeds LS2 9JT, England

Moorwood, A. F. M., University College, Gower Street, London, W.C.1, England

Müller, E., Sterrewacht Sonnenborgh, Zonnenburgh 2, Utrecht, Holland

Offermann, D., Physikalisches Institut, Nussallee 12, 53 Bonn, Germany

Owens, S. J., Imperial College of Science and Technology, 10 Prince's Gardens, London, S.W.7, England

Page, D. E., Space Science Department, ESTEC, Domeinweg, Noordwijk, Holland

Pedersen, A., Space Science Department, ESTEC, Domeinweg, Noordwijk, Holland

Peraldi, A., Engins Matra, Avenue Louis Bréguet, 78 Velizy, France

Pick, D. R., Clarendon Laboratory, Oxford OX1 3PU, England

Renard, P., CNRS-CEDEX 166-38 Grenoble Gare, France

Ring, J., Imperial College of Science and Technology, 10 Prince's Gardens, London, S.W.7, England

Sanderson, T. R., Space Science Department, ESTEC, Domeinweg, Noordwijk, Holland

Sandwell, R., Heriot-Watt University, Riccarton, Currie, Midlothian, Scotland

Schultz, G., Max Planck Institut für Radioastronomie, Argelanderstrasse 3, 53 Bonn, Germany

Seddon, H., University of Edinburgh, Royal Observatory, Edinburgh 9, Scotland

Selby, M. J., Imperial College of Science and Technology, 10 Prince's Gardens, London, S.W.7, England

Shaw, M., Space Science Department, ESTEC, Domeinweg, Noordwijk, Holland

Shivanandan, K., Naval Research Laboratory – Code 7126.3, Washington D.C. 20390, U.S.A.

Sibille, F., Observatoire de Lyon, 69 Saint Genis Laval, France

Sirou, F., Laboratoire de Météorologie Dynamique, CNRS, 92 Meudon-Bellevue, France

Slingerland, J., University of Groningen, PO Box 800, Groningen, Holland

Smith, S. D., Heriot-Watt University, Riccarton, Currie, Midlothian, Scotland

Smyth, M. J., University of Edinburgh, Royal Observatory, Edinburgh 9, Scotland

Solheim, J.-E., The Auroral Observatory, PO Box 387, 9001 Tromsø, Norway

Švestka, Z., Space Science Department, ESTEC, Domeinweg, Noordwijk, Holland

Taylor, B. G., Space Science Department, ESTEC, Domeinweg, Noordwijk, Holland

Torre, J. P., Service d'Aeronomie – CNRS, BP No. 3, 91 Verrières-le-Buisson, France

Trendelenburg, E. A., Space Science Department, ESTEC, Domeinweg, Noordwijk, Holland

Wenzel, K.-P., Space Science Department, ESTEC, Domeinweg, Noordwijk, Holland

Wijnbergen, J. J., University of Groningen, PO Box 800, Groningen, Holland

Willis, R., Space Science Department, ESTEC, Domeinweg, Noordwijk, Holland

Witteborn, F. C., Ames Research Center – Code SS, Moffett Field, Calif. 94035, U.S.A.

Zander, R., Institute of Astrophysics, University of Liège, B-4200 Cointe-Ougrée, Belgium

"Well, gentlemen, we've discarded the 'Big Bang' and 'Pulsating' theories of the origin of the universe – what do you think of 'Armonk the Sun God throwing a handful of glowing pebbles into the sky?'" *Dawes in The Christian Science Monitor © TCSPS*

INTRODUCTION

Earlier this year when possible subjects for the 1971 ESLAB/ESRIN Symposium were discussed, it was felt that, of the many areas of space physics, Infra-red Astronomy had been one of the least discussed in ESRO. The growing number of publications reporting interesting new results and proposing intriguing and puzzling problems in astronomy had shown the vitality and rapid expansion of an area which so far had been carried out from ground-based observatories or from aircraft, balloons or rockets but never from satellites. It appeared that space flights presently are excluded mainly for technological reasons, since it is difficult to keep a detector system cooled to very low temperatures for any longer periods of time. The overcoming of such technological difficulties, the improvement of systems and the finding of new concepts are the stimuli required to improve our capabilities of measuring the parameters of interest with greater accuracy and resolution.

This process might in some cases take a long time, but it can be assumed that in the near or distant future the technological problems will find a satisfactory solution. The question remained "Will IR Astronomy eventually move into the field of space when the techniques are available, and will that be a definite step forward compared to the capabilities of the presently available systems?"

During various discussions with the European scientific community it was concluded that, although much could (and had to) be done from the present observational platforms, eventually there would be a need to go higher and higher in space in order to achieve complete freedom from atmospheric effects. However, quite considerable obstacles lay in the way of this ultimate achievement, mainly the need to improve our existing techniques to their capability limit and the development of long-lasting, light-weight and low cost cryogenic systems.

The magnitude of effort required to achieve the ultimate aims of flying on a satellite appeared to be such as to require a common and co-ordinated activity among groups, well beyond the boundaries of any single nation. The indications were that a Symposium where the leading scientists in this field could freely discuss the present situation and further perspective was essential at this moment. It was only natural that ESRO, as a European organisation dealing with space research, would take one of the first steps in perhaps a long way to the full development of European IR Astronomy in space.

Considerable time was intentionally allotted to discussion in order to give the participants ample time for questions and the exchange of information. The real aim of the Symposium was to get the frontline European scientists in this field together for perhaps the first time, supported by a considerable number of highly-qualified American scientists, to inform each other of the present development and aims in the various laboratories, to discuss the prospective developments in

Manno and Ring (eds.), Infrared Detection Techniques for Space Research, 1-2. *All Rights Reserved*
Copyright © 1972 by D. Reidel Publishing Company, Dordrecht-Holland

each area and finally, supported by the evidence presented during the Symposium, to discuss the relative capabilities of the different methods and systems and to indicate the needs and the means to lead IR Astronomy higher up than balloons and into outer space.

E. A. TRENDELENBURG

1. RESULTS IN INFRARED SPACE ASTRONOMY

REVIEW OF RESULTS IN INFRARED SPACE ASTRONOMY

WILLIAM F. HOFFMANN

Goddard Institute for Space Studies, New York, N.Y., U.S.A.

1. Introduction

It is a pleasure to be at this ESLAB Symposium on Infrared Space Techniques. Over the past seven years very rapid advances have been made in the techniques of observing celestial objects in the infrared. These include the techniques that have permitted observations from mountaintop observatories in the 10 and 20 μ atmospheric windows and recent techniques for observing within the stratosphere by balloons and airplanes. A number of interesting surprises have been encountered and issues raised by these observations. I have chosen four topics to discuss at this time: (1) infrared stars, (2) diffuse infrared galactic objects, (3) infrared galaxies, and (4) cosmic radiation. In each of these topics the observed results have not been anticipated. In each of them observations above the atmosphere play a major role.

2. Infrared Stars

Visually red stars have been known for a long time. Many of these are red because of cool photospheres; others because of attenuation of the starlight by intervening dust which has a reddening effect on the light. In the past there has been no reason to expect a far-infrared or intermediate infrared emission from stars. A stellar photosphere sufficiently cool to radiate substantially in the intermediate or far-infrared would not have sufficient pressure to support itself against gravitational collapse. It was therefore a surprise when it was first discovered that some stars, in particular, T Tauri stars, have substantial infrared excesses at 10 μ [1, 2].

We now understand that the long wavelength radiation comes not from the stellar photosphere but rather from a surrounding shell of dust absorbing the stellar energy and reradiating it in the infrared at a lower temperature. Since this discovery, a large variety of stars, including supergiant stars, hot stars, and stars cooler than the sun have exhibited this kind of infrared excess.

Figure 1 shows an illustration of the difficulty of observing in the infrared. This is a schematic representation of the atmospheric transmissions from the visible through the far-infrared. The observations on infrared stars have been made in the 2, 5, 10, and 20 μ 'windows'. The 30 K Planck curve falls completely within the opaque part of the atmosphere. Thus very cool interstellar material cannot be observed from the ground, whereas circumstellar dust shells heated at 100° to 1000° radiate substantially in the transparent region from 8 to 25 μ. The locations of the 10 and 20 μ windows have turned out very fortunate for the possible identification of the circumstellar material which I will come to in a moment.

Figure 2 shows an illustration of a number of cool stars which have been observed

Manno and Ring (eds.), Infrared Detection Techniques for Space Research, 5–25. All Rights Reserved
Copyright © 1972 by D. Reidel Publishing Company, Dordrecht-Holland

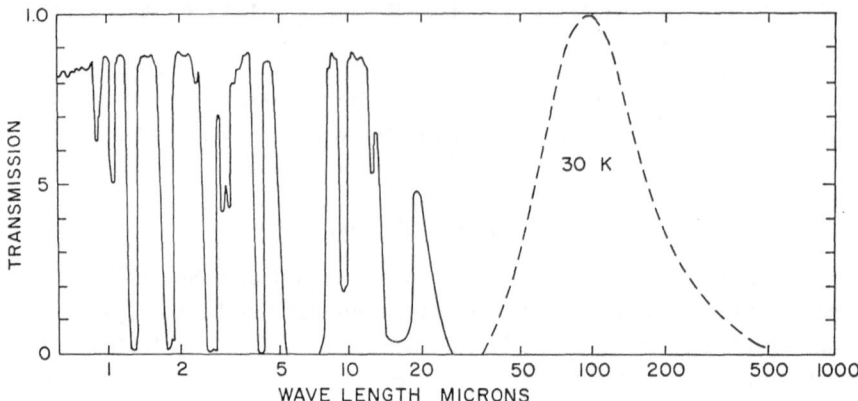

Fig. 1.   Transmission of the atmosphere from the ground in the visible and infrared regions of the spectrum. The 30 K Planck curve drawn in the opaque part of the spectrum indicates the difficulty of observing cool interstellar material from the ground.

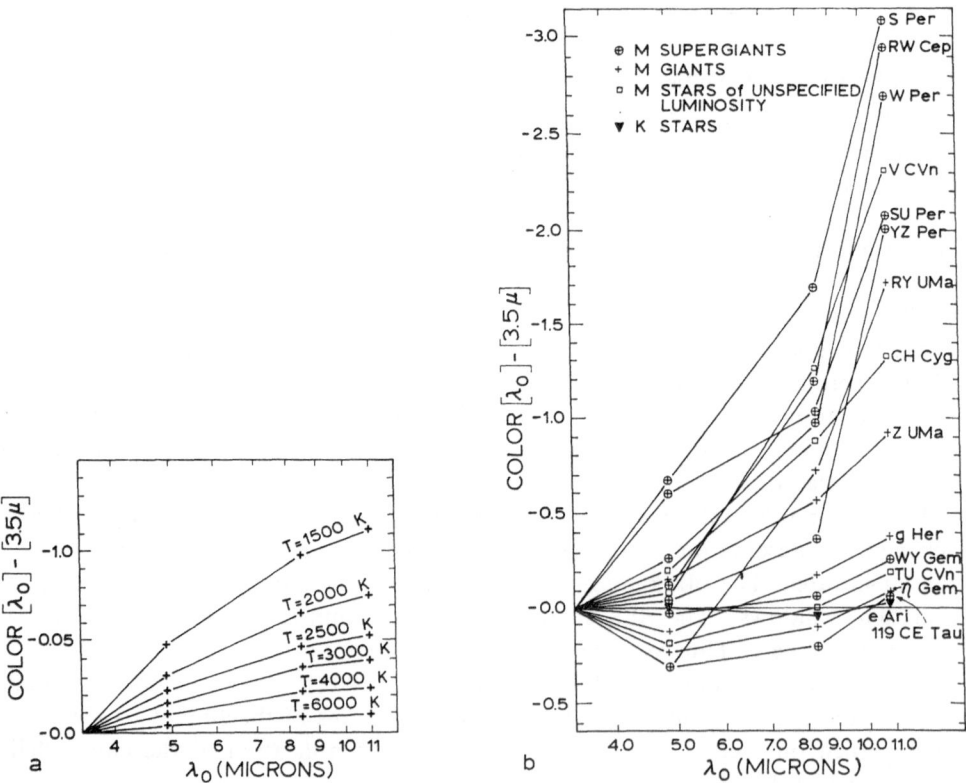

Fig. 2a–b.

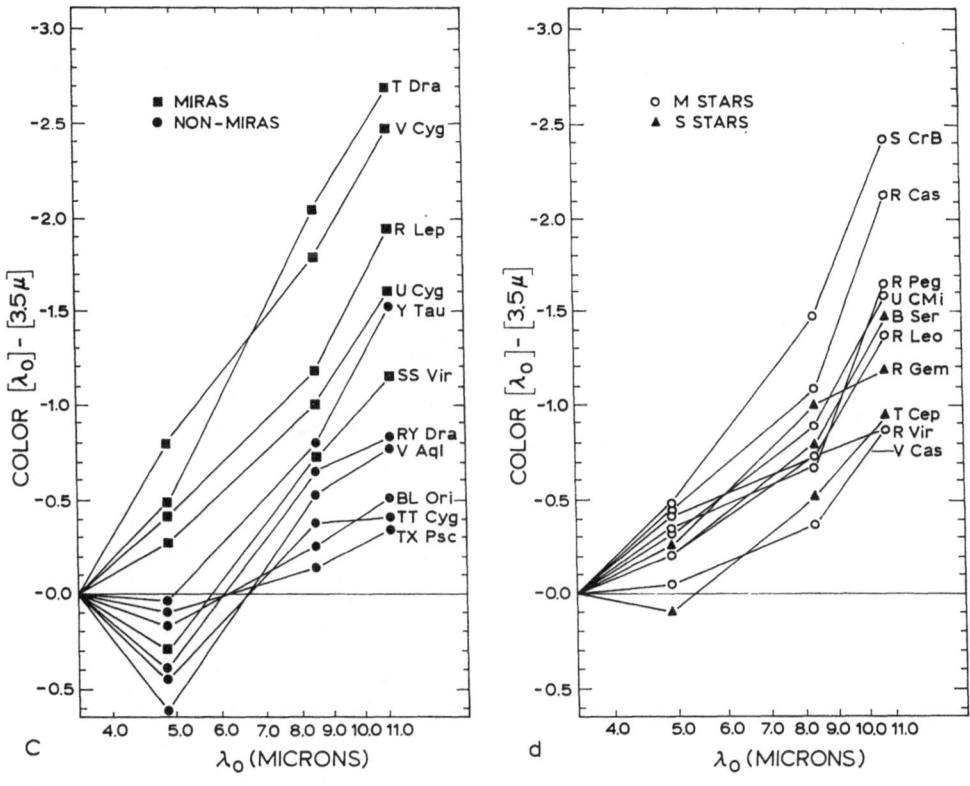

Fig. 2c–d.

Fig. 2a–d. Photometric spectra at 3.5 μ, 4.9 μ, 8.4 μ, and 11.0 μ, with color normalized at 3.5 μ plotted as a function of log λ representative of the following groups of objects: (a) black bodies at selected temperatures, (b) M giants and supergiants, (c) M Mira variables and S stars, and (d) carbon stars. (From Gillett *et al.* [14].)

from 4 to 11 μ. What is so notable about these stars is that the spectrum of each of them departs from a black-body curve with very sharply rising intensities toward 10 μ. In the case of the M supergiant stars, there is now a host of such evidence of general infrared excess. For some time there has been evidence of mass loss from supergiants obtained from doppler shift measurements indicating a gas flow away from the star. The picture now is becoming clear that part of the mass which is lost from these stars is condensed into solid particles which stay around the star and provide the excess infrared radiation.

The composition of the dust which would be formed in this way is a central question of astronomical interest. Some of the possible dust compositions discussed in the literature such as water ice and solid hydrogen cannot survive at the relatively high temperatures of the circumstellar dust. For this reason astronomers did not anticipate dust shells so close to stars. However, it is also possible for more refractory items to condense out in the atmospheres of cool stars such as carbon, iron, and, in oxygen rich stars, various kinds of silicates.

Figure 3 illustrates an identification of one of these materials around μ Cephei. This star shows a striking departure at 10 μ from a black body. Figure 4 shows the suggested interpretation of that departure. This is an absorption spectrum of enstatite,

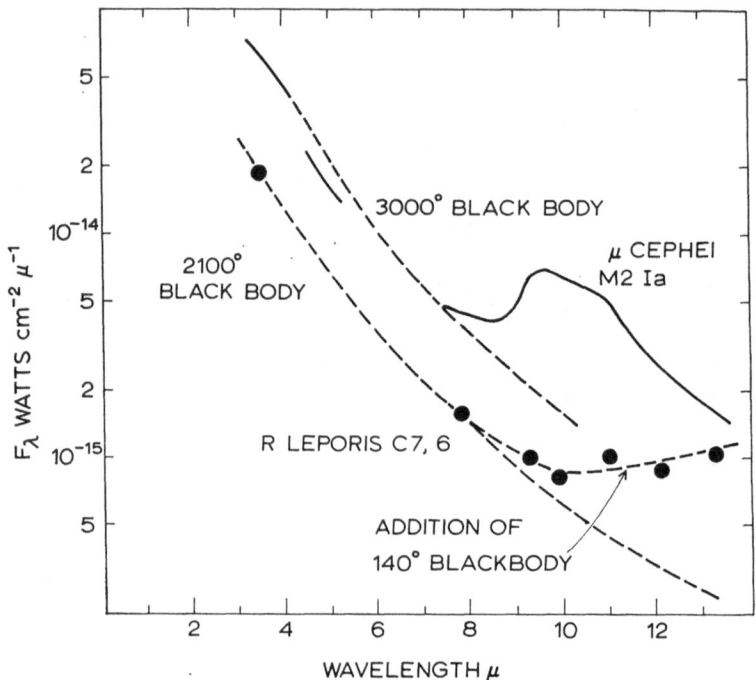

Fig. 3. Observed spectrum of μ Cep (solid lines), with the region 7.5–13.5 μ normalized to the results of broadband photometry. Points are observations of R Lep. Point at 3.5 μ is an unpublished broadband observation by D. Strecker, and the points from 7.5 to 13.5 μ were obtained with a spectrometer with 1% resolution at Lick Observatory. Dashed lines are theoretical curves for black bodies. The 140° curve for R Lep assumes an optically thin gray body.
(From Woolf and Nye [15].)

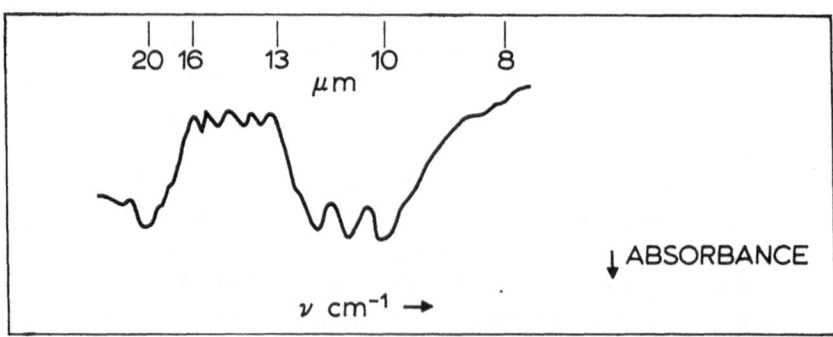

Fig. 4. Absorption spectrum of enstatite $(Mg_{0.5} Fe_{0.5}) SiO_3$. (From Hackwell *et al.* [3].)

one of the magnesium iron silicates. Enstatite, and a number of other silicates have strong absorption at both 10 and 20 μ. By good fortune these fit nicely into the two atmospheric windows. There are a number of other observations in which, either in emission or absorption, interstellar dust peaks show up both in the 10 μ and 20 μ regions. Orthopyroxene and olivene are other silicates which exhibit these same absorptions.

Olivene is a particularly exciting possibility for interstellar dust. Olivene occurs in stony meteorites. Olivene is thrown up from volcanoes. Olivene satisfies the velocity of sound expected in the earth's mantle and olivene was about 50% of the composition of Apollo 12 rocks. So, if now we find olivene around M giant stars where it has been formed in the stellar atmosphere, we encounter a very suggestive and consistent picture of the formation of materials from which solar systems are made.

Figure 5 shows an example of interstellar absorption of 10 and 20 μ. This measurement was made by Hackwell et al. [3] by observing the galactic absorption against the very bright 10 μ source at the galactic center. They estimated from the strength of the

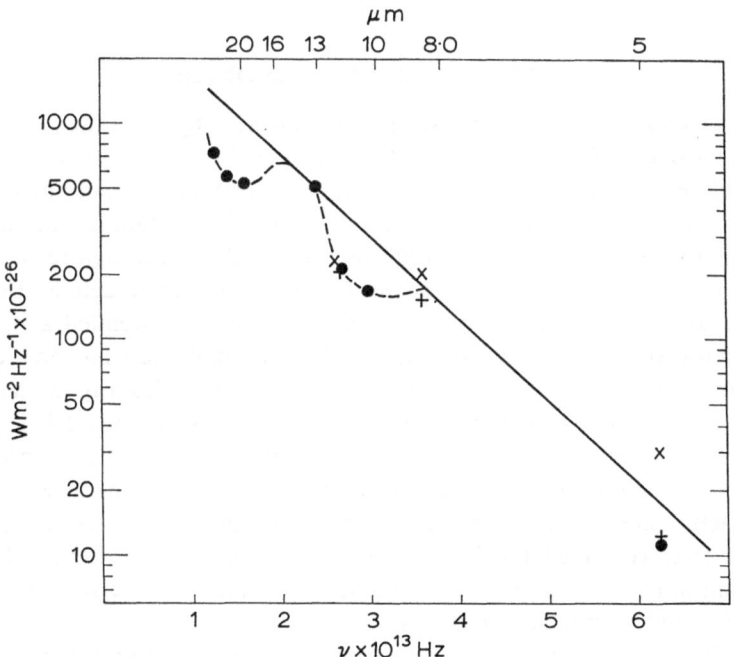

Fig. 5.   Photometric measures of the galactic center. (From Hackwell et al. [3].)

absorption that about 0.4% of the interstellar material in the line of sight toward the galactic center is silicate. This number is interesting in two respects: (1) If all the silicate of the expected cosmic abundance went into silicates, the fraction of interstellar material in silicates would be about 0.4%. (2) In general, the ratio of interstellar dust to total interstellar material is about 0.4% indicating that this kind of mineral perhaps accounts for dominant fraction of the interstellar material in the direction of the galactic center.

I have tried to indicate some of what has been learned about infrared stars. Much more knowledge can be added by space research. It is clear that there is mass loss from stars, that some of this mass loss appears in the form of solid grains, and that the equilibrium conditions in oxygen rich stellar envelopes are such that silicates can form at about 2000 °C. The radiation pressure from the stars will push these materials out. As they get larger the pressure becomes greater until the material moves out to a region where the density is insufficient to continue the accretion of material. This provides a self-limiting mechanism for determining size of the particles. Finally, the same silicates seem to be found in the interstellar medium, at least in the line of sight to the galactic center.

Considerably better material identification could be made if these spectral observations could be extended to the spectral region between 5 and 8 μ and beyond 25 μ. One of the important fields of space astronomy will be moderate resolution spectroscopy filling in these regions of the spectrum inaccessible from the ground.

3. Diffuse Infrared Galactic Objects

The most dramatic of all of the diffuse infrared galactic objects is the galactic center. This is the most intense far-infrared object outside the solar system. The galactic center cannot be observed in the visible because of interstellar extinction. The galactic center has been observed at 2 μ where the attenuation is substantial but allows some of the energy to get through. It has been observed at 10 μ with high resolution, with which one can map out on a very small scale some remarkable details. But it is extremely difficult from the ground to deal with an object as extended as the galactic center with a low surface brightness. When working through the atmospheric windows from the ground it is essential to keep the telescope beams sufficiently small so that the atmospheric radiation is kept low. This results in beam sizes of much less than an arc min. On the other hand, from the stratosphere it is possible to use large beams in searching for low surface brightness extended objects. In this manner it was discovered that the galactic center emits enormously greater radiation in the far-infrared at 100 μ than at shorter wavelengths and it emits over a very extended region [4]. In fact, the infrared luminosity indicates that a substantial fraction of the stellar output in that area is thermalized by interstellar grains.

Figure 6 shows a spectrum of the galactic center including radio, far-infrared, and near-infrared observations. The radio spectrum is for the source, Sagittarius A, which is just a few minutes of arc in size. The near-infrared spectrum, obtained by Becklin and Neugebauer [5], appears to be the reddened part of the stellar energy from this region of the Galaxy. The 10 μ radiation is a structured source about 16 arc s in size. The 100 μ radiation is from an extended source $2° \times 6°$ in size. Superimposed on this figure are thermal curves for 20 K and 30 K grains normalized to the 100 μ observation.

Figure 7 is a contour map of the galactic center region covering about 2° across the galactic plane and about 4° along it. This was obtained by a balloon experiment with a resolution of 6 minutes of arc at 100 μ. This map has a number of remarkable characteristics. The integrated 100 μ luminosity over this region is $3.4 \times 10^8 \, L_\odot$. There is considerable structure including emission peaks which coincide with the radio

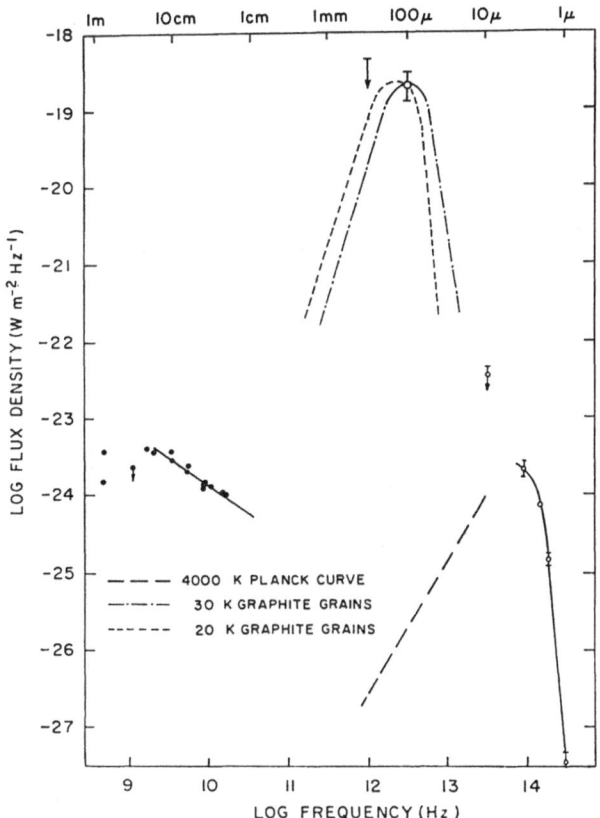

Fig. 6.   Infrared and radio spectrum of the galactic center. The 100 μ flux density and the 320 μ upper limit are for a region $2° \times 6.5°$. The radio and middle-infrared points, taken from Becklin and Neugebauer [5], are shown for reference. They are for a small source 3.5′ in diameter and hence cannot be directly compared with the 100 μ emission. The 20 and 30 K curves for graphite grains are normalized to the 100 μ observation. The 4000 K curve is normalized to the 3.4 μ observation. (From Hoffmann and Frederick [4].)

sources Sagittarius A, Sagittarius B2, KE 56, and G-0.6–0.1. This map has a remarkable similarity in the detail structure to the 2 centimeter radio maps. The main peaks are resolved. Sagittarius A is larger than the beam size and falls off rather slowly along the ridge of the galactic plane and rather sharply across it. This resolution is inade-

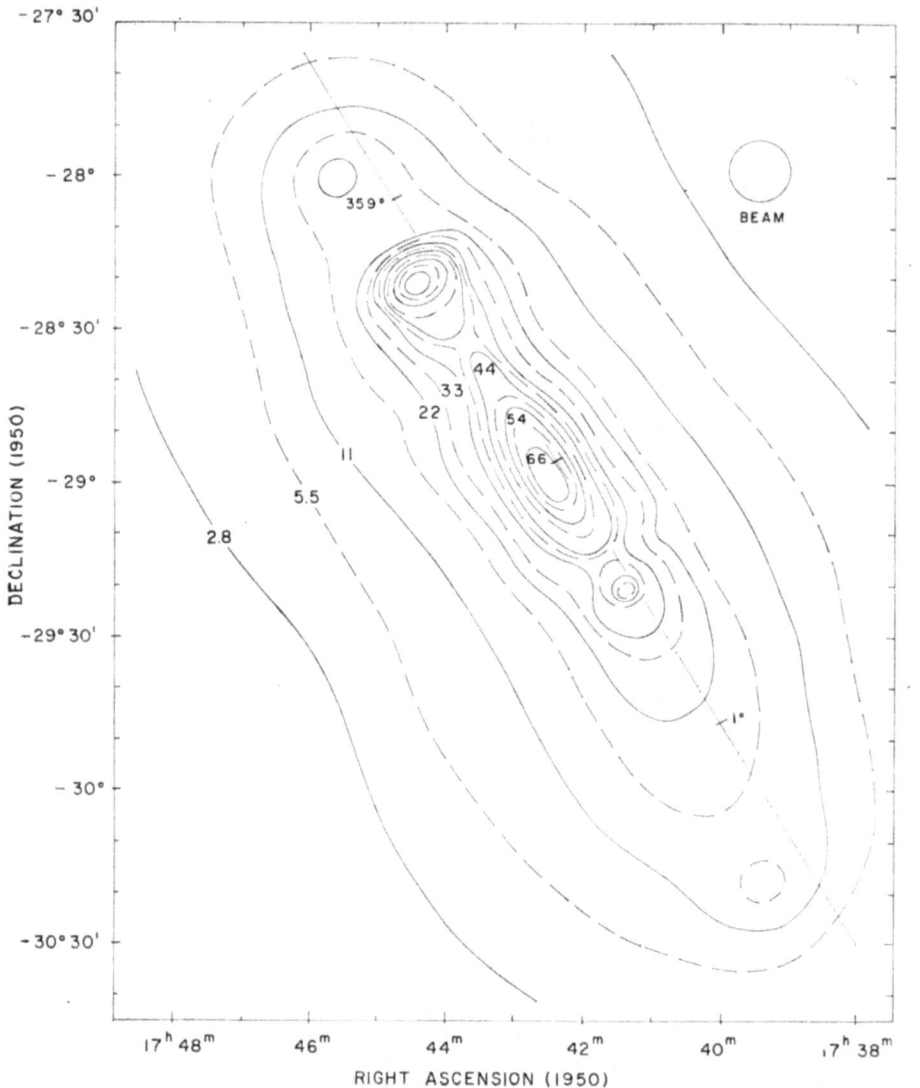

Fig. 7.   Contour map of the galactic-center region at 100 μ. The contour intervals are given in units of 10^{-4} erg cm^{-2} s^{-1} μ^{-1} sterad^{-1}. The observed coordinates have been adjusted to fit the radio position of Sgr A. The galactic equator ($b^{II}=0$) is indicated by a broken line. (Hoffmann *et al.* [16].)

quate to determine the 100 μ component of the spectrum of the small infrared object at 10 microns.

The tremendous agreement between this map and the radio maps confirms that the 100 μ emission is coming from the galactic center at a distance of about 10 kpc. Furthermore, except for a non-thermal component at Sagittarius A, the radio emission is largely thermal continuum from free-free transitions in ionized gas in an H II region. Thus it appears that the infrared energy is coming from the same region for probably two reasons. One is that the gas and dust are relatively well mixed and hence one expects the material to be available in the same area. The other is that the energy for the ionizing of the H II region is provided by very hot stars which could also be heating the dust.

We conclude then from observations of the galactic center that there is a class of diffuse galactic objects in which there is free-free radio emission from an H II region which is accompanied by interstellar dust emission. The closeness of this relationship is shown in Table I. This table gives the brightness of the four main peaks at 100 μ

TABLE I

Ratio of radio and infrared brightness in galactic center

Object	100 μ brightness	2 cm brightness	Ratio IR/radio
	10^{-3} erg cm^{-2} (s μ sterad)$^{-1}$ K		
A	7.4	28	0.26
B	7	7.4	0.95
C	5.1	3.9	1.3
D	2.2	2.5	0.9

and at 2 cm. With the exception of Sagittarius A the ratios are very nearly the same. Sagittarius A is down by about a factor of 4 in the infrared. However, we know that about three-fourths of the Sagittarius A radio emission is nonthermal. This table then substantiates the hypothesis that the energy source is the same both for the continuum radio emission and the infrared.

Figure 8 shows the Lagoon Nebula, M8, which is a particularly interesting example of a diffuse galactic infrared source. In this case, the interstellar dust is not so dense as it is toward the galactic center so that the nebula can be seen visually. Both parts of the visible nebula are observed at radio wavelengths and at 100 μ. The infrared emission from this nebula is about 10^4 solar luminosities.

There are now about 60 diffuse galactic objects observed at 100 μ. Figure 9 indicates the spectrum for four of these objects, Sagittarius A, M17, M8, and M42 (the Orion infrared nebula). These spectra must be treated with caution because different regions of the spectrum were measured with very different beam sizes. In the case of M17, the object was very similar to the size of our 12 in. telescope beam at 100 μ. The Orion nebula is a visible nebula, an H II region, a rich source of microwave molecule emission, and a region of much interstellar dust.

One of the difficulties of interpreting present far-infrared observations is the lack of adequate spatial resolution. This is required for comparing with shorter wavelength studies made with resolutions of 10 arc s or less and for determining a meaningful spectrum over the infrared and radio. In the case of Sagittarius A, for instance, the 10 μ observations are for an object 15 arc s in size whereas the far-infrared observations include emission over more than 12 arc min. As a consequence, these observations cannot be interpreted as describing the spectrum of a single object.

Figure 10 shows the work of Harper and Low [6] in which they have plotted the infrared luminosity against the 2 cm radio flux density for ten H II regions. The radio flux density is proportional to the total Lyman continuum (ionizing energy) from the

Fig. 8.   The Lagoon Nebula (M 8). (Mt. Palomar Photograph.)

heating star. The proportionality between infrared luminosity and radio flux density supports the hypothesis that the same star provides the energy sources for heating the dust for these objects, either from the Lyman α radiation from the gas or from a portion of the direct radiation from the star.

Approximately one-third of the objects in the far-infrared survey have not been identified with radio sources. These may be a new category of objects in which the source of heating is cooler stars in a very early phase of formation. Or it is possible that the radio continuum is present but too weak and too diffuse to have been detected in radio surveys.

In summary, it appears that diffuse galactic infrared objects are of two kinds: extended regions of gas and dust surrounding very hot young stars, and regions of gas and dust surrounding cooler stars. The young hot stars are probably surrounded by the dust nebula from which they were formed. The infrared objects not associated with H II regions could involve cool older stars, or perhaps protostars under formation. There is no certain evidence for this at the present time. It may be however that with observations at longer wavelengths, lower flux levels, and greater resolution, we may obtain such evidence.

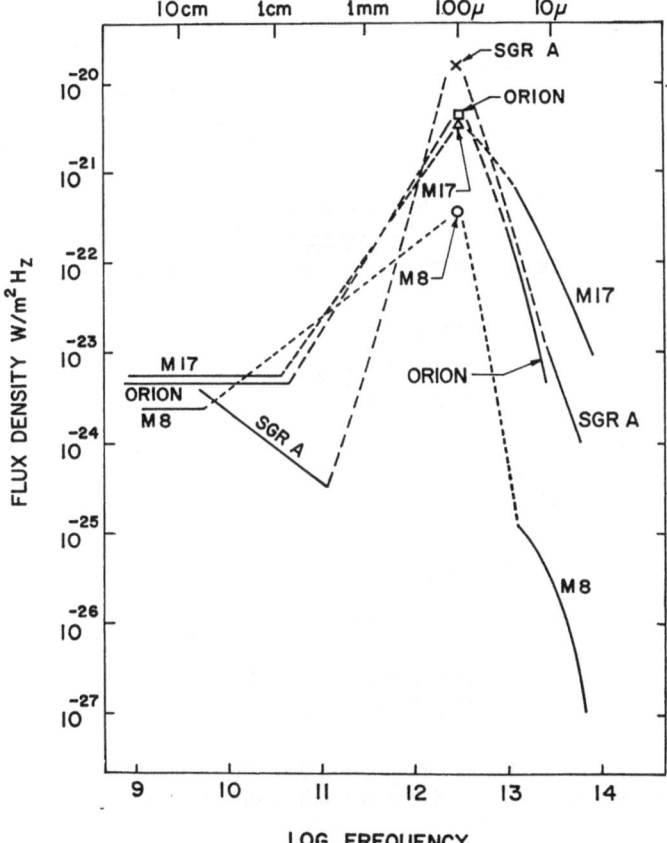

Fig. 9.   Infrared spectrum of Sagittarius A, M 17, M 8, and M 42 (Orion Infrared Nebula).

From the preliminary survey of diffuse galactic objects in the far-infrared, it is quite clear that as the sensitivity improves the galactic plane is going to be very rich with far-infrared objects. It may even be possible to map continuously the dust distribution throughout the galactic plane. If this is done with multiband photometry, then the dust temperature and some indication of the composition can be determined.

The far-infrared should be an extremely powerful method for mapping the distribution and material in the whole galaxy. The galaxy is optically then at these wavelengths. So there will be no part hidden from view.

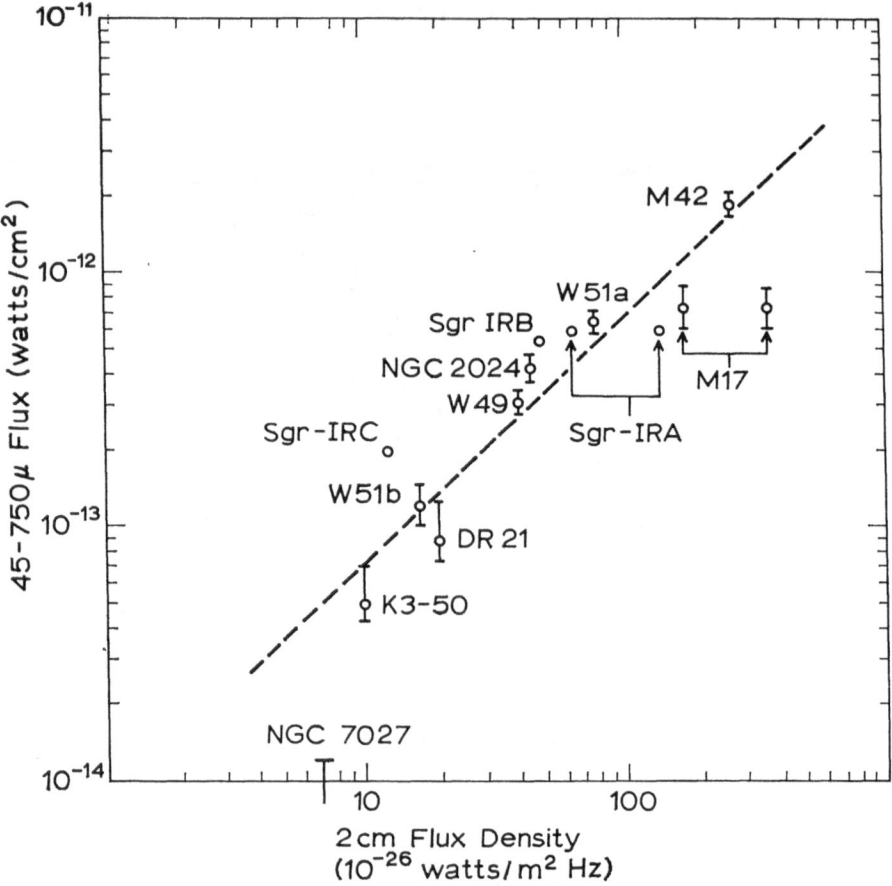

Fig. 10.   Observed far-infrared flux as a function of 2 cm flux density. Points without error bars from Low and Aumann [8]. Dashed line has unity slope. (From Harper and Low [6].)

4. Infrared Galaxies

Getting further from our immediate neighborhood, the next topic is that of infrared galaxies. Here again the infrared brought surprises. Kleinmann and Low [7] found that the nuclei of Seyfert galaxies emit an enormous excess of infrared energy. The spectrum of these galaxies turns up very sharply in the 10μ region. The infrared luminosity, in some cases, is 3 to 4 orders of magnitude greater than all other radiation from these galaxies.

Figure 11 shows the spectrum from 2 to 25 μ of five galaxies. In these five cases, the flux density is increasing quite sharply toward the long wavelengths, unlike what one would expect for a stellar-like spectrum. In two cases, NGC 1068 and M 82, flux

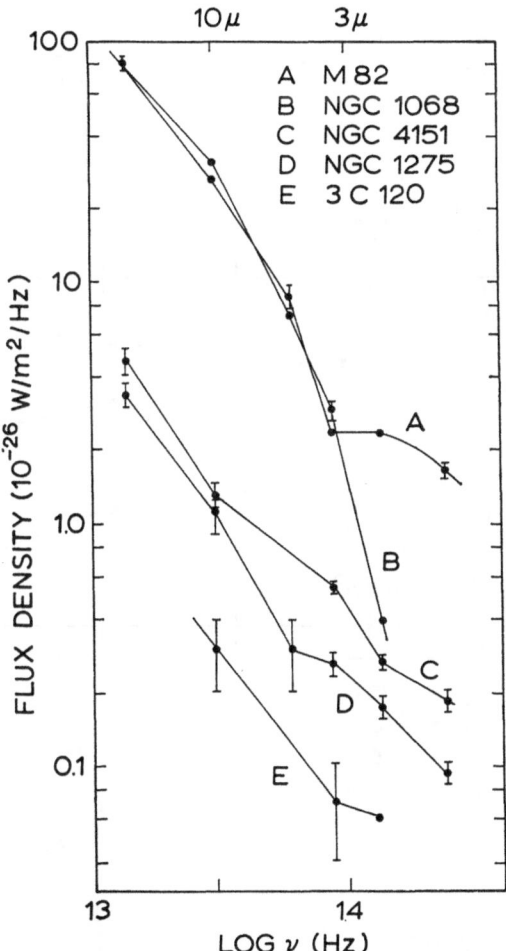

Fig. 11.   Mean flux densities of galaxies showing similarity of spectra in the far-infrared. Probable error indicated by size of data points or by error bars. (From Kleinmann and Low [7].)

density is extremely high. Low and Aumann [8] have measured a flux density for NGC 1068 from aircraft at 1.5×10^{-22} W m^{-2} Hz, indicating a very sharply peaked spectrum in the far-infrared.

Observing galaxies in the far-infrared is very difficult. The observations must be made from the stratosphere and existing telescopes for this purpose are small (30 cm). But the measurement of NGC 1068 by Low is an indication that there are a large

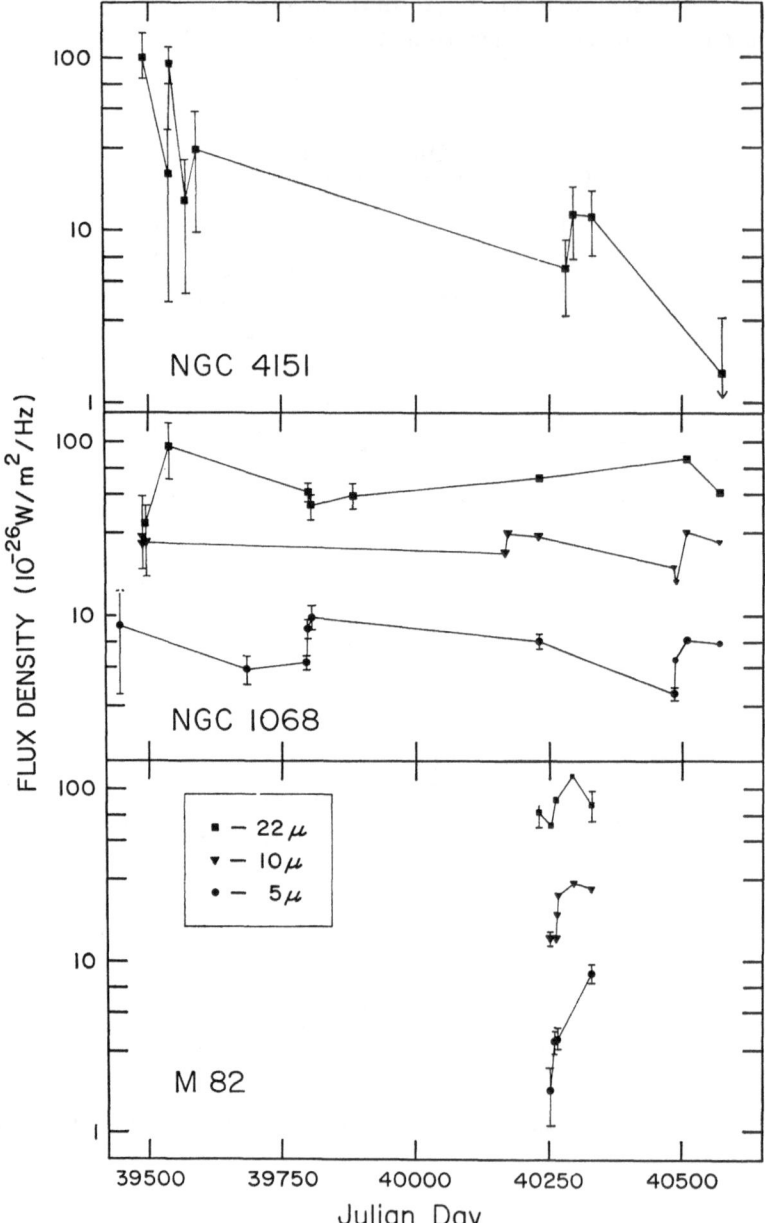

Fig. 12.   Variability of three galaxies in the far-infrared. Probable error indicated by size of data points or by error bars. (From Kleinmann and Low [7].)

number of such galaxies with enormous emissivities in the infrared sharply peaked at near 100 μ. Theoretically neither the energy production nor the emission mechanism for high infrared luminosity of Seyfert galaxies is understood. It is certainly possible that nuclear processes in the galactic nucleus can occur on a scale to produce the energy, which could then be converted to the far-infrared by absorption and reradiation of interstellar material which should abound in these galactic nuclei. Whether such thermal emission is a possible explanation depends on whether the size of the emitting region is sufficiently large for thermal emission to provide the observed luminosity.

A crucial question is then, what is the size of the emitting region of these very intense Seyfert galaxies in the infrared? The sizes are too small to be directly resolved. Present efforts to determine the size depend on looking for time variation. Assuming

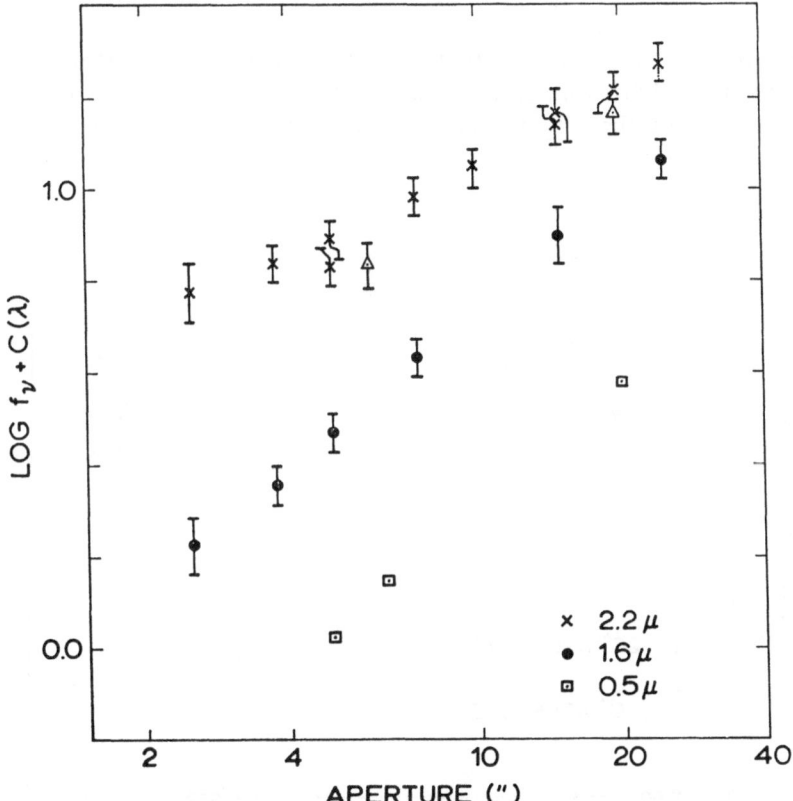

Fig. 13. Energy observed from NGC 1068, as a function of observing aperture. If f_ν is expressed in W m^{-2} Hz^{-1}, $C(0.5\ \mu) = C(2.2\ \mu) = +27.2$ and $C(1.6\ \mu) = +27.0$. Error bars represent total uncertainty in the data, including estimated setting and systematic effects. Data designated with a triangle are the average of 2.2 μ measures taken on several nights at the Lunar and Planetary Laboratory; the remaining 1.6 μ and 2.2 μ data have been obtained at the Hale Observatories. (From Neugebauer *et al.* [10].)

coherent variation of the radiation from an extended region the time scale of the variation cannot be less than the time for light to cross the region. Figure 12 provides some information about time variation of NGC 4151, NGC 1068, and M 82 at 5, 10, and 22 μ. NGC 4151 appears to have had variations over a period of about $2\frac{1}{2}$ yr. NGC 1068 appears to have variations. M 82 is difficult to observe. The apparent variations may be due to observational problems. Because NGC 1068 is the brightest of these three, its possible variation has received the most attention. It is observed to vary in the visual spectrum. Pacholczyk [9] has reported striking variations in NGC 1068 at 2.2 μ by a factor of 2 or more within one night of observing. This would indicate the 2 μ source to be extremely small.

Figure 13 shows some of the problems with determining time variations of NGC 1068. This figure shows that the 2 μ radiating region of NGC 1068 is resolvable by ground telescopes. The flux increases with aperture size out to 40 arc s. It is likely, then that the observed extreme variation at 2.2 microns was probably due to telescope setting errors when a 15 arc s beam was used to measure the flux from a much larger object.

At 10 μ, the nucleus of NGC 1068 is not resolved. Even at 3 μ there is very little variation in the total flux with aperture size. Figure 14 shows an effort by Neugebauer *et al.* [10] to put together the best information on the 10 μ time variations of NGC

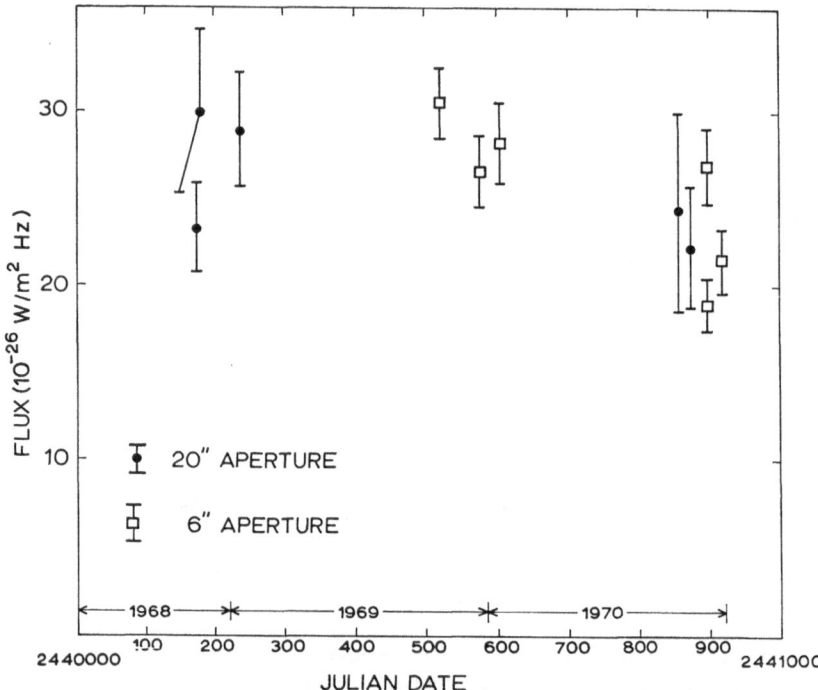

Fig. 14.   10 μ observations of NGC 1068 as a function of date and field of view. Error bars represent 1σ, determined from a statistical analysis of the data, plus a systematic error of 6%. (From Neugebauer *et al.* [10].)

1068. Low argues that these data present a sharp decrease in the 10 μ luminosity of NGC 1068 of 30% in less than a year. This implies an emitting region of 0.3 parsec. This size would be too small for a thermal source and hence would require a non-thermal mechanism such as synchrotron emission. On the other hand, these data appear to be consistent with a much slower rate of change. If a 30% change occurred over 50 yr, then the emission could be completely thermal. I think that the question of the size of the infrared nucleus of NGC 1068 remains open. Several years of good observations, particularly at the longer wavelengths of 50 to 100 μ, may be required to settle the question. Such measurements from the stratosphere are much more difficult. They require many repetitive flights with very careful relative calibration.

The present need for far-infrared observations of galaxies is for greater sensitivity to search for far-infrared emission in other Seyfert galaxies, greater resolution, and greater observing time to determine possible variations.

5. Cosmic Radiation

The cosmic black-body radiation was predicted theoretically many years ago by Gamoff and theoretically rediscovered by Dickie. The discovery of it is not a surprise in the same sense as these other items. It was experimentally discovered independently of the theoretical predictions by Penzias and Wilson in 1964, and since that time has been half well established. Figure 15 shows what is meant by half. The long

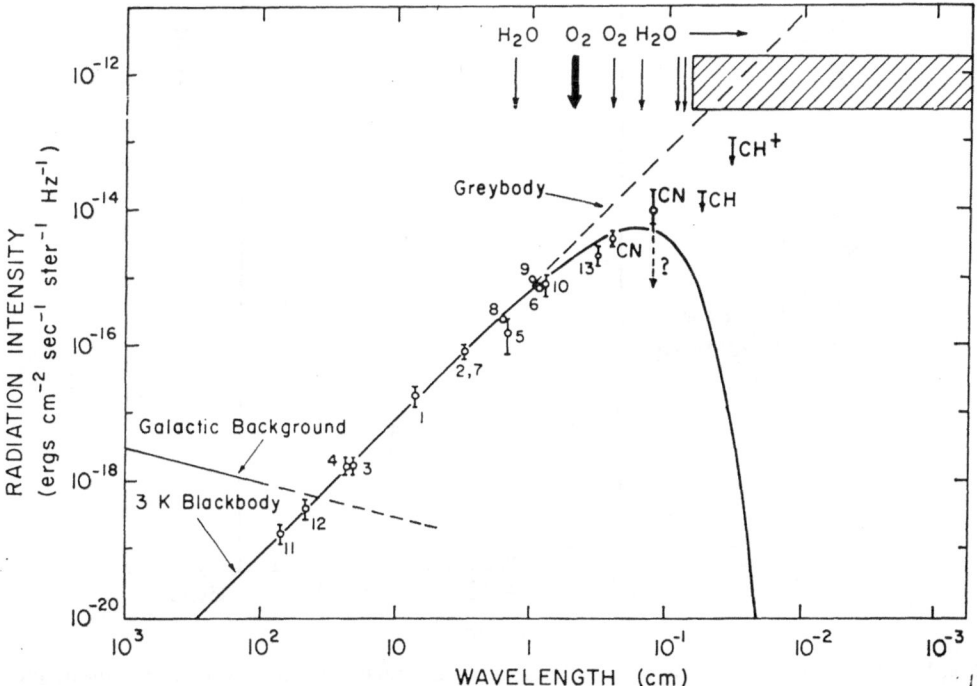

Fig. 15. Spectrum of cosmic background radiation indicating radio and molecular measurements.

wavelength half has been well established by numerous radio and interstellar molecule observations. If the radiation follows a 2.7 K Planck curve, the short wavelength side should fall off very rapidly resulting in submillimetre intensities too low to measure. Two kinds of measurements have been made at 1 mm and shorter. One is the interstellar molecule temperature measurements by Bortolot *et al.* [11]. These give upper limits. The others are the rocket measurements first made by Shivanandan *et al.* [12]. The rocket measurements indicate a flux very much above the 2.7 K curve and very broad in spectrum.

Figure 16 shows the rocket result more dramatically. More recent rocket calibrations have reduced the observed submillimetre flux by about a factor of 2. A similar

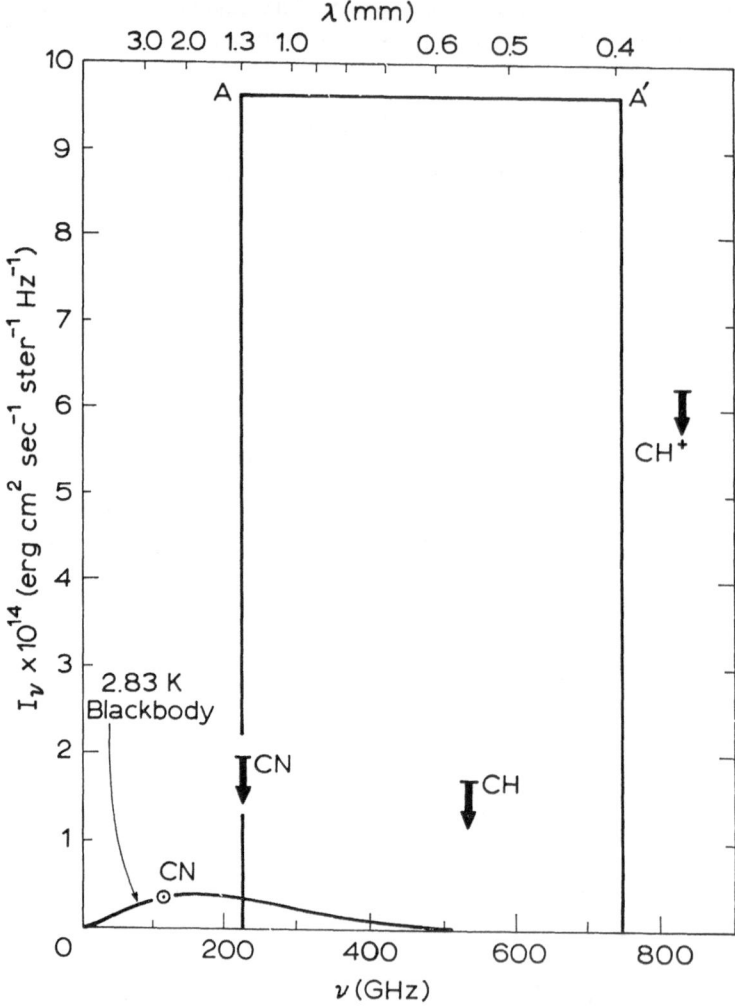

Fig. 16.   Limits on the intensity of radiation I_ν in the interstellar medium. The scale is linear, and thus the area under the line AA′ represents the flux of 5×10^{-9} W cm^{-2} sterad^{-1} observed by Shivanandan *et al.* [12]. (From Bortolot *et al.* [11].)

submillimetre flux has been observed by Muehlner and Weiss [13] by balloon measurements.

The meaning of this large submillimetre flux is one of the outstanding questions covering the background radiation. From what we have learned in the last few years about galactic objects and about other galactic nuclei, there certainly is going to be a large amount of integrated background flux in the infrared. This could conceivably account for these measurements. Observationally what is needed is both broad-band spectral measurements of meticulous care for absolute calibration and high resolution line searches in this region of the spectrum.

6. Conclusion

I have chosen these four topics partly because of their particular interest to me and partly because of how they illustrate the problems that have been encountered in infrared astronomy and the exciting questions raised.

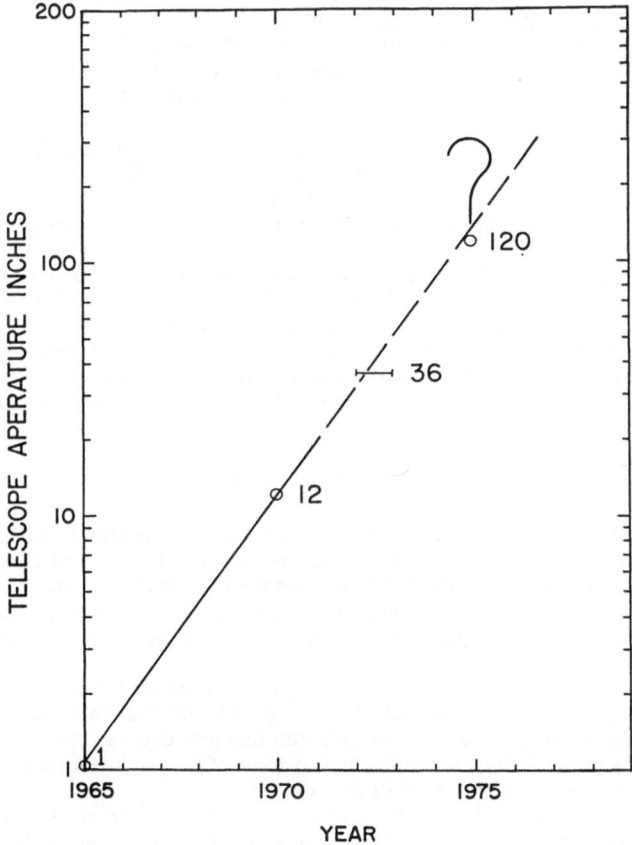

Fig. 17.   Projection of the increase in stratospheric telescope aperture with time.

Space astronomy from the stratosphere and atmosphere above can make substantial contributions to the answers through better spectral resolution, increased sensitivity and greater spatial resolution.

I have a very personal summary of the progress over the last few years in Figure 17. This shows the progression of available telescope aperture in the stratosphere. In 1965 the Goddard Institute one inch balloon-borne telescope was begun. By 1970 both balloon and aircraft telescopes of 12 in. aperture were in operation. And about 1973 we all hope to be using the NASA 36 in. airborne telescope at Ames Research Center. I look forward on the basis of this curve to the much needed 120 in. stratospheric telescope about 1975.

Acknowledgement

I wish to thank N. J. Woolf for helpful discussions about many of the topics covered in this paper.

References

[1] Mendoza, E. E.: 1966, *Astrophys. J.* **143**, 1010.
[2] Low, F. J. and Smith, B. J.: 1966, *Nature* **212**, 675.
[3] Hackwell, J. A., Gehrz, R. D., and Woolf, N. J.: 1970, *Nature* **227**, 822.
[4] Hoffmann, W. F. and Frederick, C. L.: 1969, *Astrophys. J. Letters* **155**, L9.
[5] Becklin, E. E. and Neugebauer, G.: 1968, *Astrophys. J.* **151**, 145.
[6] Harper, D. A. and Low, F. J.: 1971, *Astrophys. J. Letters* **165**, L9.
[7] Kleinmann, D. E. and Low, F. J.: 1970, *Astrophys. J. Letters* **159**, L165.
[8] Low, F. J. and Aumann, H. H.: 1970, *Astrophys. J. Letters* **162**, L79.
[9] Pacholczyk, A. G.: 1970, *Astrophys. J. Letters* **161**, L207.
[10] Neugebauer, G., Garmire, G., Rieke, G. H., and Low, F. J.: 1971, *Astrophys. J. Letters* **166**, L45.
[11] Bortolot, V. J., Jr., Clauser, J. F., and Thaddeus, P.: 1969, *Phys. Rev. Letters* **22**, 307.
[12] Shivanandan, K., Houck, J. R., and Harwit, M. O.: 1968, *Phys. Rev. Letters* **21**, 1460.
[13] Muehlner, D. and Weiss, R.: 1970, *Phys. Rev. Letters* **24**, 742.
[14] Gillett, F. C., Merrill, K. M., and Stein, W. A.: 1971, *Astrophys. J.* **164**, 83.
[15] Woolf, N. J. and Nye, E. P.: 1969, *Astrophys. J. Letters* **155**, L181.
[16] Hoffmann, W. F., Frederick, C. L., and Emery, R. J.: 1971, *Astrophys. J. Letters* **164**, L23.

DISCUSSION

B. P. Feuerbacher: When you discussed the Seyfert galaxies, you mentioned time variabilities and from that you reach a conclusion on the size of the emitters. What's the connecting link?

W. F. Hoffmann: The connecting link is a very fundamental statement that the emitting mechanism is some kind of a coherent one, where the different regions of emission would vary together by related phenomena, then the time variation must be within the transit time of light across the object.

J. Ring: In the early days of lunar infrared, people at Air Force Cambridge were very excited about the possibility of long range identification of the dust mineral composition, and then they came up against this dreadful snag that the spectrum one gets depends critically on particle size. Now we are hearing rather definite identifications of some of these minerals in the dust and I don't see too much mentioned about the effect of particle size.

W. F. Hoffmann: Particle size is certainly going to dominate the regions of the spectrum where the particles emit and generally bias the emission towards the short wavelengths since the particle size is much less than the wavelength of observation. In the case of the two emission peaks which have been used to identify the silicate picture, I think that the particle size hardly produces that kind of double resonance. In the case of dust grains of graphite or iron, nobody has sorted out the question as far as I know. I don't see right now how we can put together the enormously complicated picture derived from interstellar scattering and polarization of size, shape, and composition, but it seems to me that in a way we can bypass this problem by looking at the right wavelength for well defined absorption features.

F. Kneubühl: We know from solid state measurements that the emission spectrum and reflection spectrum of small particles are very different from the solid especially in emission. If one takes very simple crystals like alkali halides, which have a very simple reststrahlen reflection, and if one makes very small particles of 10 μ, then one gets a completely different spectrum even on the solid surface. If one wants to compare, one has to make a dust and not a solid. In small particles one has so-called surface modes which are more or less defined by the size of the particles and not by their atomic structure. One can make a simple dielectric theory which fits the emission and has not much to do with the crystal structure.

2. SYSTEMS FOR SPACE RESEARCH

THE NASA 91.5 cm APERTURE AIRBORNE TELESCOPE

MICHEL BADER* and FRED C. WITTEBORN**

Ames Research Center, NASA, Moffett Field, Calif., U.S.A.

NASA's Ames Research Center is currently operating several aircraft (Figure 1) for science and applications research. The various aircraft and much special equipment (telescopes, stabilized mounts, data systems, etc.) are available to all U.S. and other

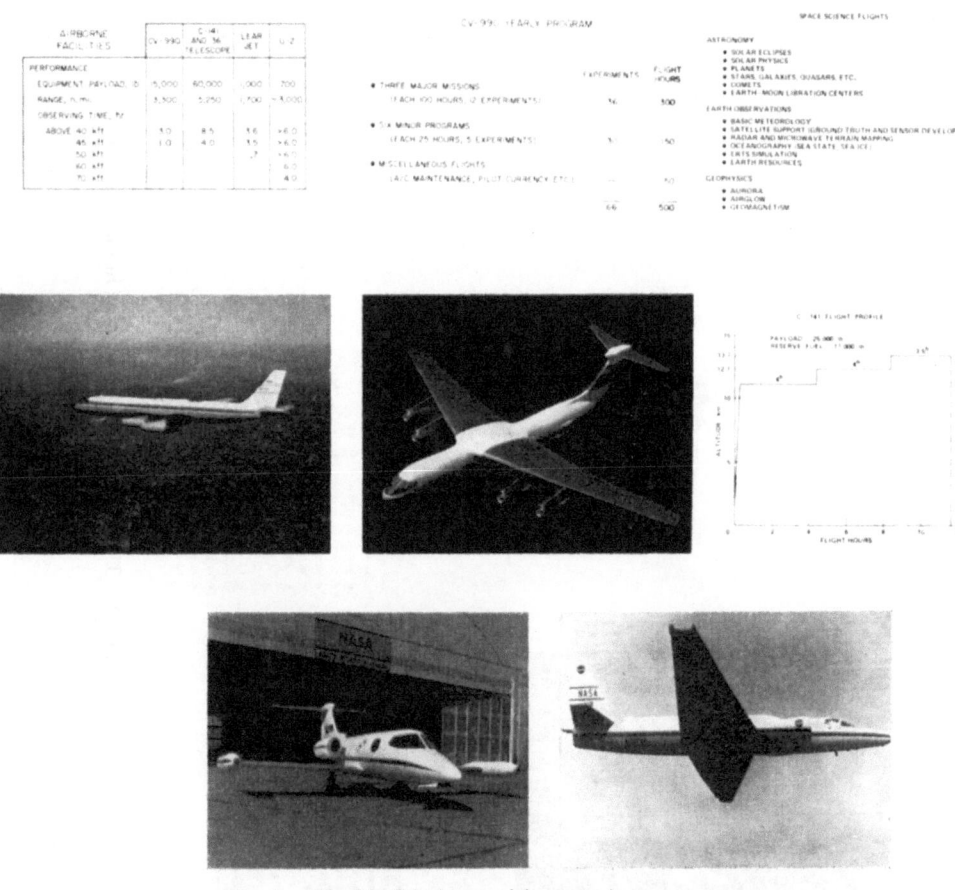

Fig. 1.   The NASA-Ames airborne science program.

experimenters submitting valid research proposals. All operational costs are borne by NASA. Most of this program has been described in Ref. [1], where further details may be found.

* Chief, Space Science Division.
** Chief, Astrophysics Branch.

Manno and Ring (eds.), Infrared Detection Techniques for Space Research, 29–31. All Rights Reserved
Copyright © 1972 by D. Reidel Publishing Company, Dordrecht-Holland

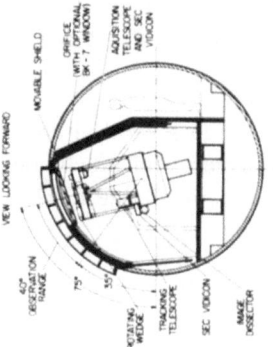

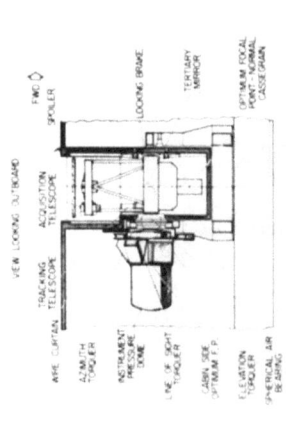

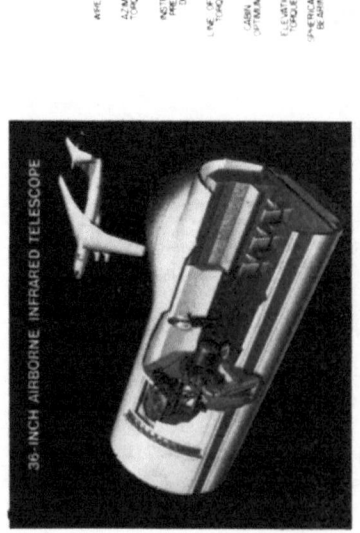

OPTICAL

- CERVIT REFLECTING OPTICS, "ZERO" THERMAL EXPANSION AT 220° K. SPECTRAL RANGE 1_μ TO 1 MM
- PARABOLIC 91.5-CM F/2 PRIMARY (f; = 1.8 M)
- 21 CM-DIAMETER HYPERBOLIC SECONDARY
- OVERALL f-RATIO = 13.5 (f; = 12.3 M)
- PLATE SCALE = 16.8 ARCSEC/MM
- MAXIMUM FIELD OF VIEW = 14 ARCMIN
- AREA OBSCURATION OF PRIMARY = 8 PERCENT
- EFFECTIVE AIRY DISC = 0.6 ARCSEC @ 5500 Å
- 80 PERCENT OF VISIBLE POINT SOURCE RADIATION IN 1 ARCSEC BLUR CIRCLE
- ALL REFLECTING ELEMENTS GOLD-COATED
- ALTERNATE SECONDARY AND TERTIARY MIRRORS ALUMINUM COATED
- 76 CM BACK FOCUS IN FOLDED MODE
- 61 CM BACK FOCUS IN NORMAL CASSEGRAIN MODE
- FULL APERTURE SCHOTT BK-7 WINDOW FOR VISIBLE AND NEAR IR OBSERVATIONS

MECHANICAL

- LENGTH OF TELESCOPE = 183 CM
- OPEN INVAR A-FRAME AND HEAD RING
- PRIMARY SUPPORTED ON BACK AND SIDES BY 21 PNEUMATIC BELLOWS
- WEIGHT OF PRIMARY = 197 KG
- DIAMETER OF INVAR AIR BEARING = 42 CM
- WEIGHT OF AIR BEARING AND SEAT = 690 KG
- WEIGHT OF TOTAL INSTALLATION IN AIRCRAFT APPROXIMATELY 5450 KG
- EXPERIMENTER'S EQUIPMENT WEIGHT
  - FOLDED MODE = 180 KG
  - NORMAL CASS. = 45 KG
- NAT FREQ OF TELESCOPE APPROXIMATELY 39 Hz
- BANDWIDTH GYRO-TORQUER SYSTEM = 6 Hz
- CUT OFF VIBRATION ISOLATION SYSTEM = 3 Hz
- ALT SECONDARY MIRROR CAN BE WOBBLED TO 140 Hz FOR SPACE FILTERING
- CAVITY AND TELESCOPE CAN BE PRECOOLED TO 220 K

TRACKING AND STABILIZATION

- TRACKING ACCURACY GOAL (OPEN PORT) = 6 ARCSEC PEAK FOR 30 MIN TIME (3 ARCSEC CLOSED PORT)
- OPTICAL AXIS DRIFT = 2 ARCSEC/HR
- 7' TRACKING RADIUS WITHOUT VIGNETTING, 5' RADIUS MECHANICAL LIMIT
- DRIFT BETWEEN TRACKERS AND MAIN TELESCOPE AXIS = 2 ARCSEC/HR
- JITTER = LESS THAN 1 ARCSEC
- 3 GAS BEARING GYROS AND D.C. SEGMENTED 35 KG TORQUERS
- 7¾ CM ACQUISITION SCOPE WITH 6' FIELD RELATED TO CONSOLE BY WESTINGHOUSE STV 606 SEC VIDICON (M, LIMIT = 4)
- 15 CM TRACKING SCOPE WITH 40 ARCMIN F O V RELAYED TO CONSOLE BY STV 606 VIDICON (M, LIMIT = +6)
- I T T F4012 IMAGE DISSECTOR FOR TRACKING POINT AND EXTENDED SOURCES (UP TO SUN AND MOON, M, RANGE = -27 TO +6)
- ROTATING OPTICAL WEDGES FOR OFFSET TRACKING OVER 6° FIELD
- OFFSET ACCURACY 1 ARCSEC
- FOCAL PLANE MONITOR FOR CONTINUOUS OR COMMAND VIEW OF MAIN FOCAL PLANE EITHER FROM TELESCOPE OR CONSOLE
- M, LIMIT AT MAIN FOCAL PLANE = 13 WITH GOLD-COATED OPTICS
- MOVABLE DOOR SLAVED TO TELESCOPE POSITION

Fig. 2.　The NASA 91.5 cm airborne infrared telescope.

The 91.5 cm airborne telescope is intended primarily for the infrared region (1 μm to 1 mm); its main characteristics are given in Figure 2. Also intended for international usage, it is expected to be operational in 1973. Further details may be found in [2].

The airborne program is also considered a precursor or prototype of future manned orbital research. Payload constraints and management structures are minimized, so that lead times are short (less than a year) and costs are little more than for ground laboratory experiments. Most of the advantages of airborne research can be carried over to orbital manned space science [3], as the payload retrievability and the presence of man make complex reliability and operational procedures unnecessary.

References

[1] Bader, M. and Wagoner, C. B.: 1970, *Appl. Opt.* **9**, 265.
[2] Cameron, R. M., Bader, M., and Mobley, R. E.: 1971, Design and Operation of the NASA 91.5 cm Airborne Telescope, to be published in *Appl. Opt.*, Sept.
[3] Bader, Michel and Farlow, Neil H.: 'Potential Reductions in Cost and Response Time for Shuttleborne Space Experiments', to be published in the Proceedings of the AIAA Space Systems Meeting held July 19–20, 1971 in Denver, Colo.

DISCUSSION

J. Ring: How much absorption or emission is left at 40 K feet at 100 μ?

M. Bader: In terms of absorption one is above more than 99% of water vapour, but about 30% of the CO_2 still remains above. This means that in the region between 1 and 15 μ, one is virtually above all the contaminants. Going up to the few hundred microns range, there are still several significant absorptions, some of them from water dimers, and some have not yet been identified. On the average the transmission is 60% or so in the several hundred microns range.

J. Ring: How does this compare with the emissivity of the optics?

F. C. Witteborn: The optic emissivity is about 0.02. The optic will transmit also, so that the overall reflection will be of the order of 95%.

W. Hoffmann: I am more optimistic than Bader about the transmission at 40 K Ft at 100 μ. Extrapolating from 100 K feet one would get that, at a 45° elevation angle, typical of the operations of a telescope from an aircraft, the emissivity would be about 10%. In terms of transmission this is completely adequate, but in terms of emissivity this will produce a background flux perhaps 10^4–10^5 greater than what one is measuring.

F. Kneubühl: What is the time variation of the residual absorption?

M. Bader: That depends very much on the location of the tropopause.

A 32-cm AIRBORNE INFRARED OBSERVATORY

P. LÉNA, N. CORON, C. DARPENTIGNY, K. HAMMAL, and
G. VANHABOST

*Université de Paris, Observatoire de Meudon et Centre National de la Recherche Scientifique,
Meudon, France*

1. Introduction

An open-port, 32 cm telescope is in current installation in an aircraft to fly at 35 000 ft. Present steps are taken to use in 1972 another aircraft climbing to 45 000 ft, where residual water vapour is expected to be of 1 μ cm^{-2} (precipitable at zenith).

The telescope is inertially guided within 1 arc min, allowing to track any point in the sky; the optics shall give ultimate performance of a few arc sec, either for short exposures, or with secondary guiding system, which may be included.

The Cassegrain focus is equipped with a Michelson interferometer, and a multi-band-pass, multi-channel helium cooled bolometer. Other equipments are in design, including fixed focus, high resolution interferometer, on-time, on-board Fourier transform system for 1024 points.

Several observations are planned with the system, which will fly in Spring 1972.

It is not the purpose of this short contribution to discuss the virtues of airborne infrared observations versus other spatial techniques such as balloon, rockets or satellites. Such a discussion may be found in other papers [1], [2] or in contributions to these proceedings.

2. Atmospheric Transmission

To briefly summarize what can be achieved from aircraft, let us examine the chief absorber between 1 and 1000 μ, namely H_2O. Figure 1 shows the vertical distribution of water (overhead precipitable amount), versus altitude. This measurement was made with a spectral hygrometer built for this purpose: it compares the solar flux in the continuum at 1.85 μ in a narrow spectral window (about 0.1 μ) to the nearby flux at 1.65 μ where the Ω band of H_2O strongly absorbs.

The distribution given in Figure 1 is fairly typical, although fluctuations will occur depending on the position of the tropopause, the season.... Very little work has been done on the stratospheric water content and its fluctuations, and it is yet uneasy to guess the optimum observing conditions.

Figure 2 summarizes the spectral transmission for amount of H_2O varying from 100 to 1 precipitable microns. The former value is certainly hard to reach from the ground, even from mountain top observatories (measured values at the Gornergrat Observatory, in Switzerland, are of the order of 500 precipitable microns). Balloon altitudes give amount of water less than 0.3 precipitable microns.

Despite the fact that the H_2O absorption is not negligible at aircraft altitude, it is easy to prove that much important and useful work is possible. To illustrate this point,

Manno and Ring (eds.), Infrared Detection Techniques for Space Research, 32–40. All Rights Reserved

Figures 3a and 3b show a theoretical H_2O spectrum, computed for various water amounts, in a spectral range where interesting lines occur. The O I line, which is strongly emitted by the high atmosphere, falls in a convenient position to be observed from aircraft. The O III line which is an important line for coronal studies and also for H II regions studies, will be fairly easy to observe from aircraft. Other lines, such as the C II line, are strongly blended by H_2O, but this is not the common case.

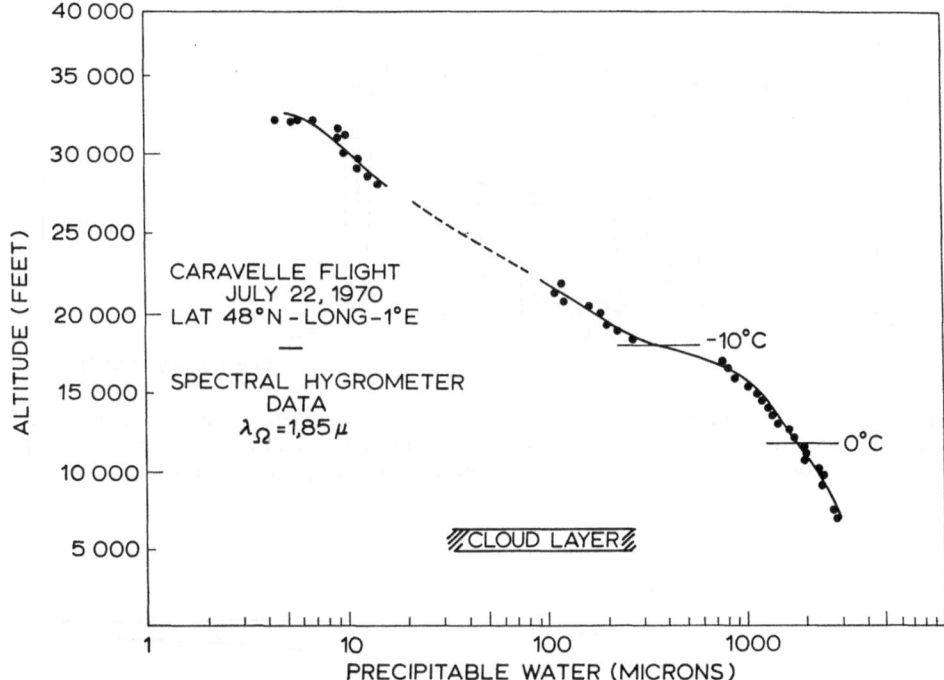

Fig. 1. Measurement of vertical distribution of precipitable water on a particular day. The absolute scale is deduced from the laboratory calibration of a spectral hygrometer.

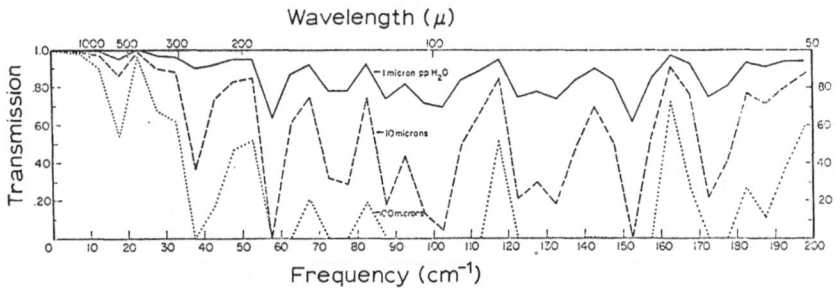

Fig. 2. Absorption of H_2O in the submillimetric and far-infrared. Amounts of precipitable water are indicated on the curves. Typical aircraft conditions vary between $10\ \mu$ and $1\ \mu$. The curve represents spectral transmission averaged over $1\ cm^{-1}$ spectral interval.

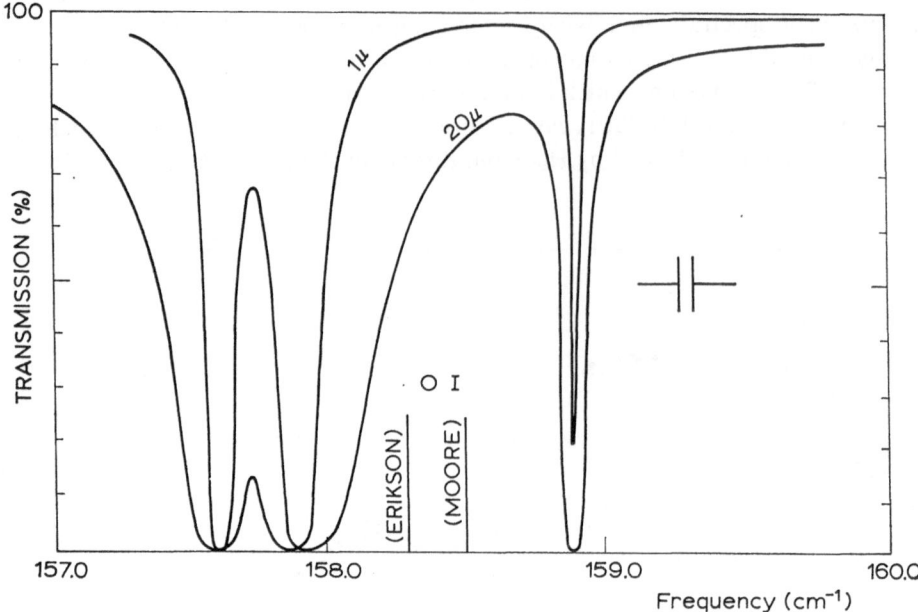

Fig. 3a.   Theoretical transmission spectrum of H_2O for two different amounts of precipitable water. The band pass of a reasonable Fourier spectrometer is indicated (0.1 cm^{-1}). The 63 μ oxygen line is shown for two computed positions by different authors.

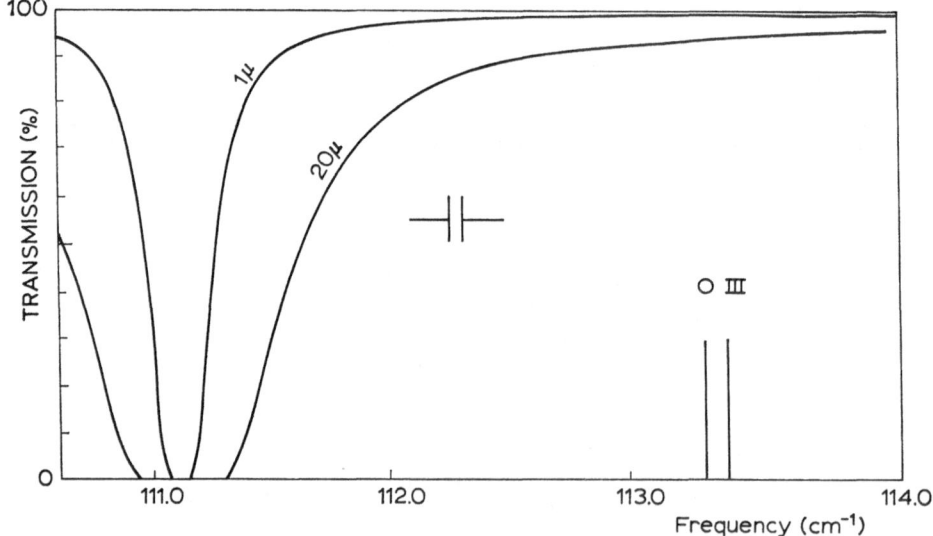

Fig. 3b.   Similar transmission spectrum as Figure 3a. Here the transmission is better and the O III line has very little blend by atmospheric transmission.

One major concern of infrared astronomy is, of course, the cancellation of spurious sky signals and the elimination of sky noise. This is an unknown field for aircraft work and shall be studied extensively in the future.

Other absorbers should not be overlooked. Minor constituents of the atmosphere may create strong blends (such as O_2, CO_2, N_2O, NO_3H,...), or become a goal by themselves in pollution studies. Continuous absorption by collision-induced dipolar moments may also become non-negligible [3].

A good high-resolution (better than 0.1 cm^{-1}) spectroscopic study of this spectral range from aircraft remains to be done and will be a valuable tool for aircraft astronomy.

3. Aircraft Conditions

3.1. Aircraft stability has been recorded by taking time-exposures of a star with a camera bound to the aircraft. The exposures were intercepted at regular intervals to give a time scale. Figure 4 illustrates the results. The *roll* is the most important

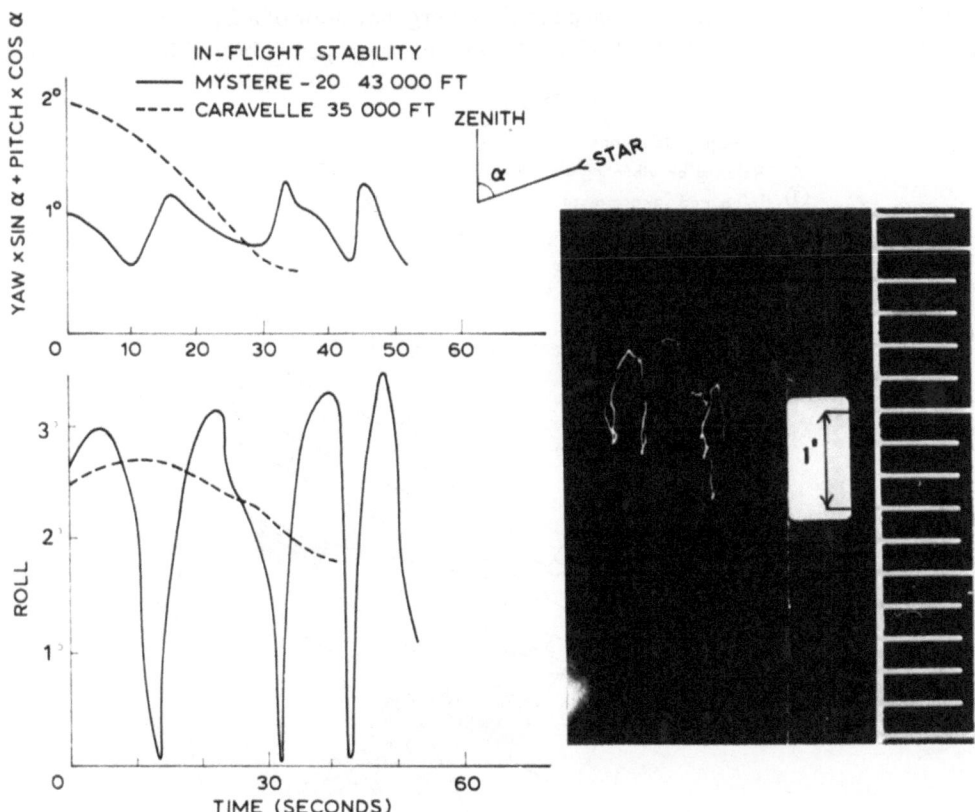

Fig. 4.   Aircraft in-flight stability. A time-exposure of a star is taken in flight, with periodic interruptions to give the time scale.

perturbation. Small aircrafts do present high roll frequencies, which places more severe constraints on the design of any servo system.

Four guiding possibilities may be considered to overcome these motions:

– *manual guiding* for limited accuracy (field of view larger than 30 arc min).

– *photoelectric guiding* (direct or offset) if the object itself is bright enough or has a bright star in its immediate vicinity: magnitude ⩾3.5.

– *infrared guiding*, which uses the infrared signal of the object itself to control the servos. This type of guiding requires sensitive and fast IR detectors, but may become of interest in the future.

– *inertial guiding*, where the aircraft rotates around the telescope, being kept fixed with respect to the inertial frame of the stars.

3.2. Aircraft turbulence may somewhat degrade image quality. This point remains to be studied.

4. The 32-cm Airborne Telescope

4.1. AVAILABLE AIRCRAFTS

The system has been built to be placed in the emergency door of a Caravelle, which has an operational ceiling altitude of 35 000 ft, where it can spend about 3 h, at a cruising

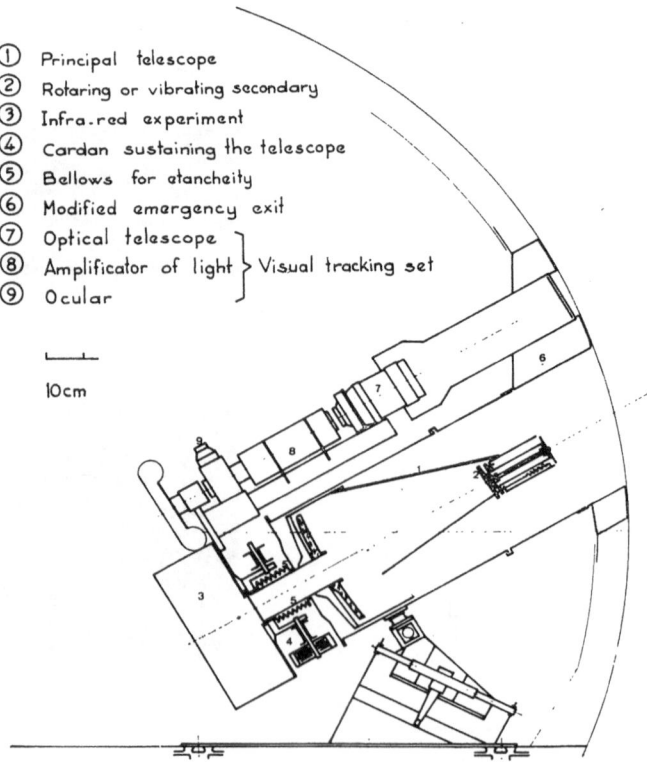

① Principal telescope
② Rotaring or vibrating secondary
③ Infra-red experiment
④ Cardan sustaining the telescope
⑤ Bellows for etancheity
⑥ Modified emergency exit
⑦ Optical telescope ⎤
⑧ Amplificator of light ⎬ Visual tracking set
⑨ Ocular ⎦

10cm

Fig. 5.   Overall design of the 'Mystère-20' telescope.

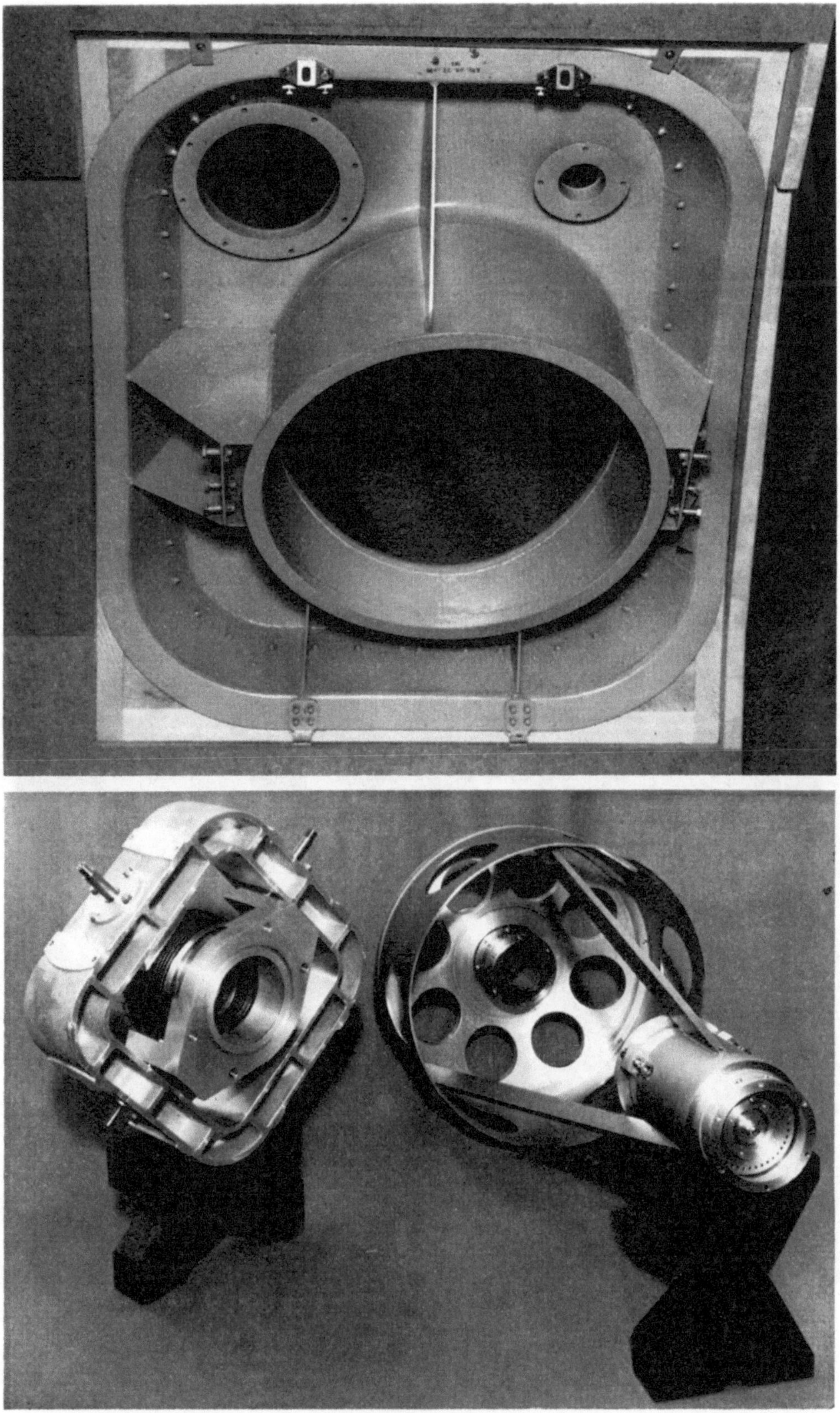

Fig. 6. Pictures of the modified emergency exit and of the telescope mount.

speed of 400 knots. The aircraft is operated from Bretigny (Essonne), but may explore in latitude the region between 28° and 65°N.

An improved version of the same telescope will be placed on a Mystère-20 (Falcon Jet by Dassault), with a ceiling altitude of 45 000 ft, where it can spend 2 h. The very good navigation equipment can give ground truth within 1 mile. This jet should be able to fly down to equatorial latitudes if necessary, and is operated from Creil (Oise).

4.2. THE SYSTEM

The telescope is a 32-cm Cassegrain, $f/7.6$ system. The primary is a Zerodur mirror, the secondary is a beryllium hyperbolic mounted on an invar tripod. Off axis rotation or vibration of the secondary are two possible operating modes.

The general schematic of the mount is shown on Figure 5. Due to the Caravelle constraints, coarse manual pointing is limited to $30° \pm 4°$ in elevation, and $\pm 4°$ in azimuth, the aircraft position providing the main alignment of the telescope with the object.

In the Mystère-20 version of the system, a hole is cut in the fuselage, and elevation manual pointing shall be possible between 30 and 60° (tentative values).

Figure 6 shows the Emergency door of the Caravelle, and parts of the telescope.

The telescope is counterweighted by the experiment chamber, which may be

Fig. 7.   View of the interferometer.

operated at the external pressure in the no-window mode, or at the internal pressure if a thin window is placed near the focus. The guided part also includes a tracking set: small 10 cm-telescope, image intensifier, eye-piece, camera or TV camera, connected to a monitor. All this system is gimbal-mounted and pressure tightening is achieved with low-torque bellows. Two gyros provide the error signals for the torque motors, and long term (1 minute or more) drifts of the gyros may be manually compensated by the operator.

The experiment chamber may support about 15 kg of equipment, the overall weight of the guided system being 60 kg. The present equipment is made of a Michelson interferometer (Figure 7), with a 1 cm^{-1} resolution capability. The interferometer is a fast-scanning one (scan duration between 1 s and 100 s), and scan addition is made on-board in a 20 bits, 1024-words memory; they are also tape recorded.

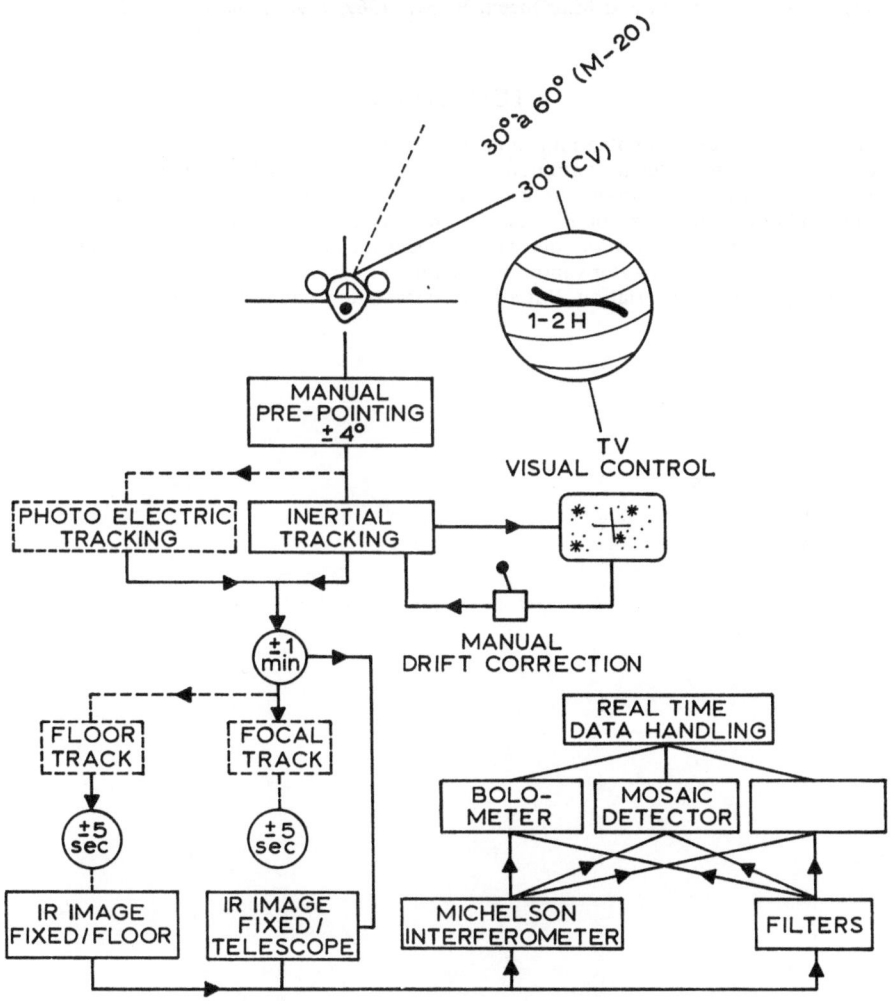

Fig. 8.   Overall block-diagram of the Caravelle and/or Mystère-20 system.

The interferometer is coupled to a Germanium bolometer, which has various adjustable cooled filters to deal with background radiation and optimize its photons shot noise.

Figure 8 describes the various performances which may be expected from the system, both in its present and future versions. Improved tracking may be useful in the near infrared around 20 μ. At 100 μ, the diffraction limit of the telescope is 1 arc min, and a tracking accuracy of this magnitude is at least required.

Flights are expected to start in the fall 1972, and the system should be completed by middle 1973.

References

[1] Bader, M.: 1970, *Appl. Opt.* **9**, 265.
[2] Léna, P.: 1970, *Space Sci. Rev.* **11**, 131.
[3] Eddy, J. A., Lèna, P. J., and MacQueen, R. M.: 1969, *J. Atmospheric Sci.* **26**, 1318.

DISCUSSION

R. A. Hanel: There may be a potential problem with aircraft observation of this nature. The water vapour or whatever provides a continuum may not be horizontally homogeneous and may be stratified. So as the aircraft moves, one is scanning the horizontal distribution and this may give rise to fluctuations of the background. This may become important for very weak sources.

P. Léna: This is true. I may quote one measurement that we made with the Convair 990 when we found that the gradient of water vapour with path was approximately 10^{-4} km^{-1}. However, we did not see any short time variation. But a more systematic study of this effect should be made.

FAR-INFRARED OBSERVATIONS OF
CELESTIAL OBJECTS BY BALLOON-BORNE TELESCOPE

WILLIAM F. HOFFMANN

Goddard Institute for Space Studies, New York, N.Y., U.S.A.

Abstract. A newly developed balloon gondola with 12 in. telescope has been flown a number of times in 1970 and 1971 making far-infrared observations of celestical sources. These include the planets Jupiter and Venus, the galactic center, a number of bright H II regions, and some dark dust clouds. The technical aspects of the balloon gondola and telescope system and some of these results will be discussed.

I would like to talk about some of the technical aspects of the Goddard Institute for Space Studies 30 cm balloon-borne infrared telescope. This instrument was developed by Carl Frederick, Roger Emery, and myself after it had been discovered that the galactic center was very luminous in the infrared and that it was worth investing some trouble in building an instrument of greater sensitivity and resolution (Hoffmann and Frederick [1]). The guiding philosophy behind this instrument was to attain a high level of reliability and simplicity and a rapid turnaround. This was achieved to a large extent by the efforts of Carl Frederick who was largely responsible for the mechanical design.

Figure 1 shows the telescope, which is an $f/5$ Newtonian telescope with a parabolic pyrex primary mirror. The telescope tube is 1.5 m long including a 61 cm baffle to provide shielding from off-axis objects such as the Earth, the Moon, and the Sun.

The $f/5$ beam is reflected off the secondary diagonal into liquid helium dewar at the side through a polyethylene window. The field optics consists of a 5 mm diameter

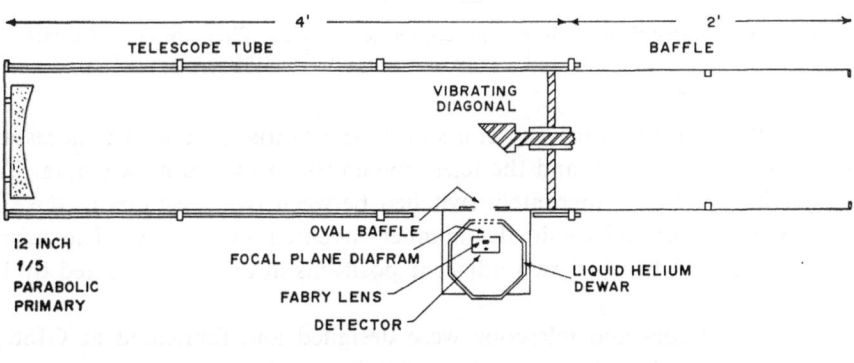

Fig. 1. 30 cm telescope used for balloon-borne far-infrared observations.

Manno and Ring (eds.), Infrared Detection Techniques for Space Research, 41–52. All Rights Reserved
Copyright © 1972 by D. Reidel Publishing Company, Dordrecht-Holland

crystal quartz lens which images the primary on the detector. The field of view in the sky is determined by a 6 mm aperture stop in front of the lens which gives a field of view of 12 arc min.

The filtering system which defines the 100 μ passband consists of the 2 mm thick quartz lens, the 1.6 mm thick white polyethylene window, two layers of black polyethylene, 0.15 mm thick each, a 1.6 mm thick plate of cold teflon, and a No. 300 electroformed mesh. Figure 2 shows the spectral passband of the filter system. The half power points are approximately at 80 μ and 125 μ. The overall peak transmission is between 10 and 20%.

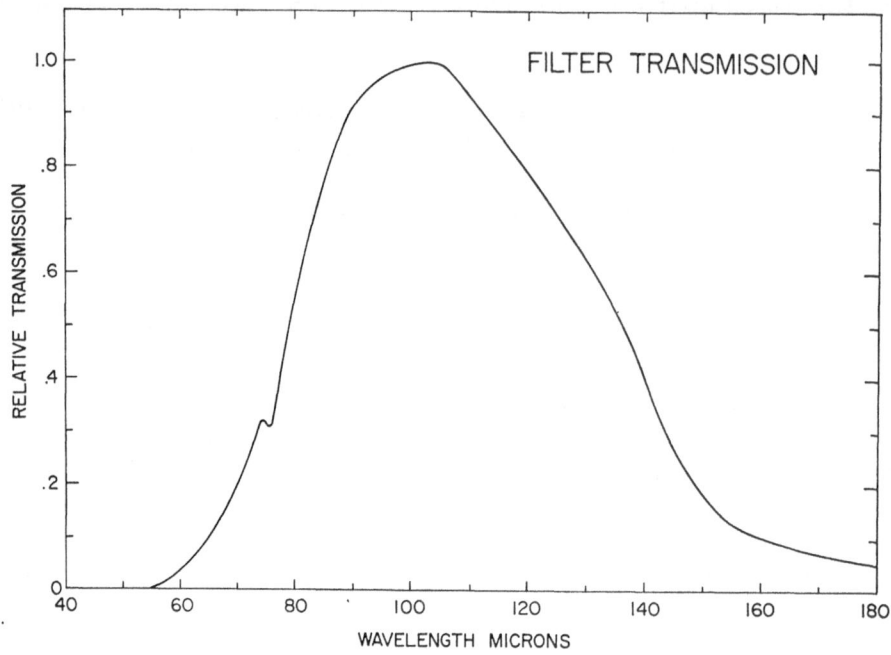

Fig. 2.   100 μ filter transmission curve. The components of the filter are crystal quartz, black polyethylene, white polyethylene, teflon and electroformed mesh.

Even at balloon altitudes the sky emission is substantial so that it is necessary to subtract off the sky emission and the telescope emission by beam switching. This is done by having the beam alternately switched between two positions in the sky by vibrating the secondary mirror in the manner invented by F. Low. The mirror is percussively switched between two adjacent positions in the sky separated by 18 arc min.

The balloon gondola and telescope were designed and fabricated at GISS with prime consideration given to reliability, durability, and very rapid turnaround between flights. For this reason, an elevation-azimuth axis system was chosen with the elevation referred to the vertical as determined by Earth's gravity and the azimuth referred to the Earth's magnetic field. This system has been proven to satisfy the reliability requirement, to have a turnaround time of only six days, and to have achieved a stabilization and pointing accuracy of 3 arc min. The liquid helium dewar containing the detector and field optics has survived about eight landings with this and a previous gondola with no damage.

Figure 3 illustrates the main features of the present balloon gondola. The overall height is 5 m. The weight is 2200 kg. The telescope, complete with cryogenically cooled detector, is counterbalanced by the insulated electronic unit mounted on the opposite side of the central mechanical bearing and drive assembly. The main horizontal shaft can be commanded to any elevation angle in half degree steps. It can also be commanded to scan up and down about a central position through an arc 2.5° to 20° in size.

Alongside the main vertical support rod is a magnetometer mounted on a post above a motor controlled table which can by command orient the magnetometer in half degree steps relative to the gondola. The magnetometer is operated in a null servo system which drives the gondola in azimuth against the inertia wheel hung immediately below the main bearing. A tachometer on the inertia wheel controls a bias in a dynamic bearing mounted at the top of the vertical shaft in order to transfer excess angular momentum of the inertia wheel to the balloon. The gondola can be commanded to scan in azimuth over a range 2.5° to 10°.

The fundamental limit of magnetometer stabilization is the penduluming of the balloon itself and the constancy of the Earth's magnetic field. We find that the rms deviation of the orientation stabilization of multiple observations of the same object over one half hour is 1.6 arc min. The zeros of the elevation and azimuth settings are determined by observations of known objects. The long term accuracy from one part of the sky to another is 3 to 6 arc min.

Hanging directly below the inertia wheel is the telemetry transmitter and ballast hopper supplied by the National Center for Atmospheric Research (NCAR). Upon flight termination these two units are dropped onto a 20 ft long cord and the telescope is automatically stowed in a horizontal position so that it will land on the crush pads mounted directly onto the telescope and electronics structure. In addition, the vertical support rod is separated just above the bearing assembly in order to provide a flexible coupling at landing. This prevents the vertical support rod from providing leverage for the wind driven parachute to upset the gondola on the ground.

Figure 4 shows that even the best laid plans can come to naught. But in this case, the practical design did work out even though the gondola landed upside down. The crush pads were re-usable. We only had to clean the mirror off and reassemble the gondola after it was returned to Palestine in order to fly it again in 6 days. I think that without a major catastrophic landing, like a lake, a marsh, a very high wind, this gondola does indeed satisfy the requirement of rapid turnaround.

Figure 5 illustrates the signal obtained while scanning in azimuth. This is a strip chart recording of the telemetered signal obtained during a flight. The lower channel is the magnetometer signal. Full scale represents 10°. The scan rate is 0.4°/s. The bolometer signal is in the upper channel. This is first negative, then positive as the object passes through the two telescope beams. This signal is from the galactic center. This is

Fig. 3.   Far infrared telescope and balloon gondola.

Fig. 4. The gondola after a landing in a wooded portion of east Texas.

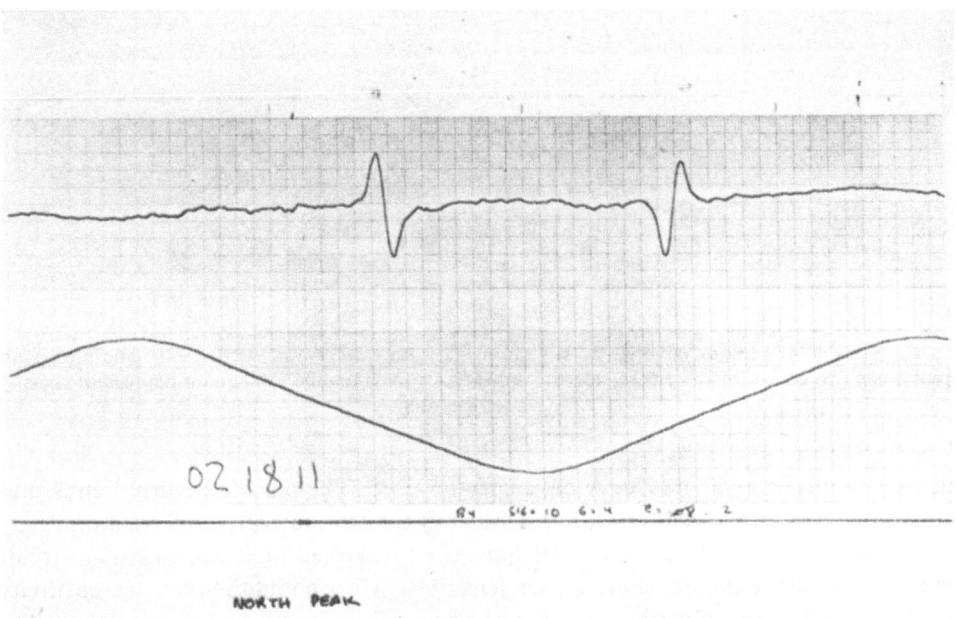

Fig. 5. Strip chart recording of telemetry signal and orientation channels during scans of the galactic center. The upper trace shows the photometer signal as the source passes first through the negative then the positive beam. The lower channel gives the scanning magnetometer signal with 10 degrees full scale.

the brightest 100 μ object in the sky outside the solar system. The weakest signals which we can identify on single scans are about $\frac{1}{50}$ of that.

With our choice of scanning speed and beam size we are assured of a minimum of 3 to 4 and sometimes 6, successive overlapping scans. This is necessary for distinguishing celestial objects from noise. Also we have the option of rescanning over the same region an arbitrary number of times and by later computation adding these to produce a much better sensitivity. This signal in Figure 5 is produced with a filter time constant of 0.08 s. This represents an extremely short integrating time for a telescope photometer operation. By summation of multiple scans over a specific object for an hour, we can achieve an effective accumulation of data of 5 to 10 min and considerably better signal to noise.

Figure 6 shows the regions of the sky covered during the last three flights. Approximately 750 square degrees of the sky have been covered including 100 degrees along the galactic plane. Also regions away from the galactic plane such as M 42 (the Orion

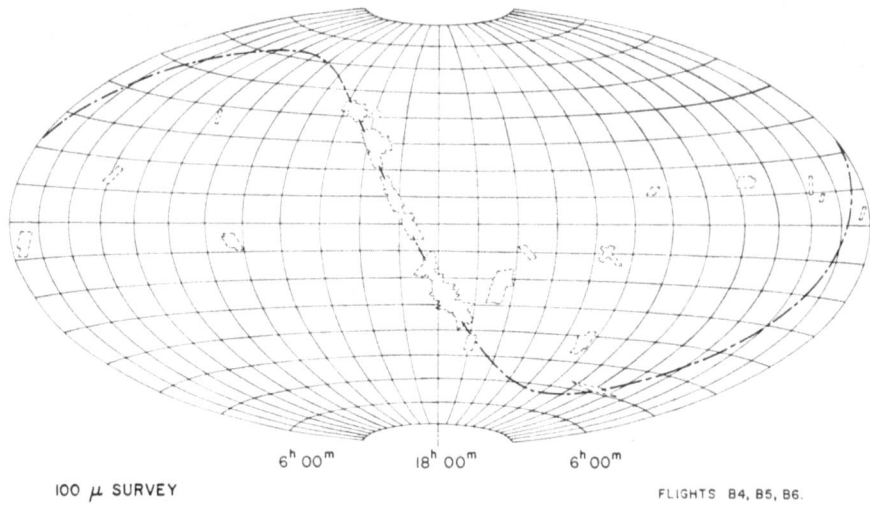

100 μ SURVEY   FLIGHTS B4, B5, B6.

Fig. 6.   Outline of regions covered by sky surveys of June 1970 and April 1971. The plot is an equal areas projection of the entire celestial sphere. Approximately 750 square degrees of the sky have been covered.

infrared nebula) and Centaurus A have been covered. The planets Saturn, Venus, and Jupiter are used for calibration of flux sensitivity and for correction of position determinations. A total of about 60 objects have been observed at 100 μ. The majority of them are just above the threshold of our sensitivity. If we should reduce the sensitivity by a factor of 5, most of them would not be observed. It is apparent that if we increase the sensitivity by a factor of 5, the number of sources will increase dramatically. We may obtain a continuous map of large regions of the galactic plane, similar to that already obtained for the galactic center region (Hoffmann et al. [2]).

We have also learned from these flights about the 100 μ transmission of the atmosphere at 100 000 ft. Figure 7 shows the transmission of the atmosphere as a function of elevation angle. The points plotted were obtained by observing the Orion infrared nebula over a range of 10° to −2° in elevation angle. The horizon at 100 000 ft is depressed by 6°. At −2° the transmission is 50%. This depression of the horizon and

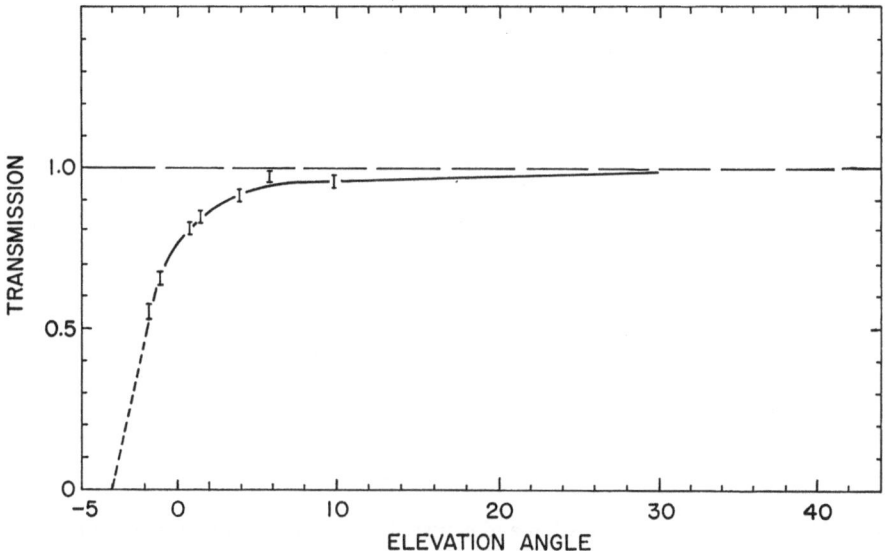

Fig. 7.   100 μ atmospheric transmission at 90 000 ft versus telescope elevation angle.

high transmissivity of the atmosphere permit observations from Palestine, Texas as far south as −60° declination. The curve is a theoretical one based on a model assuming constant scale height with altitude for the absorbing material (water vapor). This model yields a zenith transmission of 99.3%. Hence the zenith emissivity is only 0.7% at 100 μ.

In order to compare atmospheric transmission and emission at various altitudes, I have extrapolated this data to lower altitudes, as shown in Figure 8. The extrapolation is made with the simplifying assumption of constant scale height and constant temperature. Hence this model does not take into consideration the slight decrease in temperature from −65° to 100 000 ft to −85° at the tropopause (40 000–50 000 ft). Figure 8 shows curves of transmissions both for the zenith and for 45° elevation, a typical angle for observation from balloons and aircraft. These curves are for the atmospheric properties integrated over an interval of wavelength 75 to 125 μ.

Even at the very high transmission obtained at balloon altitudes, the residual atmospheric emissivity can be a problem. In our observations the flux from the sky entering the telescope beam is 100 times higher than the smallest astronomical signals we observe. This is not at present the primary limit to our sensitivity since the instrumental radiation is substantially greater than that of the atmosphere. However, the

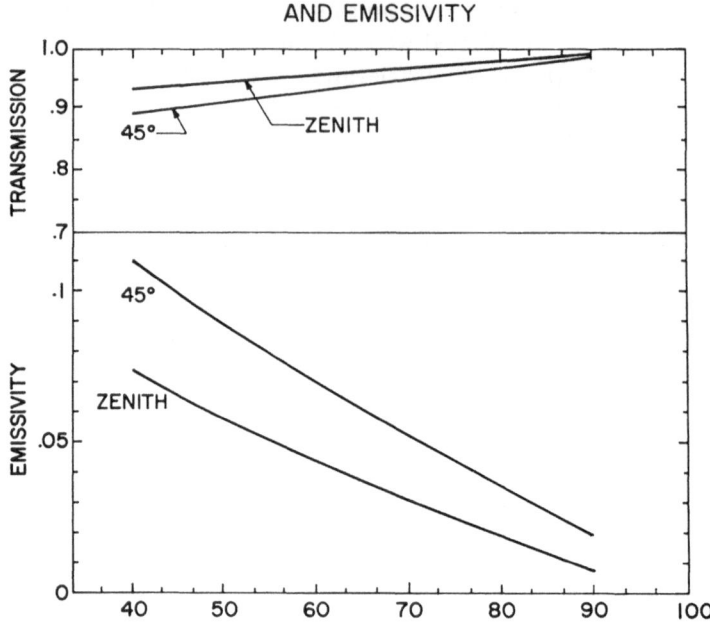

Fig. 8.   100 μ atmospheric transmission and emissivity as a function of altitude. Curves for both zenith and 45° elevation angle are shown.

atmospheric emission will become the fundamental limit to sensitivity from balloon altitudes if we use a cryogenically cooled telescope to reduce the instrumental emissions, as we plan to do in the future.

References

[1] Hoffmann, W. F. and Frederick, C. L.: 1969, *Astrophys. J. Letters* **155**, L9.
[2] Hoffmann, W. F., Frederick, C. L., and Emery, R. J.: 1971, *Astrophys. J. Letters* **164**, L23.

DISCUSSION

F. Melchiorri: You have used an open telescope: in this case, a condensation of water vapour may occur on mirrors or metallic parts during the ascent, mainly at the traversal of the inversion layer: in which way do you take this possibility into account?

W. F. Hoffmann: There is an enormous lag in the temperature of the optics of the telescope and in fact in the case of the pyrex mirror it never, over a period of twenty hours, reaches ambient temperature. In the case of the secondary mirror, it never quite reaches ambient temperature because there is a heat input, from the vibrating secondary solenoids, so that condensation directly on the mirrors is unlikely. In fact we have no evidence that we've had any problem from that. The window, which is polyethylene and hence a poor heat conductor, falls considerably below the ambient, because one side of it is in a very cold radiation field, and that certainly can frost up and does on the ground. So we use a double window. There is a second very thin polyethylene window outside of it and there is a desiccant container in between which keeps it dry on the ground. Our experience is that aloft, at the low vapour pressures at 100 K ft, the ice which does accumulate on the way up will sublimate.

F. Melchiorri: Which is the rejection factor of your filters in the visible and near infrared, where the Moon and the planets emit most of their radiation?

W. F. Hoffmann: I don't have in my mind right now the rejection factor in calibration. It is extremely great, in fact quartz is very absorbing. I seem to remember it to be about 10^{-5}. We measure it by putting in a filter which blocks, extremely effectively, the pass band, and see what gets through at low wavelength. We believe that in no case, do we have emission coming through, anywhere short of 40 μ.

Contribution from Dr R. E. Jennings (University College, London): A BALLOON-BORNE TELESCOPE SYSTEM.

The balloon-borne telescope which has been constructed by the Physics Engineering Group at University College under the direction of the late Mr H. S. Tomlinson is shown in Figure 1 (during construction). The $f/5.5$ cassegrain (Dall-Kirkham) telescope has a primary mirror of 39 cm diameter. Guidance onto a particular part of the sky is achieved first by coarse guidance using a magnetometer to bring the guide star (mag 4 or brighter) into the field of view of the star-tracker, and then by the star-tracker, which can be seen to the left of Figure 1. The pointing accuracy to the guide star should be ± 1 min arc peak to peak. The telescope can be offset from the star-tracker by $\pm 5°$ in elevation and cross elevation and facilities are provided for a vaster scan of the sky around the guide star. Alternatively the telescope can be set in any direction within the 10° square patch of sky.

The optics of the telescope are shown in more detail in Figure 2. Chopping is achieved by rocking the secondary mirror (the mechanism is not shown) and the beam is brought out through one of the main bearings. By means of a dichroic mirror, the infrared beam is reflected up into the cryostat which has the filters, condensing lens and germanium bolometer mounted on the base plate of the inner container in the usual manner, while the visible beam passes to a photomultiplier. In this way the infrared scan can be 'over laid' on the visible scan.

The appearance of the system shortly before flight is shown in Figure 3. The gondola has been fitted with roll bars to prevent damage on landing, the telescope being latched into the vertical position prior to cut down.

Fig. 1.

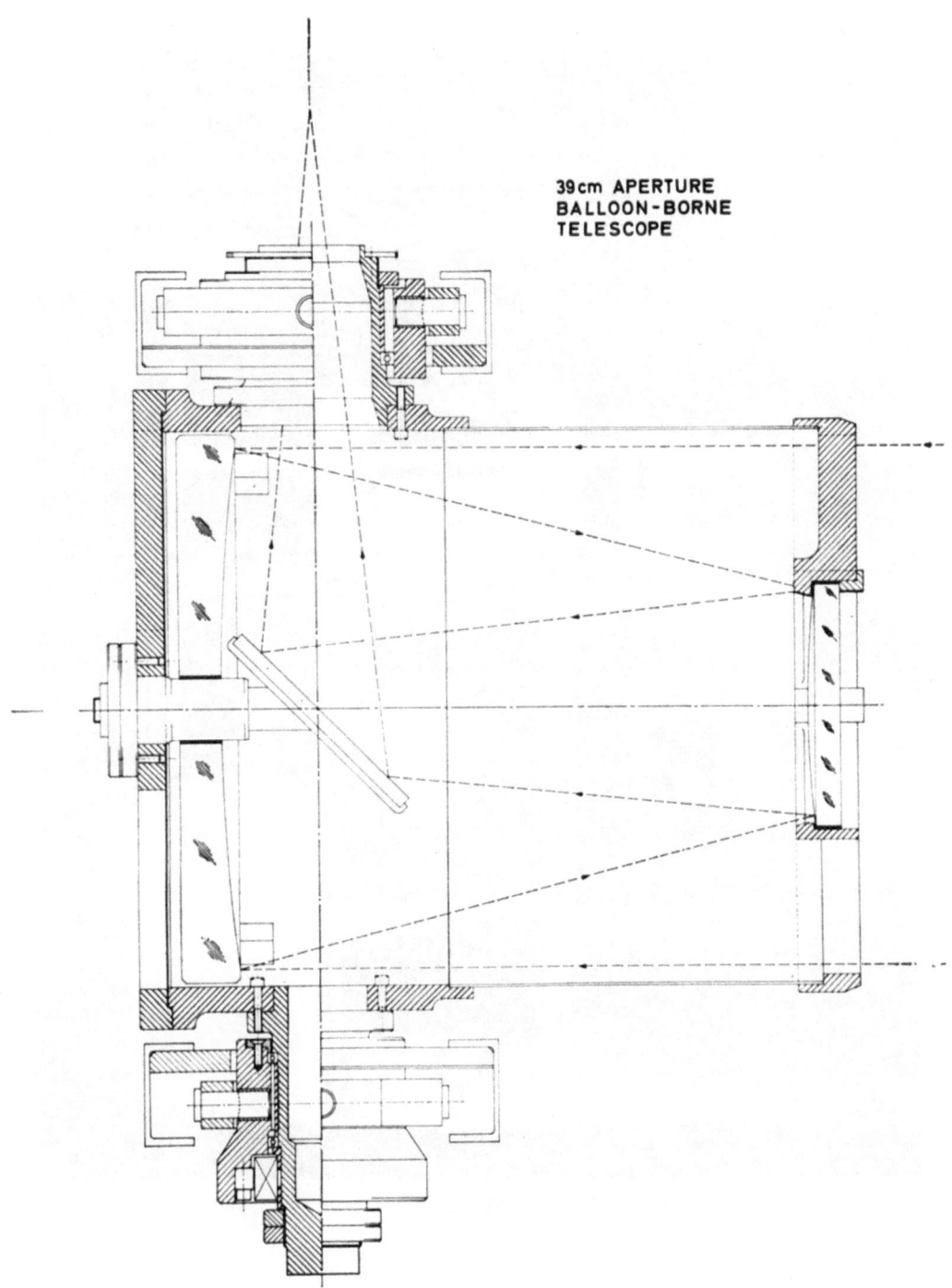

39 cm APERTURE BALLOON-BORNE TELESCOPE

Fig. 2.

Fig. 3.

AN INFRARED PHOTOMETER FOR THE BALLOON-BORNE
TELESCOPE THISBE

D. LEMKE

Max-Planck-Institut für Astronomie, Heidelberg-Königstuhl, G.F.R.

After two flights of our balloon-borne telescope Thisbe in 1970 for measurements in the visible and near ultraviolet wavelength region we hope to launch in October 1971 our first balloon-borne infrared photometer. By this experiment we will make a surface photometry of the night sky at 2.4 μ. I will describe in the following the main components of our project: The balloon gondola Thisbe and the 2.4 μ photometer. Finally I will give some details about our future plans in the field of extraterrestrial infrared experiments, these are (i) the feasibility study of a highly stabilized balloon-borne telescope, (ii) the application of a bolometer, (iii) the planning of an infrared satellite experiment.

First Thisbe: In 1967 we started the planning phase of a balloon-borne telescope. According to this plan we are realizing the project in two steps. The first step was the construction of a coarsely stabilized gondola. With this gondola we can get some experiences in the field of balloon-borne experiments. After that we can realize the highly stabilized gondola in the second stop.

This method has further advantages: since we got our coarse stabilized gondola after only one year of construction, we can quickly perform some astronomical observations. The stabilization accuracy of only $\pm\frac{1}{2}°$ is not so hard for us because we are interested in some programmes we can realize with this stabilization, these are mainly surface photometries.

The first observational flights of Thisbe were launched last October from Palestine, Texas. The purpose was to measure intensity and polarization of the zodiacal light and of the Milky Way. Figure 1 shows Thisbe at the hook of Tiny Tin, the big lauching machine at Palestine. In the centre of the gondola you see the elevation axis with a circular frame for fixing the payload. In this case the payload consisted of four small telescopes for measurements of the zodiacal light. For attitude control of azimuth the whole gondola rotates around the azimuth bearing by contrary rotations of the reaction wheel. The whole observational programme is stored on a punched tape on board the gondola. All observational and housekeeping data are transmitted by a PCM-telemetry system which has 18 channels with a transmission accuracy of better than $\pm0.5\%$.

The weight of the whole gondola including the payload is 180 kg. For absorption of the landing shock there are 12 crash pads made of paper. They protected the gondola at the three flights made up to now very reliably. Figure 2 shows a landing point in a very dense hickory forest in East Texas. As one can see, the gondola stands upright and there was no damage of the gondola nor of the payload.

The first flights gave some results concerning the intensity of the zodiacal light at

Manno and Ring (eds.), Infrared Detection Techniques for Space Research, 53–59. All Rights Reserved
Copyright © 1972 by D. Reidel Publishing Company, Dordrecht-Holland

Fig. 1. The balloon-borne telescope Thisbe at the hook of the launching machine at Palestine, Texas.

small elongation angles from the Sun and about dust concentration in the upper atmosphere [1]. There were also some initial difficulties with the gondola's attitude control system which are removed now.

One of the next balloon-borne experiments we will launch this fall is the 2.4 μ-surface photometry [2]. First I will give some information about the existing measurements of infrared flux from the sky and then describe our instrument. Figure 3 shows

Fig. 2.   Landing point of Thisbe near Center, Texas.

a diagram summarizing all existing infrared flux measurements. Peebles [3] has published the data recently in a paper concerning cosmology and infrared astronomy. There is given the surface brightness of the sky background versus wavelength. In the near infrared, at some microns, there exists up to now only a rocket measurement of Harwit *et al.* [4]. They could give an upper limit of the flux in the region 1–3 μ. An experiment prepared for the ESRO-satellite TD1 in nearly the same wavelength-region was cancelled [5].

The disadvantages of a rocket experiment, the short observation and pointing time and the limited payload are not present in a balloon-borne experiment. But there are other limits: for this kind of absolute flux measurements the thermal radiation of the remaining atmosphere overhead the balloon makes severe restrictions in the intermediate and far infrared. In the near infrared, where we shall measure, the night airglow is orders of magnitude brighter than the flux from the sky. The airglow radiation is produced mainly by the hydroxyl-radical. But in the emission spectrum of OH exists a gap near 2.4 μ with a width of $\Delta\lambda \approx 0.3\ \mu$ [6]. This gap was verified by some airglow investigations [7]. Outside the gap there is no chance to see even the brightest regions of the Milky Way. This fact is the reason for the choice of our wavelength region.

Figure 4 shows a sketch of the photometer. The main components are: (i) the photometer box, (ii) the container for dry ice, (iii) a large baffle system, against stray light from the Earth and the balloon.

The 15 cm objective lens is made of Infrasil, an infrared transmitting fused silica. The chopper is an oscillating plane mirror driven by an electronic motor. Next in the

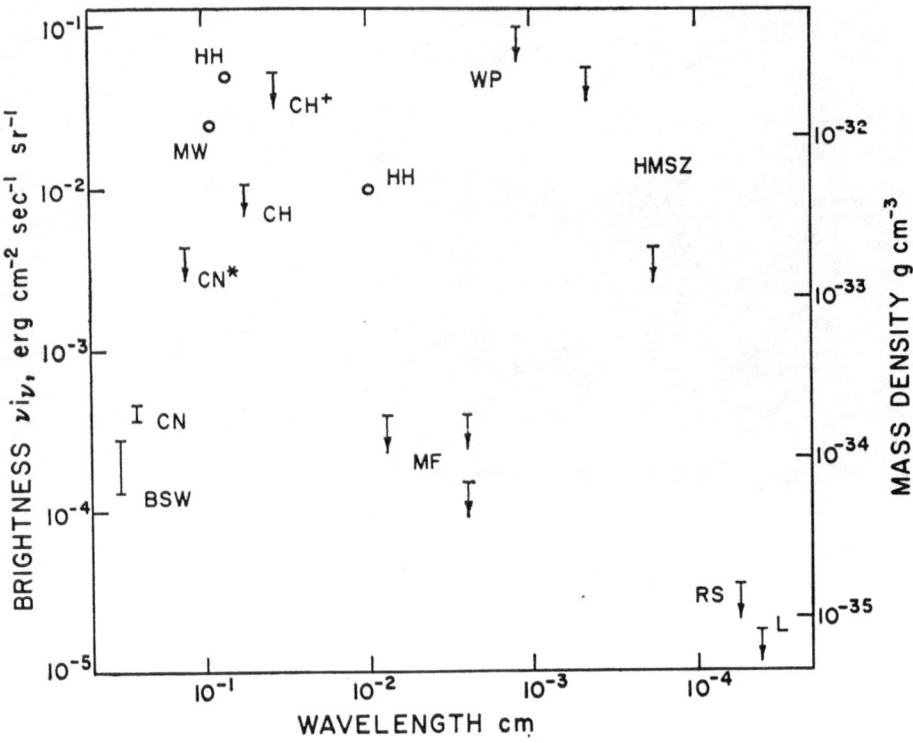

Fig. 3.   Measurements of infrared background. For details see [3].

beam are the filter wheel, the field lens, and the lead sulphide detector. The detector has an area of 4 by 4 mm².

At 35 km – floating altitude the dry ice box will get a temperature of −120 °C. The reason for this temperature drop is the low environment pressure of about 6 mb. That means the thick metallic plate in the centre of the photometer has a homogeneous temperature of nearly −120 °C during the observation time. This temperature is sufficient for sensitive operation of the detector. Moreover it is low enough that the left part of the plate can act as an internal reference for the photometer. In the first half period of a chopping cycle the chopper images this internal reference, emitting a flux nearly 'zero' on the detector. In the second half cycle it images the region of the sky under observation on the detector.

This method requires that all parts of the photometer have sufficiently low temperatures. Especially the temperature of the lens must be lower than −80 °C. This value will be reached since the environment temperature at floating altitude is below −50 °C. The thermal behaviour of the photometer were studied with a dummy in a test chamber simulating the environmental conditions at floating altitude. For a flight duration of 8 h we need 5 kg dry ice.

What will be the sensitivity of the photometer? We calculated it for a detectivity $D*$ of the lead sulphide detector: $D*$ (2.4, 25, 1) $\approx 10^{12}$ cm Hz$^{1/2}$ W^{-1}. The diameter

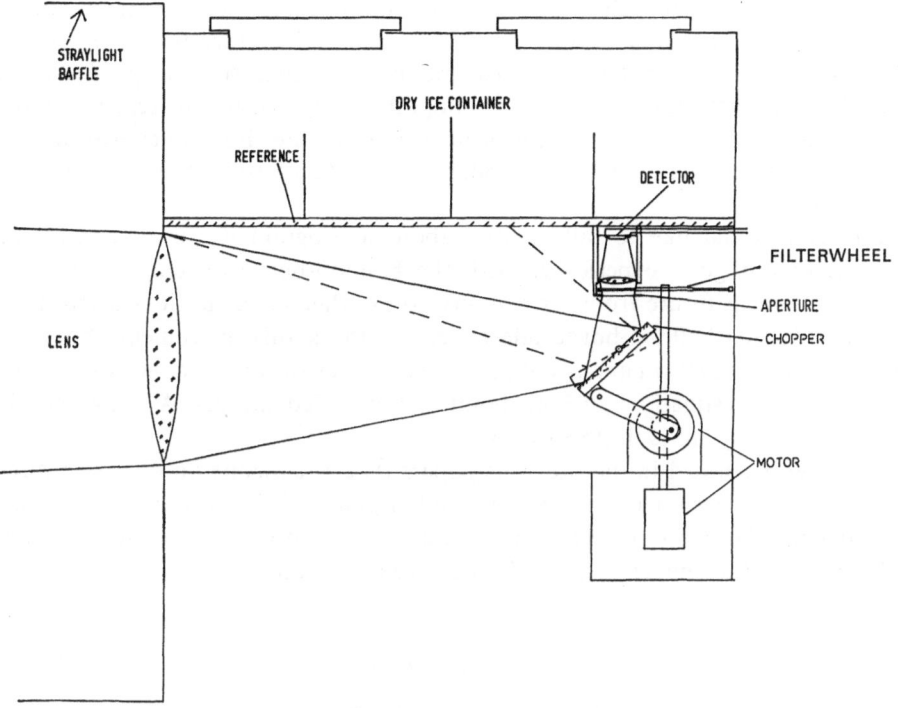

Fig. 4. Sketch of the 2.4 μ photometer.

of the sky field to be observed is 2°. Then we are able to measure at $\lambda = 2.4$ μ a flux of 10^{-11} W cm^{-2} sr^{-1} μ^{-1} with an accuracy of $\pm 20\%$ and an integration time of 400 s. The expected flux of the Milky Way for galactic latitudes smaller than $\pm 10°$ will be in the same order of magnitude. This flux was calculated from the known colour and intensity of the Milky Way in the visible region. It is quite possible that the infrared flux will be much larger because of the reduced interstellar extinction. The infrared measurements will be supplemented by simultaneous measurements in the near ultraviolet. Therefore we hope to derive some information about the galactic structure and the interstellar extinction. We also hope to obtain additional information about the interplanetary dust, especially about the temperature of the grains and the density of the dust.

Infrared flux measurements have a further very interesting aspect: Partridge and Peebles [8] have shown that the spectrum of young galaxies in the early universe – and therefore to observe at large distances – should be highly redshifted. This should result in a detectable infrared background radiation. The best chance to detect this radiation by a surface photometry should be between 5 and 10 μ. At shorter wavelengths the integrated star light of the Milky Way and the zodiacal light are probably brighter than the background radiation. Our 2.4 μ photometry could give better knowledge about this foreground components.

A measurement of the cosmic infrared radiation will be possible only with a satellite experiment. From such measurements one can hope to derive information about cosmology and the early development of galaxies. Therefore we have proposed an infrared surface photometry for an astronomical satellite. But with passive cooling of the telescope, that means radiation cooling, it is only possible to reach a wavelength shorter than 4 or 5 μ on a 500 km orbit of the satellite. For larger wavelengths one needs an active cooling system on board, for instance solid coolants or a thermo-electrical system.

Finally I would like to talk a little about a program we have recently started: because of our good experiences with the balloon-borne telescope Thisbe we will learn in the near future from a feasibility study whether or not it is useful to have a highly stabilized balloon-borne telescope. We think this instrument should have a diameter of about 80 cm and a stabilization accuracy of some seconds of arc. We plan to use it for investigations of both point and extended sources. A bolometer for this far infrared program is in preparation.

Although we heard in the previous papers that a telescope on board an aircraft is a very useful instrument, there are good arguments for a similar balloon-borne instrument. The main advantage is, in our special situation, that we can handle a balloon project by ourselves at the Heidelberg Observatory.

References

[1] Gabsdil, W.: 1971, Thesis, University of Heidelberg.
[2] Hofmann, W., Lemke, D., and Thum, C.: 1971, *Forschungsber. des BMBW*, F B W 71–31.
[3] Peebles, P. J. E.: 1971, *Comments Astrophys. Space Phys.* 3, 20.
[4] Harwit, M., McNutt, D. P., Shivanandan, K., and Zajac, B. J.: 1968, *Infrared Astronomy*; Gordon & Breach, New York, p. 91.
[5] Delbouille, L.: 1969, *Phil. Trans. Roy. Soc. London*, A **264**, 319.
[6] Chamberlain, J. W. and Smith, C. A.: 1959, *J. Geophys. Res.* **64**, 611.
[7] Moroz, W. I.: 1959, *Dokl. Akad. Nauk S.S.S.R.* **126**, 983.
[8] Partridge, R. B. and Peebles, P. J. E.: 1967, *Astrophys. J.* **148**, 377.

DISCUSSION

M. Douma: What is the time scale for completion of the highly stabilized gondola?

D. Lemke: The project will be completed in 3–4 years if it is financed.

W. Hoffmann: For a number of years there have been comments about balloons carrying up some kind of moisture, which is something to which as far as I know no experiment has been specifically devoted to measuring with any detailed accuracy. The experience we had with that is rather favourable. It seems to us that there is no water vapour from the balloon that has caused any problem to the experiments. In particular, what we have done is to scan our telescope to the balloon. Doing this without chopping allows us to measure the gradient of the atmosphere emissivity as we go towards zenith. If, in fact there was a halo of water vapour around the balloon, contributing anything more than the content of water vapour above the balloon itself, it would show up as an aberration in the slope. In our experience there is no such evidence.

J. Ring: Summarizing what has been said so far, it looks as though there is a factor of 14 increase in sky radiation at the aircraft altitude over what we might get at balloon altitude (0.1–0.007 emissivity). The sky background radiation at aircraft altitude is about what one would expect from the uncooled mirrors so that the comparison between aircraft and balloon, as far as signal to noise ratio is concerned, seems to depend very much on (a) how well one chops at long wavelength particularly on balloons and (b) how well one can cool the mirrors, perhaps in the two cases.

W. Hoffmann: Yes, whether the balloon can take advantage of the lower emissivity depends critically on these two questions.

K. Shivanandan: Is Dr Lemke firmly convinced that one needs a spacecraft for a sky survey at 2.4 μ?

D. Lemke: It all depends on the influence of the airglow at these wavelengths. At 3 or 4 μ the thermal radiation from the Earth is too large, so that at these and longer wavelengths one would need a satellite.

W. Hoffmann: But you can certainly differentiate the airglow from the Milky Way just because of the spatial difference in their distribution.

MEASUREMENTS OF THE ATMOSPHERIC BACKGROUND EMISSION BETWEEN 10 μ AND 100 μ

A. F. M. MOORWOOD

Dept. of Physics, University College, London, U.K.

1. Introduction

Measurements of atmospheric emission in the interval 1000 cm^{-1} to 100 cm^{-1} (10μ to 100μ) have been made to a height of 38 km in connection with the infrared astronomy programme at University College London. The instrument used was a Michelson interferometer, flown at night on a balloon from Palestine, Texas during August 1969. Details of the instrument and the methods of data reduction and calibration are given elsewhere [1].

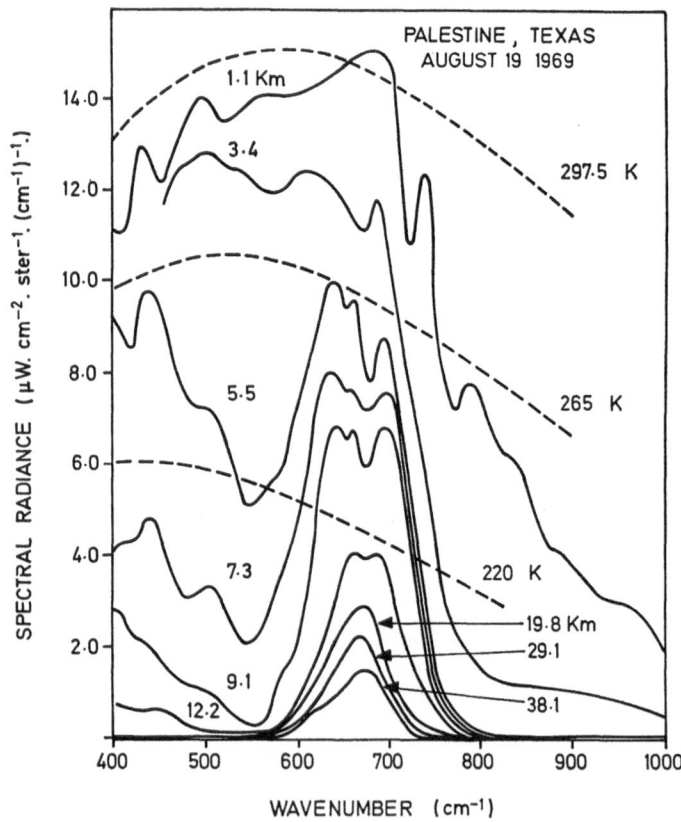

Fig. 1. 400 cm^{-1} to 1000 cm^{-1} portion of atmospheric emission spectra recorded at an elevation of 40°. The black-body curves correspond to altitudes of 1.1 km, 5.5 km and 12.2 km.

Manno and Ring (eds.), Infrared Detection Techniques for Space Research, 60–62. All Rights Reserved

2. Results

Spectra obtained at an elevation of 40° and with a mean spectral resolution of around 34 cm^{-1} are shown in Figures 1 and 2. The accuracy of the absolute calibration is about 5% while an analysis of the float altitude spectra gives a figure $\simeq 0.15$ μ W cm^{-2} ster^{-1} (cm^{-1})$^{-1}$ for the mean noise equivalent radiance through the range. The spectral region between 310 cm^{-1} and 400 cm^{-1} is less reliable due to low instrument transmission and has been omitted.

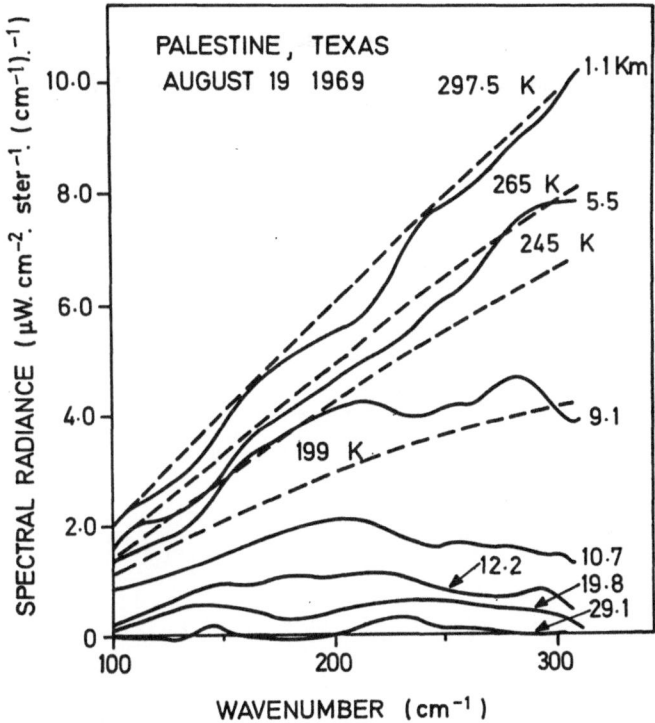

Fig. 2. 100 cm^{-1} to 300 cm^{-1} portion of atmospheric emission spectra recorded at an elevation of 40°. The black-body curves correspond to altitudes of 1.1 km, 5.5 km, 9.1 km, and 15 km.

Figure 1 shows the region from 400 cm^{-1} to 1000 cm^{-1}, characterized by the atmospheric windows around 10 μ (1000 cm^{-1}) and 20 μ (500 cm^{-1}) and the strong emission due to CO_2 centred at 15 μ (667 cm^{-1}). The black-body curves included for comparison correspond to the measured air temperatures at 1.1 km, 5.5 km, and 12.2 km.

The region between 100 cm^{-1} and 300 cm^{-1} shown in Figure 2 is dominated by emission from the pure rotational band of water vapour. Comparison black-body curves are included for altitudes of 1.1 km, 5.5 km, 9.1 km, and 15 km.

At the lower altitudes, effects due to self absorption in the unevacuated path between chopper and detector can be identified in the low wavenumber region and at the centre of the CO_2 band. Of principal interest however is the altitude range above

10 km where aircraft and balloons are being used increasingly as platforms for astronomical measurements in the mid and far infrared. It will be observed that around 12 km, typical of aircraft operation, considerable emission is still present in the 30 μ to 100 μ region and the peak CO_2 emission is still almost black at the local air temperature. Although the problem is less severe at balloon altitudes, some emission due to water vapour is still present above 29 km and the CO_2 emission is still prominent even at 38 km. The residual emission at balloon altitudes may have quite important consequences for measurements which require high spectral resolution.

References

[1] Jennings, R. E. and Moorwood, A. F. M.: 1971, *Appl. Opt.* **10**, 2311.

A BALLOON-BORNE HELIUM-COOLED
INTERFEROMETER FOR INVESTIGATION OF THE
ISOTROPIC SUBMILLIMETRE BACKGROUND

J. E. BECKMAN and E. I. ROBSON

Dept. of Physics, Queen Mary College, London, U.K.

1. Introduction

Isotropic radiation at microwave frequencies was first detected in the classic observation of Penzias and Wilson [1] in 1965. The intensity distribution of this radiation has been followed into the centimetre and higher millimetre wavelength regions by a number of observers [2, 3, 4] confirming that it appeared to comprise the spectrum of thermal equilibrium radiation from a black body at a temperature close to 3 K. Indirect evidence setting upper limits to any such spectral intensity in the low millimetre region and submillimetre range was provided from measurements on interstellar lines by Field and Hitchcock [5], also by Thaddeus and Clauser [6]. However, the observational situation was then complicated by a series of broad-band measurements near the peak of a 3 K spectrum, i.e. close to 1 mm in wavelength, made from rocket altitudes by the Cornell/Naval Observatory group [7, 8] indicating an equivalent black-body temperature in the range 0.7 to 1.0 mm which is close to 8.3 K. These estimates have been confirmed in a balloon-borne broad-band radiometric observation by Mühlner and Weiss [9], giving an estimated temperature of 7 K in a passband between 0.8 mm and 1 mm.

Attempts to achieve higher resolution around the one millimetre wavelength region are severely hampered in any ground-based work by the strong absorption and re-emission of atmospheric water-vapour. It is this which has led to two currently contradictory measurements from high-altitude ground-based sites [10, 11] one of which indicates the presence of a spectral feature at 0.85 mm wavelength, containing apparently sufficient energy to account for the submillimetre excess described above, whilst the other shows no such feature. We have designed an interferometer to perform a spectral scan of the background radiation between 0.3 mm and 3 mm wavelength from a balloon altitude of 35 km, where the water vapour above the observer is typically 10^{-5} that at even high-altitude sites.

2. Optics

The need to provide maximum energy throughput, and a resolution better than 1 cm^{-1} throughout the spectral range led to the following optical specifications for the interferometer, whose layout is illustrated in Figure 1. A polarizing grid, of grating interval 100 μm produces from the incident sky radiation a plane polarized beam, which is chopped mechanically then deflected into the instrument by a plane mirror, and collimated by a 45° off-axis paraboloid. The paraboloid acts as a field stop,

Manno and Ring (eds.), Infrared Detection Techniques for Space Research, 63–72. All Rights Reserved

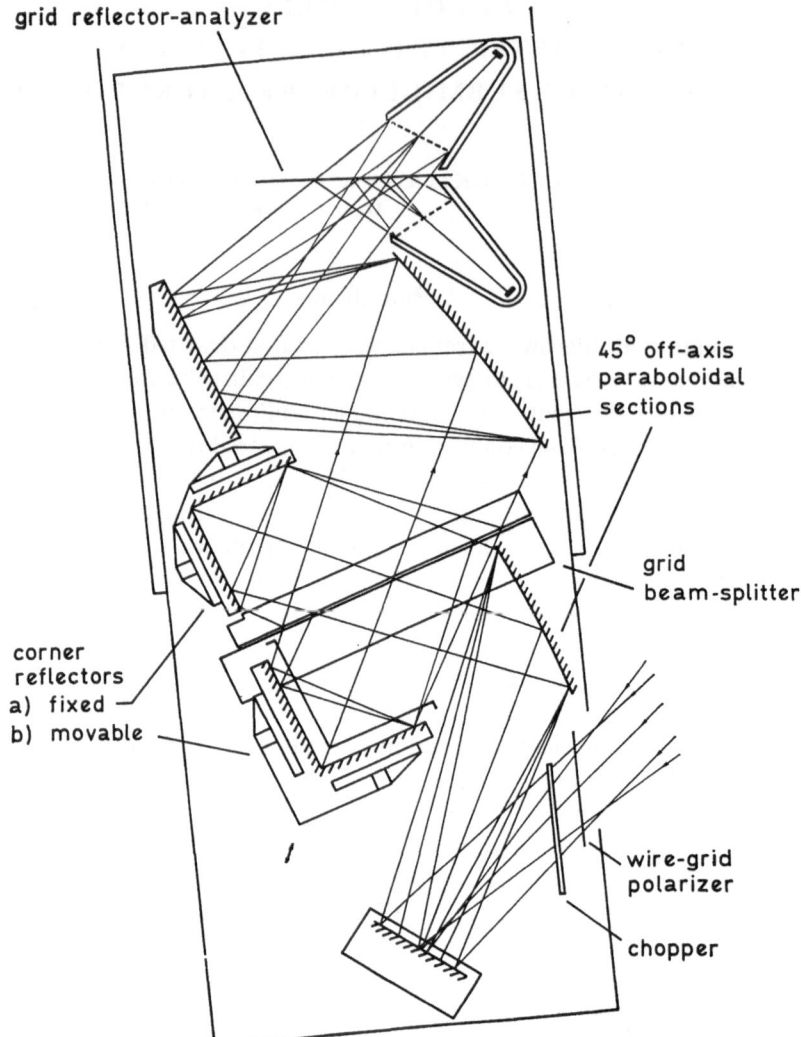

grid reflector-analyzer

45° off-axis
paraboloidal
sections

grid
beam-splitter

corner
reflectors
a) fixed
b) movable

wire-grid
polarizer

chopper

Fig. 1.   Optical layout of 2 K interferometer.

limiting the cone-angle of the beam to 3° unvignetted, which amounts to an average 8° for the total vignetted beam on the sky. The mean diameter of a cylinder of parallel radiation entering the interferometer is 56 mm, giving an area-solid angle product of 0.21 cm² sterad. The beamsplitter comprises a second polarizing grid, with wires making a projected angle of 45° to the incident beam, and with its plane at 45° to the beam. This is shown in Figure 2. Two components polarized perpendicularly are reflected from the 90° 'roof' reflectors, and are transmitted and reflected by the beam-splitter respectively without substantial loss, thus recombining to form a single output beam. The lower of the two roof reflectors is mobile, and scanning is performed by moving it through 1 cm to either side of the zero path difference position.

Output radiation from the interferometer is focused with another 45° mirror, and after deflection of the converging beam by a plane mirror, it is analyzed with a third wire grid producing two interferometric output images, each of which is modulated and contains all the spectral information required. The detectors are each at the apex of a conical optic, which transmits radiation from the 4 cm diameter image plane to the 5 mm diameter detector crystal itself back by a polished hemisphere. The whole optical train is folded inside a blackened copper chamber, cooled to liquid helium temperature by a helium jacket.

Fig. 2.   Wire-grid polarizer in stainless steel stretching and mounting frame.

Advantages of such a system compared with a conventional Michelson interferometer are: (1) The key advantage of using a two-beam system; subtracting the two interferograms leaves a modulated output and eliminates any level fluctuations due to changes in atmospheric transparency, temperature of the system, (which could lead to a variety of electronic and optical level changes) or indeed any non-isotropic radiative input. Of course each interferogram is available separately also. (2) There are no dielectric films in the path of the interfering beams, which could lead to a limitation in the spectral range of the instrument due to hooping [12] in the interferogram. (3) Unpolarized radiation finding its way into the instrument is not modulated and therefore adds only to fluctuations in the zero intensity level of a random nature, and not to the signal level itself; this leads in particular to an equivalent chopper temperature of 0 K, even though the actual chopper temperature is unlikely to be less than 10 K even at balloon altitudes. The chopper, and all the optical components are in any case of polished copper or aluminium for minimum emissivity, and strapped to a jacket

Fig. 3.   Flight interferometer with twin detector cones.

at helium temperature. The optical train, which is shown in Figure 3, is enclosed in an evacuated space within a helium cooled vessel, which at the prevailing pressures at balloon altitude will be pumped to 1.8 K.

3. Operation and Performance of the Interferometer

Interferometric scanning is carried out step-wise with steps in path difference of 80 μm over a path difference up to 2 cm, providing a possible high frequency cut-off at 100 cm^{-1} (and a resolution limit of 0.5 cm^{-1}). These limits are imposed by telemetry timing considerations and the size of the operating chamber respectively. The upper frequency limit will in practice be determined by the grating interval of the polarizer and the response curve of the detector, which gives a half-power cut-off between 40 and 50 cm^{-1}. Given an estimated NEP [13] for our detectors, (InSb crystals operated in the Rollin [14] mode) of 2×10^{-12} W Hz$^{-1/2}$ and a 2.7 K background spectrum throughout the range, one can estimate a signal to noise ratio of 30 over the whole pass-band in the 16 min taken to obtain a two-winged interferogram. This is equivalent to a ratio of 3:1 per data channel over continuous observations during a three-hour flight, the minimum set by the worst case of helium boil-off, and over 5:1 for the estimated flight time of 10 h. If the energy falling on the instrument is that corresponding to 8 K, the above figures are low by a factor of 50. An advantage of the interferometric method is that it is always possible to trade off resolution for improved signal-to-noise ratio during the data reduction; a rapid estimate of the effective temperature of the radiation may be made by examining the peak of the modulated signal; an arbitrary resolution interval may then be chosen, down to 0.5 cm^{-1} by including the appropriate portion of the interferogram, in order to look for spectral features.

4. Radiative Input

In designing the package, it is of course important to have an estimate of the contributions to radiation from unwanted sources which tend to reduce the precision of the background measurement. To separate the vacuum space containing the interferometer and detectors from the ambient atmosphere a thin mylar window, thickness $12.5 \ \mu\text{m}$ is employed, which is exposed to the atmosphere only at balloon float height by removing a metal cover-plate. The average optical thickness of this window over the beam acceptance cone is $\tau_w = 10^{-3}$ giving a net temperature contribution of less than 0.3 K with known spectral response. The only other catadioptric element is a black absorption filter covering the cone-mouth, and hence at helium temperature. This cuts off below $50 \ \mu\text{m}$ wavelength, preventing any detection of atmospheric emission in the 5–$7 \ \mu\text{m}$ region by intrinsic electronic excitation of the InSb; between $200 \ \mu\text{m}$ and 2 mm its measured absorptivity is $\tau_f = 0.01$, giving a contribution of 2×10^{-2} K. As described above, the polarizing interferometer cannot modulate any contributions from the vessel, the polished optics, or the chopper blade, but it will certainly detect any residual atmospheric water vapour, and any scattered light from the balloon or terrestrial emission which is travelling within the acceptance cone. Using the paper by Gay [15] in which the residual water vapour above 28.4 km was determined, we can estimate that at a minimum flight altitude of 30 km there will be no more than $0.01 \ \mu\text{m}$ of precipitable water vapour above the instrument; allowing for the secant of the $45°$ zenith angle of observation an estimated 0.5 K equivalent mean emission temperature is added to the beam. This of course is concentrated in the rotational emission lines of the H_2O molecule, which are, however thoroughly observed spectroscopically both in the laboratory and atmospherically. Our resolution should be enough to distinguish particularly as large a feature as is needed to explain the observed submillimetre excesses. Finally, the rayleigh scattered input from Earth and balloon has as a generous upper limit a contribution of 0.1 K, whilst that from the atmosphere itself no more than 0.01 K can be expected during a night flight. The total for all these sources gives a value at 30 km of less than 1 K, and we expect to fly the instrument always above 32 km during observations.

5. Thermal Control and Calibration

The cryostat containing the interferometer chamber is shown in Figure 4. Heat can leak into the vessel, reducing the helium hold-time, by several routes. The significant ones are: radiation through the sky-aperture and from the nitrogen shield, conduction through electrical leads, down the neck of the helium container, and along the shaft of the interferometer stepping motor, and dissipation both electromagnetic and frictional in the chopper assembly, shown in Figure 5. Estimated magnitudes of these quantities are: radiative, through the window 0.3 W, from the shield 0.2 W; conductive, down the electrical leads 0.2 W, down the neck 50 mW and along the drive motor

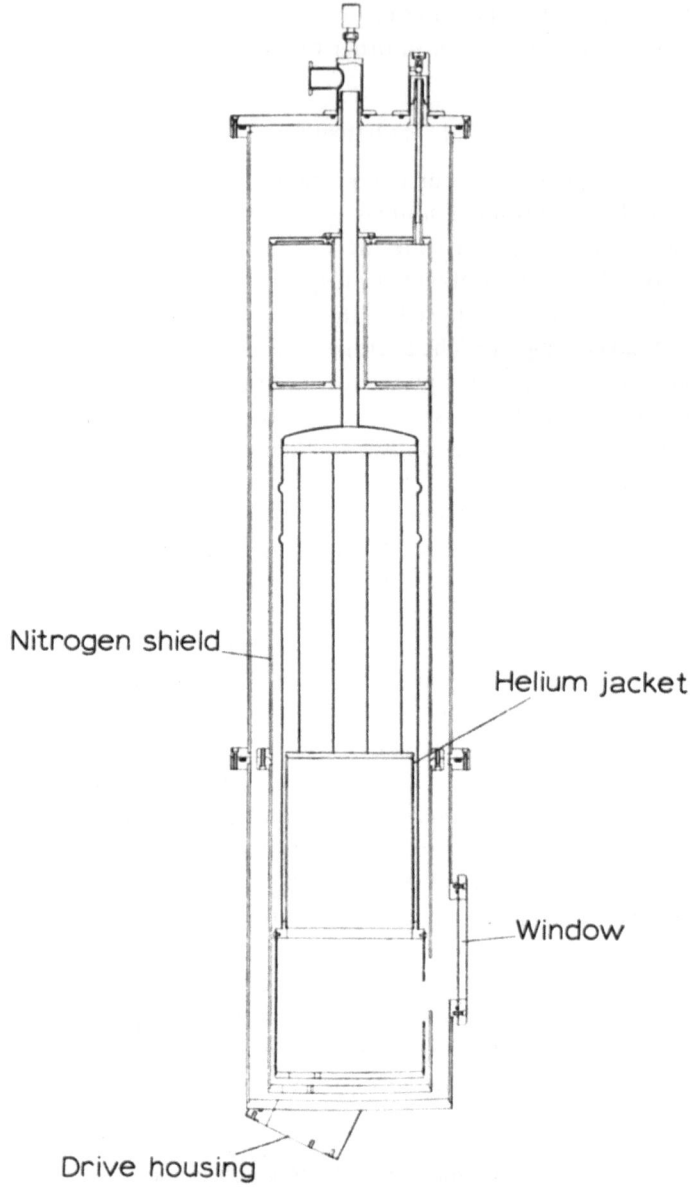

Fig. 4.   Cryostat showing interferometer chamber, blow-off window, and motor housing.

shaft 0.1 W; dissipative, in the chopper bearing 0.1 W. The total figure for these upper estimates is very close to 1 W which leads to an estimated hold time for the 17 litre vessel of 12 h. This is a realistic figure for in-flight purposes, for although the estimates for each inflow of heat are high, the boil-off rate on ascent of the balloon will reduce the maximum observing time to around nine or ten hours. Laboratory tests have

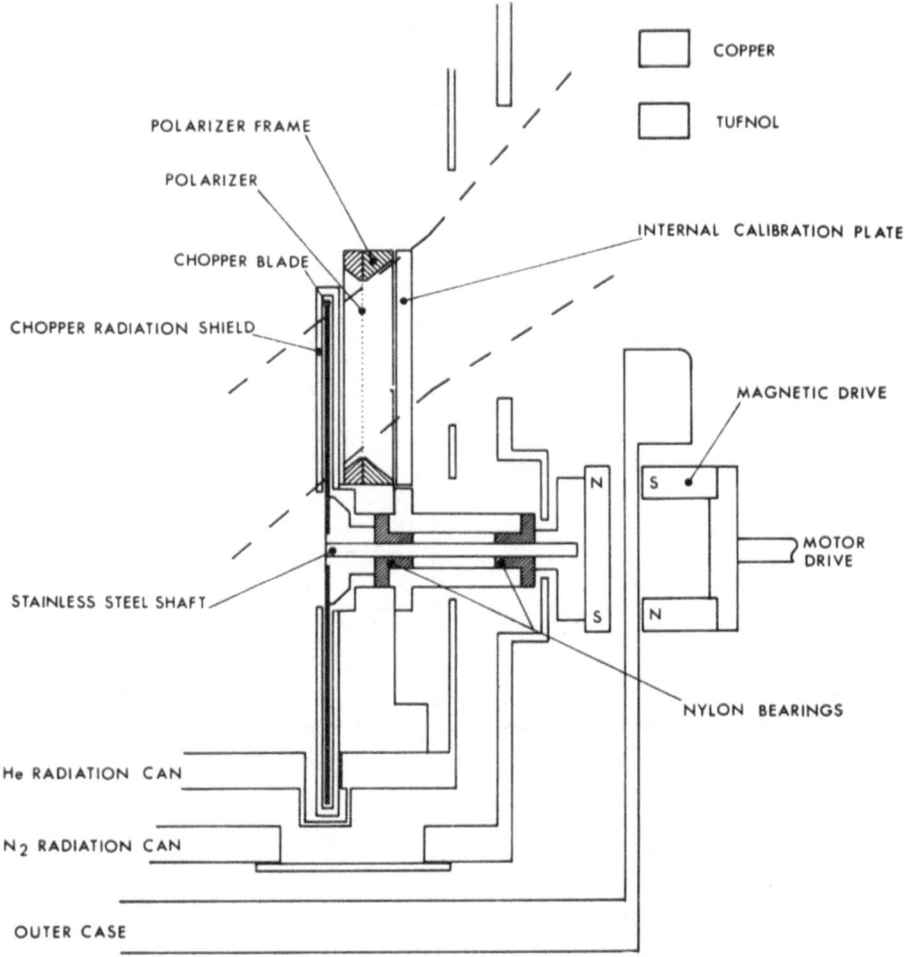

Fig. 5.   Calibration, polarizer, and chopper assembly.

borne out the hold time estimates; no limitation is imposed by the 10 litre nitrogen vessel, for which an estimated boil-off rate of 4.5 h per litre has also been borne out by experiment.

Temperature drifts in flight, in particular of the detector crystals are monitored with a series of resistance thermometers. In-flight calibration is however provided by a copper plate shown in Figure 5 just in front of the first polarizing grid. A check on the performance of the interferometer is provided by closing the two jaws of the plate, which then provides a source at a temperature close to 10 K. An interferogram of this source can be monitored once for every six or eight sky spectra. The energy distribution of the calibrator is itself compared in the laboratory with a 'black body' which has been tested with the apparatus shown in Figure 6. We use a blackened cone of semi-angle 30° maintained at constant temperature either by a water bath, or using a nitrogen or helium dewar in its place. The dewar clips onto the front window of the

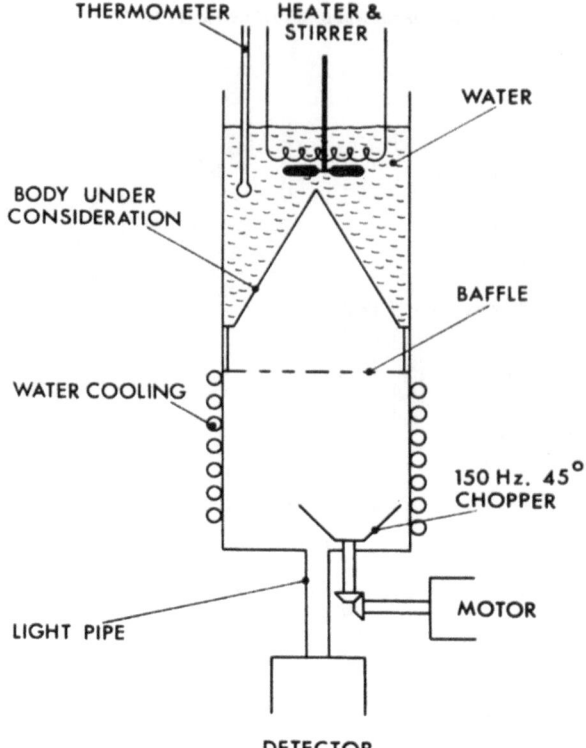

Fig. 6. Experiment to measure radiation from test black bodies.

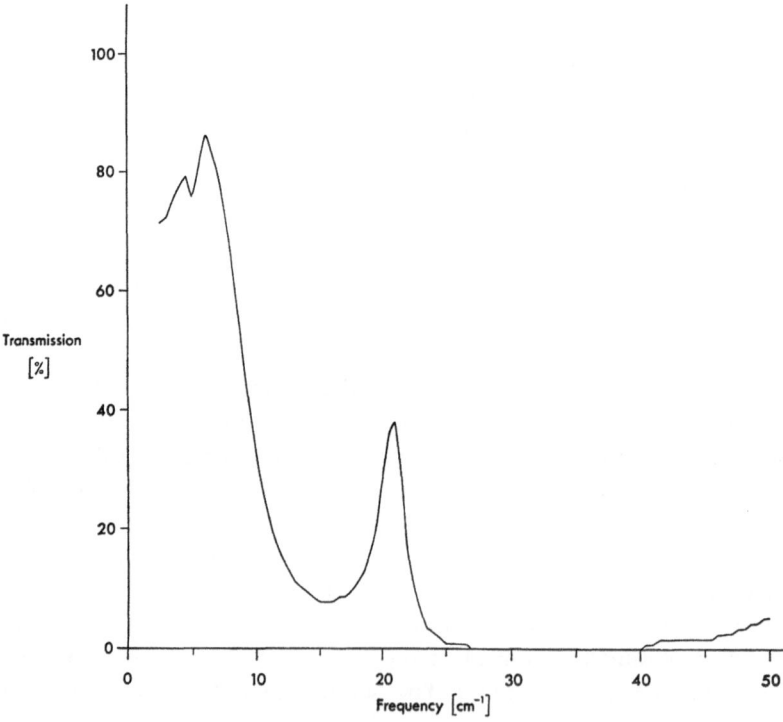

Fig. 7. Spectrum of metal mesh filter, made with polarizing interferometer.

balloon cryostat so that the calibration body fills the input beam. Spectra at each temperature are then compared with that from the in-flight calibrator in a simulated flight situation. Finally we show the spectrum of a metal mesh transmission filter (Figure 7) taken in the laboratory with a polarizing interferometer due to Martin and Puplett [16] showing the effectiveness of the instrument and InSb detector combination in producing high-resolution in the region of the spectrum between 3 mm and 0.3 mm, (3 cm^{-1} to 30 cm^{-1}), for which our instrument is to be used.

Acknowledgements

We should like to thank Prof. D. H. Martin and Mr E. Puplett for their extensive contributions to the design of polarizing interferometers, of which we have made free use throughout. Mr D. G. Vickers was in large part responsible for the execution of the design programme; Mr J. Weston and Mr K. Nickels gave us their own services and those of their workshop colleagues at Queen Mary College, whose craftsmanship produced a beautiful and functional instrument. Many other technical staff have contributed to our work. The instrument was built with funding from the Science Research Council, which we acknowledge with gratitude.

References

[1] Penzias, A. A. and Wilson, R. W.: 1965, *Astrophys. J.* **142**, 419.
[2] Roll, P. G. and Wilkinson, D. T.: 1966, *Phys. Rev. Letters* **16**, 405.
[3] Stokes, R. A., Partridge, R. B., and Wilkinson, D. T.: 1967, *Phys. Rev. Letters* **19**, 1199.
[4] Ewing, M. S., Burke, B. F., and Staelin, D. H.: 1967, *Phys. Rev. Letters* **19**, 1251.
[5] Field, G. B. and Hitchcock, J. L.: 1966, *Astrophys. J.* **164**, 1.
[6] Thaddeus, P. and Clauser, J. F.: 1966, *Phys. Rev. Letters* **16**, 819.
[7] Shivanandan, K., Houck, J. R., and Harwit, M.: 1968, *Phys. Rev. Letters* **21**, 1460.
[8] Houck, J. R. and Harwit, M.: 1969, *Astrophys. J. Letters* **157**, L45.
[9] Mühlner, D. and Weiss, R.: 1970, *Phys. Rev. Letters* **24**, 742.
[10] Beery, J. G., Martin, T. Z., Nolt, I. G., and Wood, C. W.: 1971, *Nature Phys. Sci.* **230**, 36.
[11] Richards, P. L. and Werner, M. W.: 1971, *Proceedings of XVII Astrophysical Symposium*, Liège, to be published.
[12] Richards, P. L.: 1964, *J. Opt. Soc. Am.* **54**, 1474.
[13] Clegg, P. E. and Huizinga, J. S.: 1971, this volume, p. 132.
[14] Kinch, M. A. and Rollin, B. V.: 1963, *Brit. J. Appl. Phys.* **14**, 672.
[15] Gay, J.: 1970, *Astron. Astrophys.* **6**, 327.
[16] Martin, D. H. and Puplett, E.: 1969, *Infra-Red Phys.* **10**, 105.

DISCUSSION

F. Melchiorri: In your interferometer there is a fixed polarizer in front of the instrument. Let me suppose that the infrared radiation is polarized: this fact may occur, for example, in the emission of interstellar grains. In these conditions you cannot be sure of measuring the exact value of the incoming radiation. Is it possible to rotate your polarizer?

J. E. Beckman: No, we cannot rotate at present. Your point is valid, but there is nothing we can do about it now, though in fact we can make several flights with different orientations of the polarizer.

F. Kneubühl: Why did you use stainless steel in this grid polarizer? About 6 years ago we made similar wire polarizers with distances between the wires of about 15 μ. We checked them with a submillimetre laser and found that if you use copper wire instead of stainless steel or nickel one gets much better efficiency with regard to residual absorption by the wires.

J. E. Beckman: We don't think that in this system this is a very major problem. We are not throwing away by this as much energy as by the diffraction effect.

F. Kneubühl: How do you cope with the residual emission of the upper atmosphere?

J. E. Beckman: We will measure its spectrum and, if it turns out to be a sharp line, we will have a better chance of distinguishing this than if we did not have such a good resolution. Apart from that we are not able to do anything about measuring whether or not it is isotropic which may throw some light on the actual size of the emission.

INFRARED ROCKET ASTRONOMY

KANDIAH SHIVANANDAN

E. O. Hulburt Center for Space Research, Naval Research Laboratory,
Washington, D.C., U.S.A.

1. Introduction

The extension of astronomy in the wavelength region of 1 μ to 1 mm has in the past been limited by lack of detectors of high sensitivity and by the obscuration of the Earth's atmosphere. Ground-based near infrared astronomy could be carried out within certain partial windows: i.e. from 2.1 μ to 2.4 μ, 3.5 μ to 4.0 μ, 4.6 μ to 4.7 μ, 8–13 μ, and 20–25 μ. The region from 25 μ to 2 mm is completely opaque and measurements have to be carried out from above the atmosphere using aircraft, balloons, or rocket experiments.

At maximum balloon altitudes (30 km) one expects to find about 5 μ or precipitable water. This leads to high opacity, hence, high emissivity in many parts of the far infrared spectrum. Balloon telescopes do not permit one to obtain absolute background measurement because these telescopes require a 'hot' window to prevent the residual H_2O from condensing on the cooled detectors. The presence of the balloon itself also makes it difficult to minimize stray thermal radiation off within the telescope. Hence, telescopes that are completely cryogenically cooled without any window at the telescope entrance and which could be pointed in a direction well away from the horizon (i.e. carry out observations at small zenith angles) are necessary to do observations in the far infrared, at altitudes above 100 km using sounding rockets.

2. Infrared Rocket Telescopes

The first infrared rocket experiment [1] was a liquid nitrogen cooled telescope for observations of the diffused background in the wavelength range of 1 μ to 7.5 μ and was flown in an Aerobee rocket. To extend observations into the far infrared, liquid helium cooled telescopes have been designed and flown by various experimenters [2]. In this paper, some of the design and performance characteristics of a liquid helium cooled rocket-borne telescope built at Naval Research Laboratory to measure infrared background radiation in the wavelength region of 10–1000 μ will be discussed.

3. Instrumentation

The telescope, as indicated in Figure 1, consists of a cassegrain optical system inside a cryogenic dewar of 15 in. diameter, the cryogenic system consisting of two storage systems for liquid helium, one for the telescope and one for the cover, the latter is ejected before observation.

The optical system consists of a paraboloidal primary mirror and a secondary

Manno and Ring (eds.), Infrared Detection Techniques for Space Research, 73–76. All Rights Reserved
Copyright © 1972 by D. Reidel Publishing Company, Dordrecht-Holland

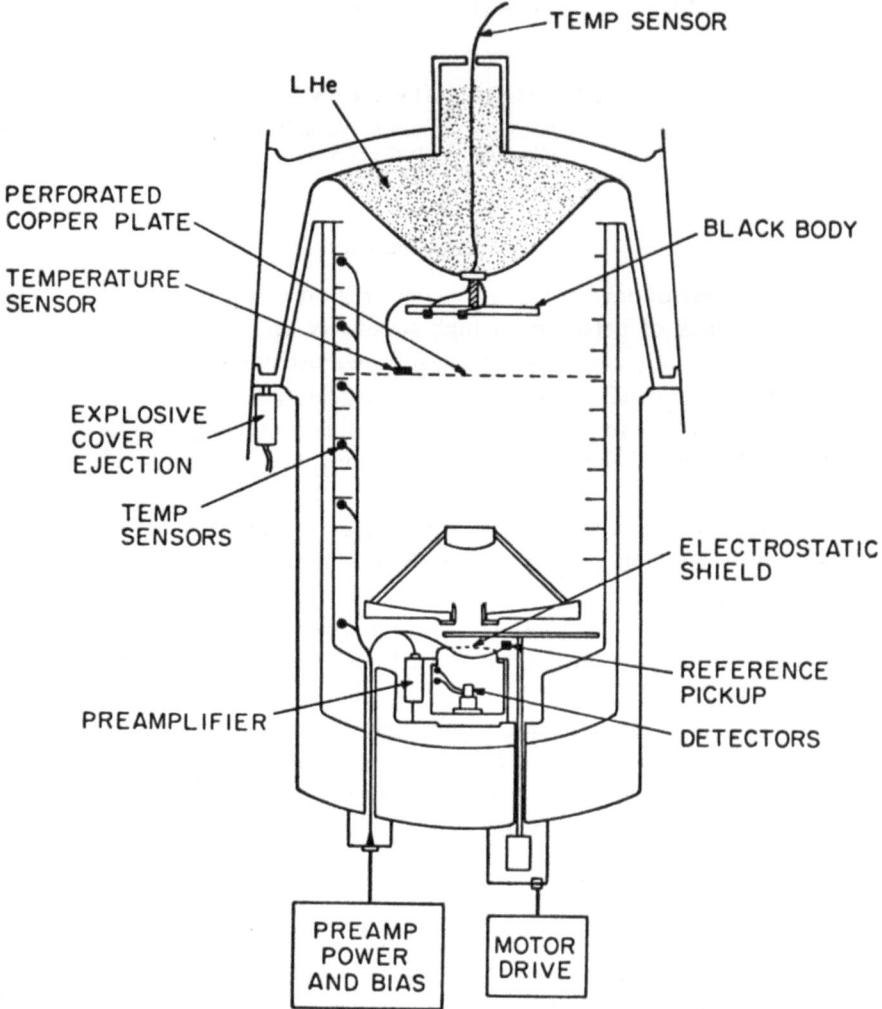

Fig. 1. NRL infrared telescope.

hyperboloidal mirror defining a three degree circular field of view. The detectors are located at the focus of an ellipsoidal condensing mirror placed suitably with the other focus defined by a field stop in front of the primary mirror. A rotating mechanical chopper was used to modulate the incoming light at a rate of 267 Hz.

Five detectors were used for infrared detection, the spectral ranges are shown in Figure 2. To eliminate intrinsic absorption bands of germanium at 1.3 μ and indium antimonide at 5 μ, two sheets of black polyethylene were used as filters between the chopper and the detectors, which attenuated these wavelengths by a factor of 100.

A wire mesh was used to cut off long wavelength radiation at 1.3 mm. A series of 1/10 W 100 Ω carbon resistors were used as temperature sensors along the profile of

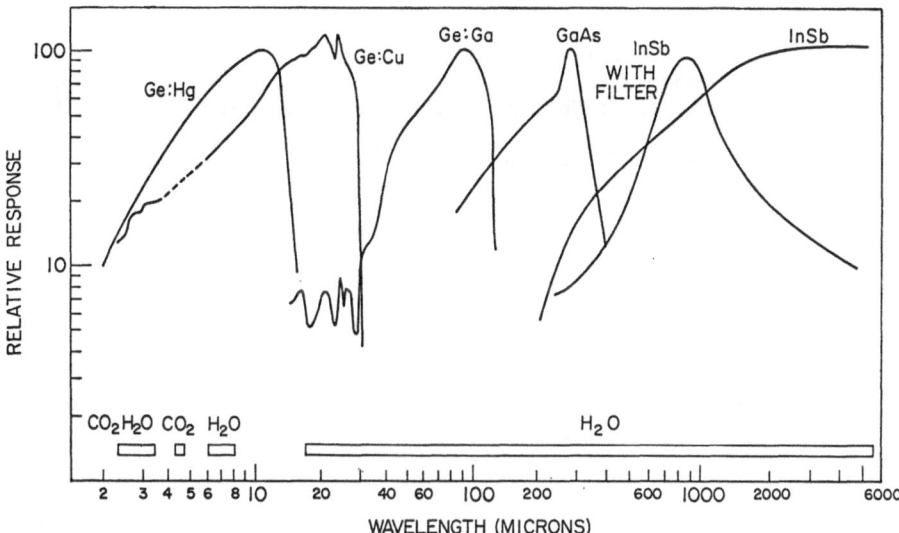

Fig. 2. Spectral response curves of infrared detectors.

the telescope and preamplifiers operating at 4.2 K were mounted on a copper block which housed the ellipsoid and the detectors.

The signal from the detectors have post amplification using tuned amplifiers to the chopping rate and pulse code modulation telemetry using an effective bandwidth of 30 Hz is utilized in the output of the detector channels. An energy absorbing damper which oscillates with the rocket motion removes energy from the system without changing the angular momentum and causes the rocket to open up in a precession cone, providing a spiral scan for observation. A gyroscopic aspect sensor, a magnetometer and an optical aspect sensor provided data for the identification of stellar objects.

The calibration of the detectors is carried out using a blackbody attached to the bottom of the helium reservoir of the flight cover. Temperature of the blackbody was monitored using carbon sensor and a qualitative response of the detectors were obtained with reference to their spectral range using various blackbody temperatures.

Successful flights have been accomplished using Aerobee and Black Brant V-B rockets and the scientific payload was recovered using water recovering systems. Preliminary astronomical results have been obtained [3] and future sky scans are being planned using spatial and spectral techniques in the far infrared.

Acknowledgement

A great many people have assisted in the successful design and operation of the infrared telescope. The system is a brainchild of Dr Douglas P. McNutt, and Dr Paul Feldman developed the detector technology in the near infrared. This work was performed under the auspices of the Office of Naval Research.

References

[1] Harwit, M., McNutt, D. P., Shivanandan, K., and Zajac, B. J.: 1966, *Appl. Opt.* **5**, 1732.
[2] McNutt, D. P., Shivanandan, K., and Feldman, P. D.: 1969, *Appl. Opt.* **8**, 2199; Harwit, M., Houck, J., and Fuhrmann, K.: 1969, *Appl. Opt.* **8**, 473; Blair, A. J. *et al.*: *Appl. Opt.* **10** (1971), 1043.
[3] Feldman, P. D., McNutt, D. P., and Shivanandan, K.: 1968, *Astrophys. J.* **154**, 131; Feldman, P. D., and McNutt, D. P.: 1969, *J. Geophys. Res.* **74**, 4791.

AN INTERFERENCE FILTER RADIOMETER
WITH COOLED OPTICS AND A COOLED PbS
DETECTOR FOR ROCKET APPLICATION

W. BANGERT, E. KRIEG, and R. SCHEIDLE

Meteorologisches Institut der Universität München, München, G.F.R.

Abstract. A filter radiometer for emission measurements at 2.45 μm and 2.80 μm from aboard a sounding rocket has been developed. Details of construction concerning the cooling of the optics and the detector during flight are given. Some experiences with the instrumentation in the laboratory and the behaviour of some critical components are discussed.

1. Basic Problem of the Experiment

The instrument which shall be described here has been developed for measurement of the OH night glow at 2.8 μm from aboard an Aerobee rocket. Because of the low intensity of the incoming radiation there would be considerable disturbances of the detected signal by the thermal emission of the optical components if these would have ambient temperature. The emission of the optical components at 295 K and with an emissivity of 5% would be 3.5 times stronger than the maximum expected OH radiation of 2×10^{-8} W cm^{-2} sr^{-1} μm^{-1}. Therefore the whole instrument is cooled to $-80\,^\circ$C which is considered on one side to be a temperature low enough to suit the problem and on the other side to be still practicable with not too much technical effort.

2. General Description and Constructional Details of the Radiometer

In the following a brief description shall be given of the principles of the radiometer and some of its constructional details. Figure 1 shows the optical layout of the instrument. The entrance optics is a focusing Cassegrainian type system consisting of two spherical mirrors (OS1, OS2). Since only the emission of an extended source is to be measured there is no need for an accurate imaging optics. A chopper (OC) with two mirrored 90° sector blades reflects the incoming radiation under an angle of 90° to a field stop (OB) in the focal plane of the Cassegrainian system. The chopper is driven by a synchronous motor giving a chopping frequency of 112 Hz. The radiation then passes through a rotating filter (OF) which was specially fabricated for this experiment and consists of two 180° sectors one carrying the 2.80 μm interference filter for the measuring channel and the other the 2.45 μm filter for the reference channel. It is rotated by a second synchronous motor by means of a worm gear at a speed of 1 Hz. Then an image reducing optics consisting of a plane mirror (OS3) and a spherical one (OS4) focuses the radiation on the sensitive area of the detector (OD) a lead sulfide cell in a dewar which is cooled by means of a miniature cryostat operating on Freon 23 gas. When the chopper blades are not in the optical path then the detector is looking

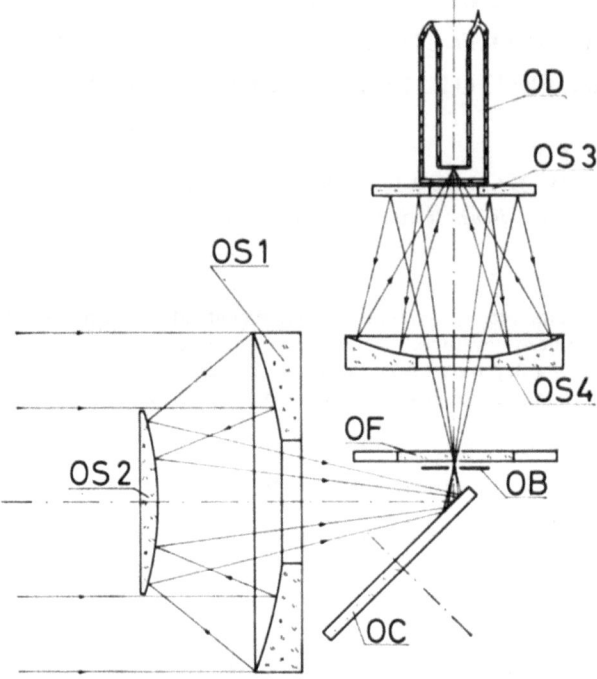

Fig. 1.   Optical layout of the radiometer.

at the black body, a small V-groove plate also cooled to −80 °C. The basic instrumental data are shown in Table I.

TABLE I
Instrumental Data

Diameter of primary mirror	100 mm
Focal length of Cassegrainian optics	210 mm
Diameter of field stop	3 mm
Reducing factor of condenser optics	2.2
Field of view	1°
Peak wavelength of filter 1	2.80 μm
Peak wavelength of filter 2	2.45 μm
Half bandwidth of filters	0.12 μm
Sensitive area of PbS cell	1.4×1.4 mm^2
Chopping frequency	112 Hz
Bandwidth of electronics	1 Hz

The whole instrument consists of five easy dismountable subunits (Figure 2), namely the entrance optics housing, the chopper block, the filter box, the condenser optics housing, and the detector housing. Figure 3 shows a front view of the radiometer i.e. the entrance optics housing. The latter is a hollow double wall container with a cylindrical cross-section inside and a rectangular one outside, thus leaving in-between space for receiving the liquid coolant. These spaces have connection to further

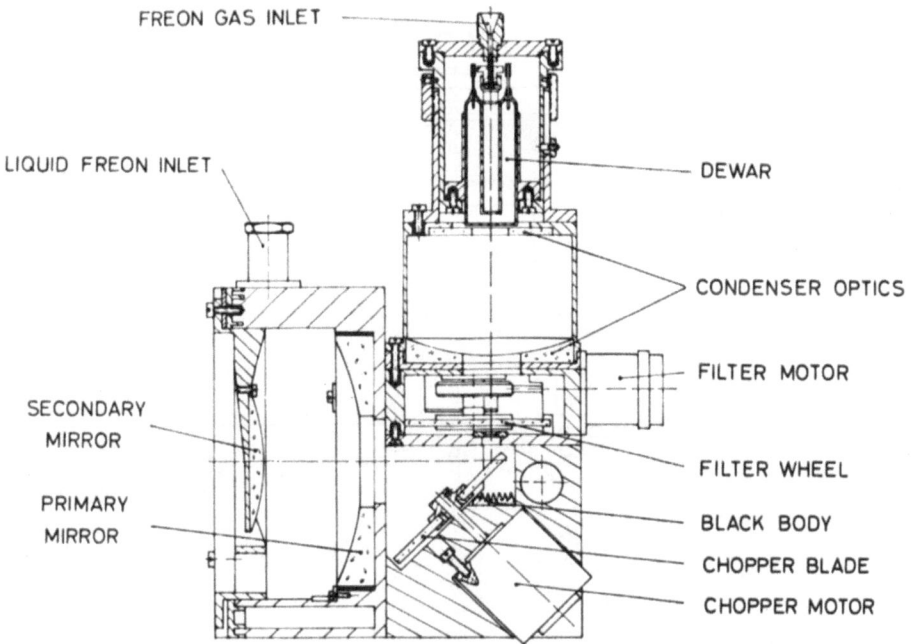

Fig. 2. Cross-section of the radiometer.

holes in the chopper block in such a way that a closed coolant cycle through the instrument is achieved. As coolant liquified Freon 23 is used which is filled in through the right pipe on the top of the housing, flows through the right half of the entrance optics housing, then through the chopper block, and through the left half of the optics housing and leaves through the left pipe. This procedure is used for precooling prior to launch. During flight the evaporating Freon gas has to pass a constant pressure relief valve which keeps the vapor pressure inside the radiometer at 760 torr $\pm 5\%$. Thus a constant boiling point temperature of (-82 ± 1) °C is maintained as long as there is liquid Freon inside. The storage capacity is appr. 0.5 kg which gives a cooling time of about 1 h.

The secondary mirror suspension is milled out of one piece and is basically a circular flange with three thin arms which hold the plate carrying the mirror. The arms have such a shape that the optical intersection is as small as possible but still have enough mechanical strength and thermal conductivity. The flange is screwed to the front wall at three points, thus enabling adjustment of the Cassegrainian optics by putting washers of different thickness between flange and housing. The washers are made of copper to give appropriate heat transfer from the mirror suspension to the housing. Wherever possible heat conduction grease is used.

The primary mirror (Figure 2) is simply clamped to the rear wall of the housing, thus giving good thermal contact. For protection against breaking due to inequal thermal expansion of the glass and the metal of the housing the mirror is centred in it holding by an elastic corrugated metal foil.

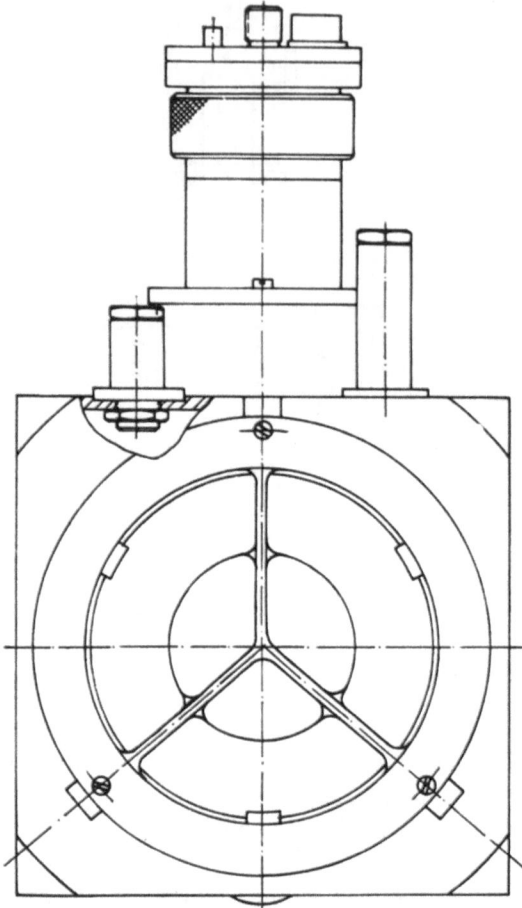

Fig. 3.   Front view of the radiometer.

The second subunit is the chopper block, containing the black body, the assembly for generating the reference signal for the synchronous rectifier, and the chopper. As mentioned above, the chopper block is cooled directly by the liquid Freon. Thus the enclosure in which the V-groove plate made of anodized aluminium is mounted is at −82 °C constantly and yields as viewed by the detector through the fieldstop, very good black-body characteristics. The temperature of this black body as well as that of the two mirrors of the Cassegrainian optics is monitored by thermistors. A critical point is the temperature of the chopper since this cannot be monitored in operation. Its only means of cooling is by radiation where against heat transfer occurs along the motor shaft. But this is negligible since most of the heat produced by the motor (appr. 2.5 W) is directly transferred to the chopper block, i.e. the coolant. In addition the motor shaft is made of stainless steel so that thermal conductivity is as small as possible.

The next unit is the filter box which is like the following units not cooled directly but only by heat conduction. Since they are optically behind the chopper their thermal emission cannot affect the electrical AC-signal.

The last subunit is the detector housing shown in Figure 4. It is basically a closed cylindrical chamber (6) which can be moved up and down in its mounting for adjustment. The dewar (19) is glued into a small tube (11) by a shock absorbing silicone resin. The high pressure Freon 23 gas enters through the inlet (2) is then liquified inside the dewar and evaporates into the chamber. Since the chamber is hermetically sealed by PTFE O-rings (5, 16) the inside pressure can be controlled by the same constant pressure relief valve which has been mentioned earlier. Thus a constant

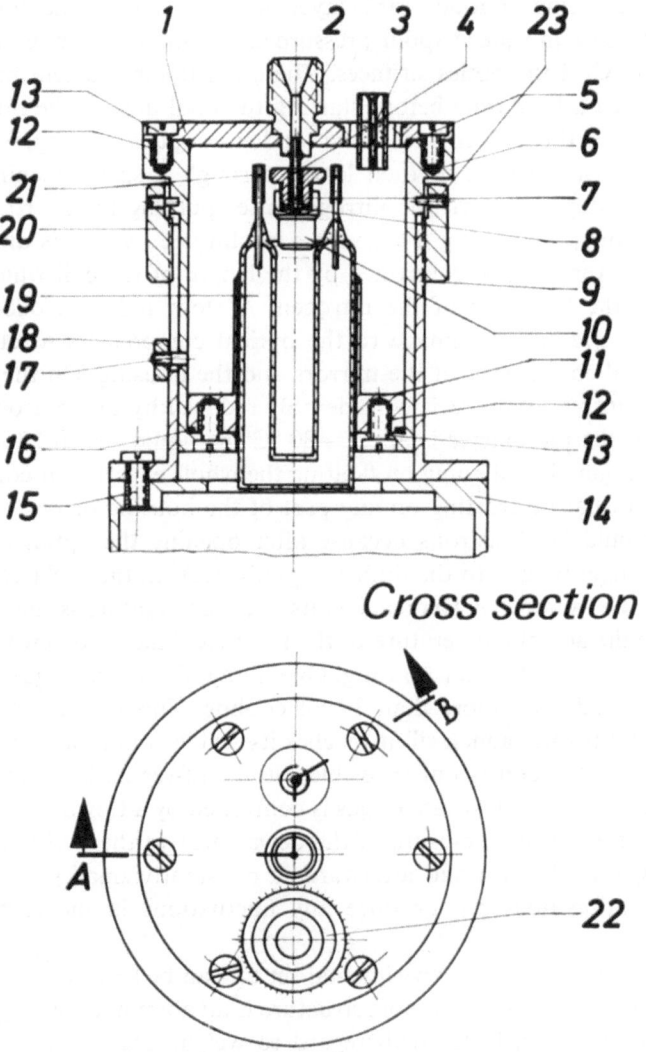

Fig. 4. Cross-section of the detector housing.

boiling point of the liquid can be maintained so that no changes in the detector responsivity during flight are expected.

The assembled radiometer is entirely insulated by polyurethane foam of 20 mm thickness and is enclosed in a cubic housing of $178 \times 178 \times 178$ mm. The opening for the optics is covered by a door which is opened after nose cone ejection.

3. Handling of the Instrument

A major problem in handling the radiometer is the protection of the optics against frosting. Calibration for example has to be done in the vacuum. But even at pressures below 10^{-2} torr the water vapour content in the vacuum tank is high enough to get under special conditions a good visible layer of ice crystals on the mirrors due to the enormous difference in water vapour pressure at the ambient temperature of the tank walls and -80 °C of the optics surfaces. Therefore the instrument has to be purged with dry nitrogen a long time before starting the cool down. The nitrogen is blown into the radiometer through an inlet in the condenser optics housing, flows through the interior of the instrument, and then leaves through a small opening in the door. During the actual count-down the nitrogen after purging for a proper time will be cooled down slowly by passing it through a cooling coil which is dipped into liquid air. The temperature can be controlled by the length of the coil which dips into the liquid air and the flow rate of the nitrogen. A slow and controlled cool down is necessary since else severe damage to the optical components would occur due to different thermal contraction of the mirrors and the housing. On the other hand the loss of the expensive Freon 23 is considerably reduced by this precooling. When the temperature of the radiometer is appr. -80 °C the liquid Freon 23 is filled in while the nitrogen purging is still going on floating the whole experiment compartment with dry nitrogen so that no frosting on any part of the housing or on the couplings can occur. This would be disastrous because after opening the optics cover, the water vapour would migrate due to the diffusion gradient from these places to the mirrors.

Only when the radiometer has reached its final temperature is the detector cooling initiated since the actual temperature of the sensitive flake is somewhat dependent on the temperature of the detector housing. One filling of the Freon gas storage bottle is 90 g at 50 atm and gives more than 3 h of cooling. This is long enough to run the detector cooling before launch till it reaches its stable condition. The cooling period of 3 h from 90 g of Freon is only possible with a self-regulating cryostat. The nozzle of the latter and thus the flow rate of gas is controlled by a tiny needle valve according to the temperature in the enclosure of the dewar near to the PbS flake. Thus a comparatively quick cool down and afterwards a constantly small gas flow is achieved. The self regulation does not produce any fluctuations in the responsivity of the detector.

On Figure 5 the integrated experimental set up can be seen which is to be flown in an Aerobee rocket. The basic support structure is an aluminium honeycomb sandwich plate which proved excellently with regard to weight and vibration characteristics. The radiometer is mounted in the center with the door being closed. On the top to the

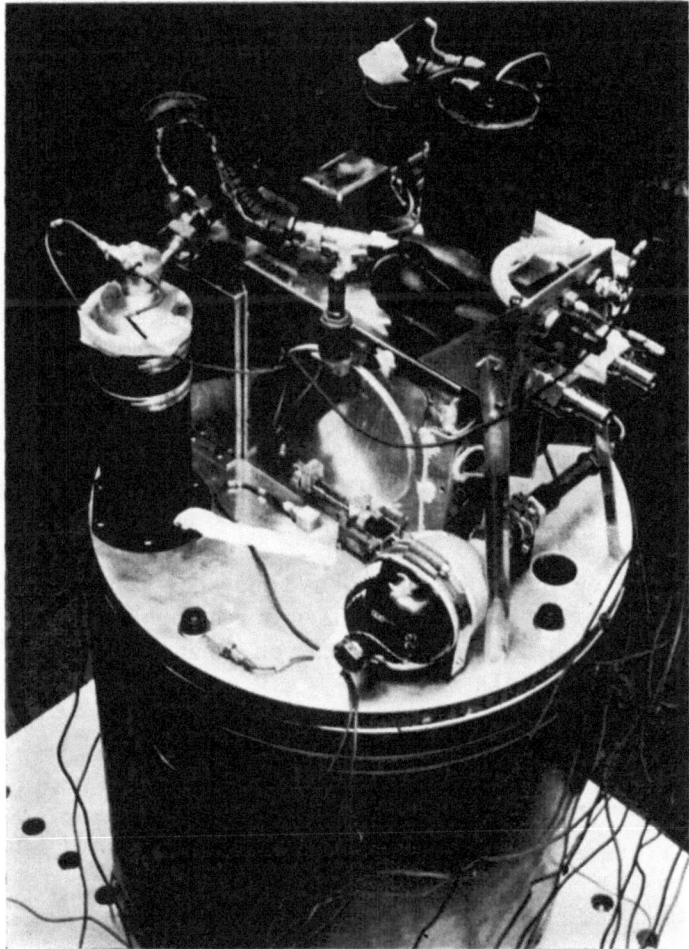

Fig. 5.   View of the integrated experiment.

right are the connectors for the supply lines which are quick disconnect shut-off couplings which can be snatched off rapidly shortly before launch. On the right side is the storage bottle for the high pressure Freon 23 gas. On the left side is the support for the constant pressure relief valve of which only the upper part with one pipe leading to the radiometer outlet and another to the detector outlet at the rear end of the instrument housing is visible. The electronics box is situated behind the relief valve.

INFRARED INTERFEROMETRY FROM SATELLITES

B. J. CONRATH and R. A. HANEL

Goddard Space Flight Center, NASA, Greenbelt, Md., U.S.A.

The infrared spectroscopy experiments (IRIS) on Nimbus 3 and 4 measure the thermal emission spectrum of the Earth between 400 and 1600 cm^{-1} with a resolution of 5 cm^{-1} and 2.8 cm^{-1}, respectively. Spectral radiances are interpreted to derive the vertical profiles of temperature, humidity and ozone, as well as for studies of radiative transfer in the atmosphere and of the emissive properties of the Earth's surface. A precision of approximately 0.5 to 1 erg cm^{-2} s^{-1} ster/cm^{-1} has been achieved. The interferometer on Nimbus 3 operated $3\frac{1}{2}$ months continuously and the instrument on Nimbus 4 recorded more than one million spectra during the first year in orbit.

The results of IRIS are a vivid demonstration of the power of Fourier spectroscopy. The multiplex advantages (Fellgett) and the étendue advantage (Jaquinot) of the Michelson interferometer were exploited fully.

The design of the interferometers on Nimbus 3 and 4 has been given in the literature [1, 2]. Scientific results from the experiments have been published also [3, 4, 5, 6, 7]. Plans for the use of a Michelson interferometer on a planetary mission, Mariner '71, have also appeared in print [1].

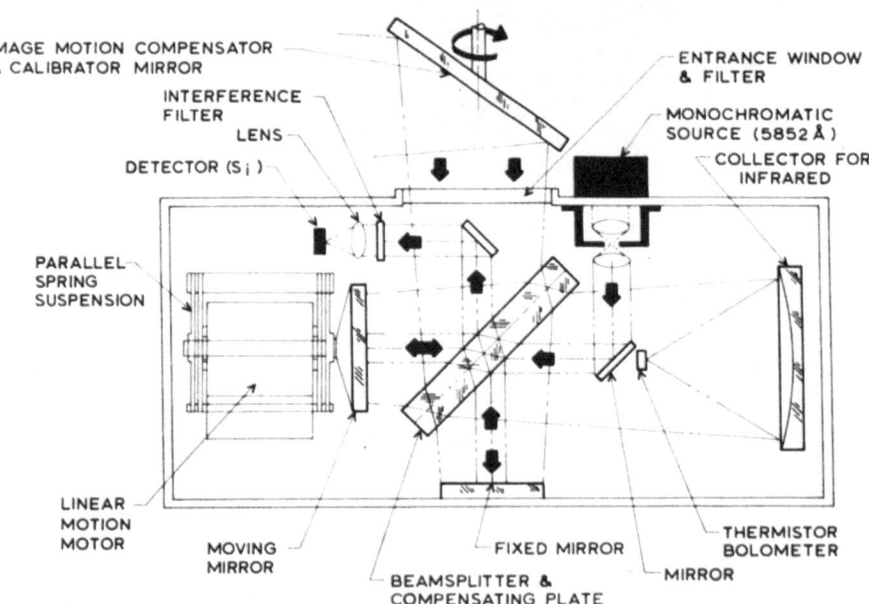

Fig. 1.   Simplified diagram of the Michelson interferometer flown on Nimbus 3 and 4. The image motion compensation and calibration mirror can be oriented to see Earth, deep space, and an onboard black body for calibration. The beamsplitter uses potassium bromide as a substrate.

Manno and Ring (eds.), Infrared Detection Techniques for Space Research, 84–92. All Rights Reserved
Copyright © 1972 by D. Reidel Publishing Company, Dordrecht-Holland

Fig. 2. External view of the Nimbus 4 interferometer (IRIS-D). The upward-looking port is for viewing Earth, and the other port is for viewing deep space. The white surfaces are exposed to space to cool the whole instrument to approximately 250 K. Maximum dimensions from Michelson mirror drive to bolometer assembly are approximately 42 cm.

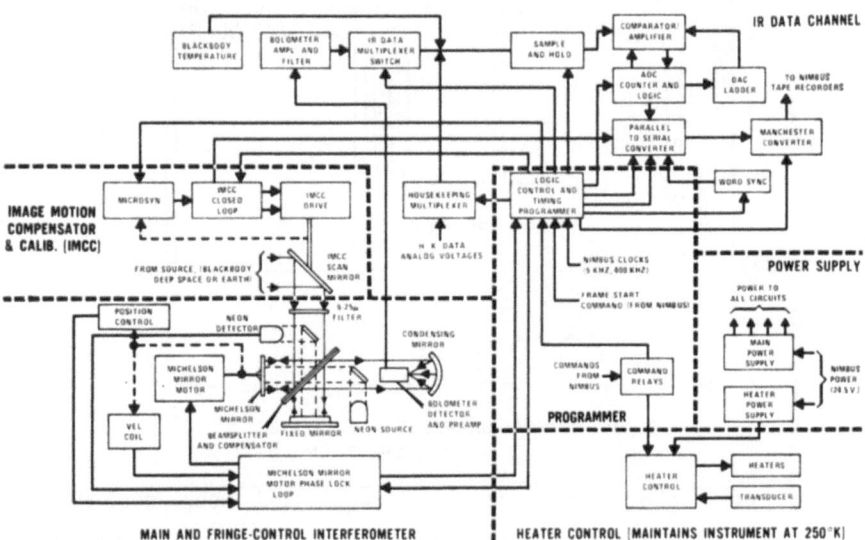

Fig. 3. Block diagram of the Nimbus 4 interferometer. Circuitry to the spacecraft telemetry subsystem to record housekeeping data is omitted. The Michelson mirror drive uses velocity feedback and a wideband phase lock loop to slow the motion of the mirror to a spacecraft clock frequency.

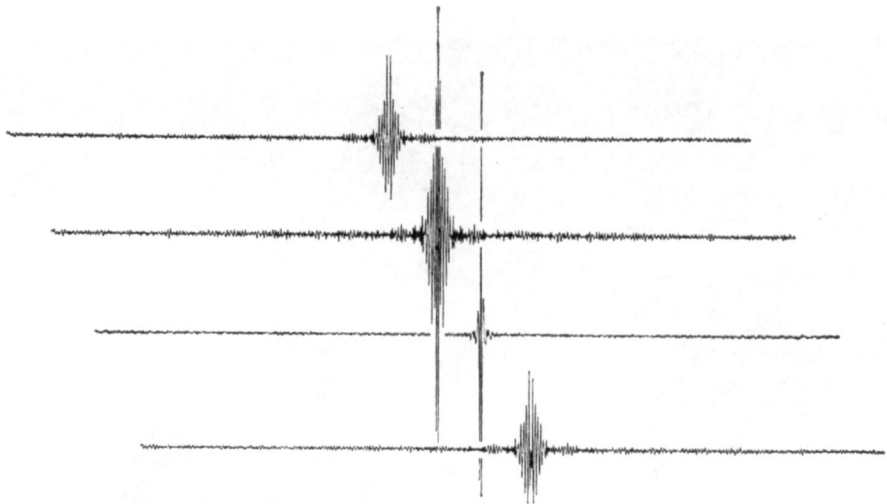

Fig. 4. Typical interferograms recorded in orbit. Interferograms 1, 2, and 4 were taken while viewing Earth; interferogram 3 was taken while viewing the warm black body. Numbers 1 and 4 are arctic cases and number 2 is a hot desert case.

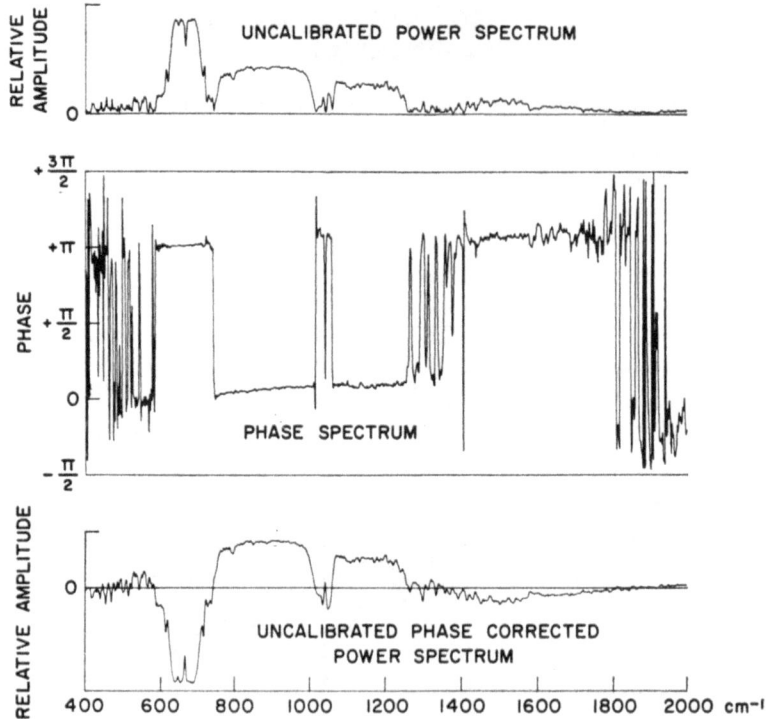

Fig. 5. An uncalibrated power spectrum from Nimbus 3 and the associated phase spectrum shown in the upper and middle part are the result of the Fourier transformation of one interferogram. The lower part of the figure shows the phase corrected power spectrum before calibration.

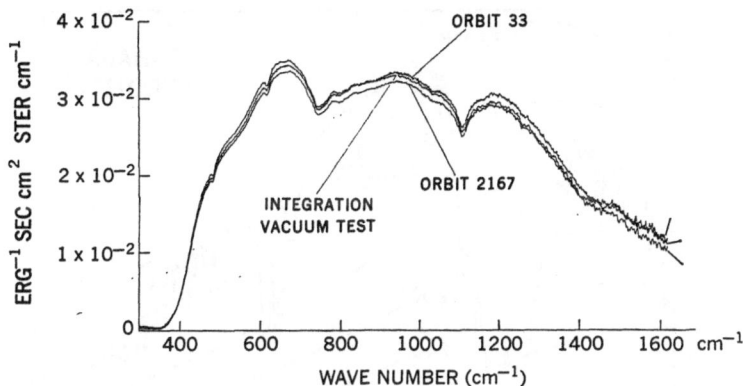

Fig. 6.   Spectral responsivity of the Nimbus 4 interferometer during the thermal vacuum test of the spacecraft about two months before launch, shortly after the instrument was turned on in Earth orbit (orbit 33) and after approximately 5½ months (orbit 2167) of continuous operation in space.   Responsivity after a whole year in orbit is essentially the same as the curves shown.

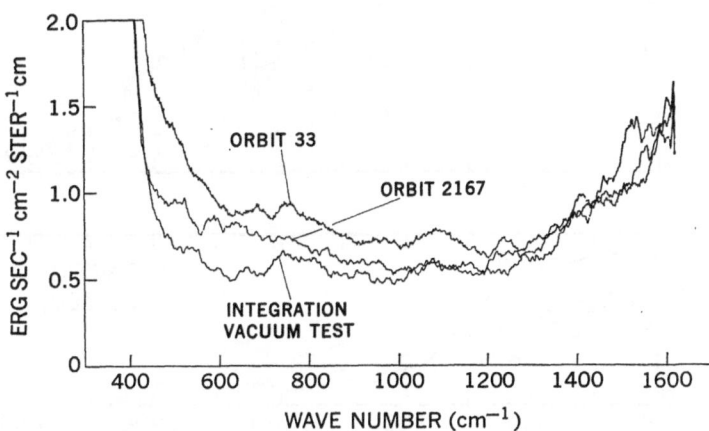

Fig. 7.   Noise Equivalent Radiance (NER) of the Nimbus 4 interferometer calculated from the standard deviation of individual responsivity measurements. The NER curves have been smoothed for display purposes by averaging over 25 cm⁻¹. The NER values obtained while in orbit contain also systematic variations due to orbital temperature changes which will be removed in the final reduction process.

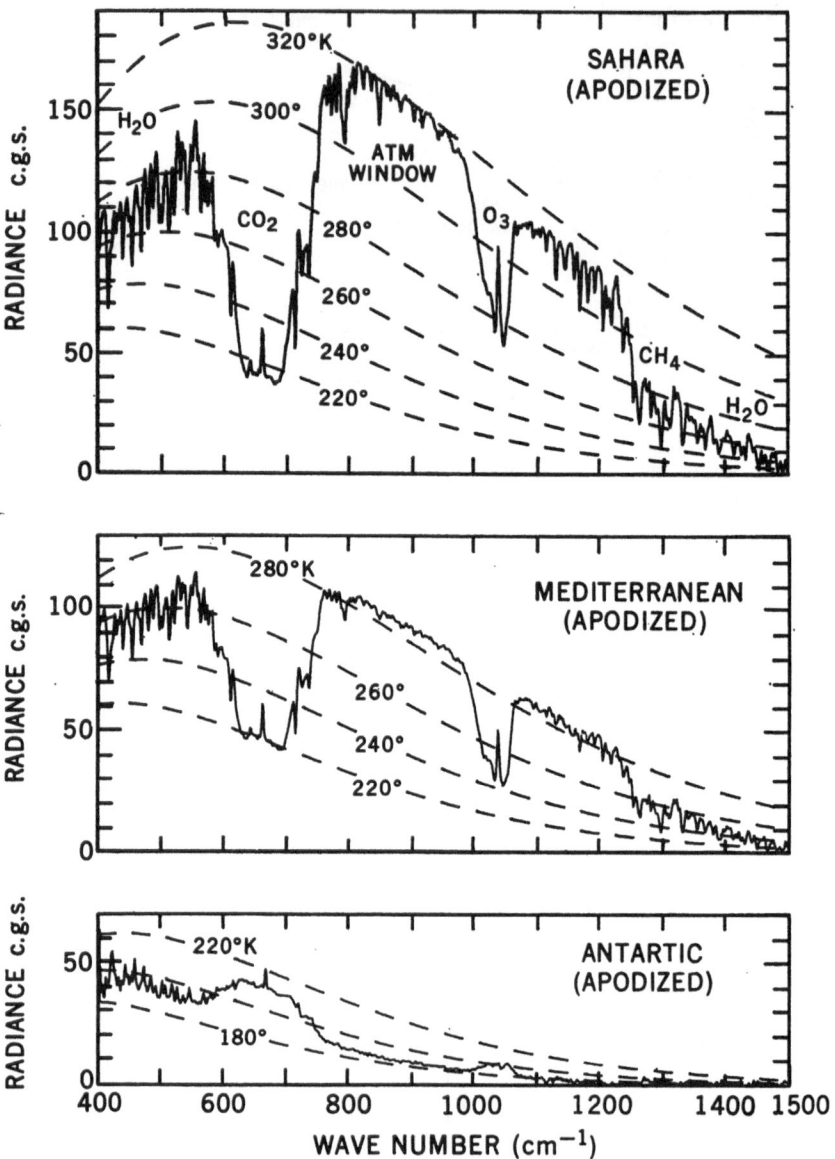

Fig. 8.   Calibrated thermal emission spectra recorded by the interferometer on Nimbus 4. The apodized spectra have a spectral resolution between 2.8 cm^{-1} and 3 cm^{-1}. A hot desert case, an intermediate case over water, and an extremely cold spectrum recorded over the Antarctic are shown. Radiances of black bodies at several temperatures are superimposed.

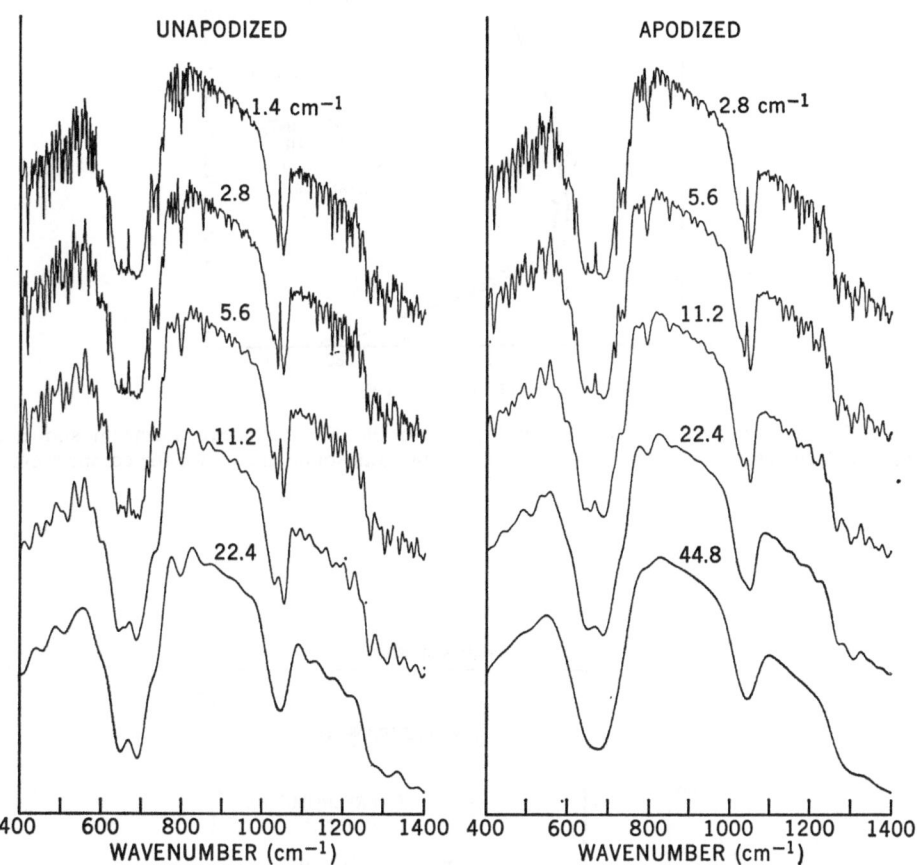

Fig. 9. A desert spectrum using all 4096 words available in an interferogram and plots of the same spectrum computed successively with one half of the words used in the previous step. The spectrum displayed lowest uses only 256 words which corresponds to a resolved spectral element size of 22.4 cm^{-1} and 44.8 cm^{-1} for the unapodized and apodized mode of data reduction.

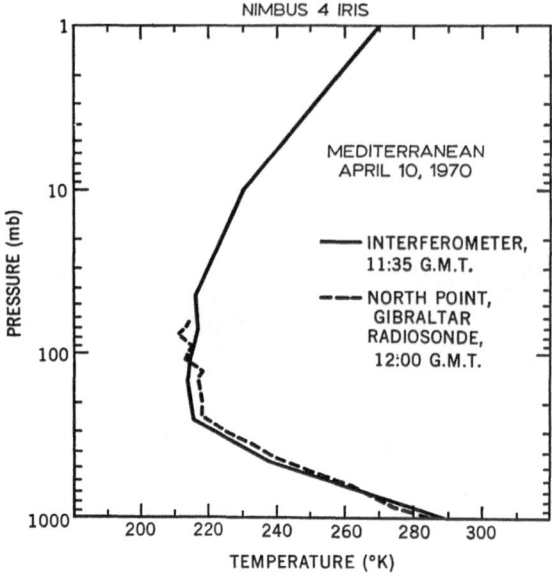

Fig. 10. Vertical temperature profile obtained from the Mediterranean spectrum shown in Figure 8. Temperature measurements from a nearby radiosonde are shown for comparison.

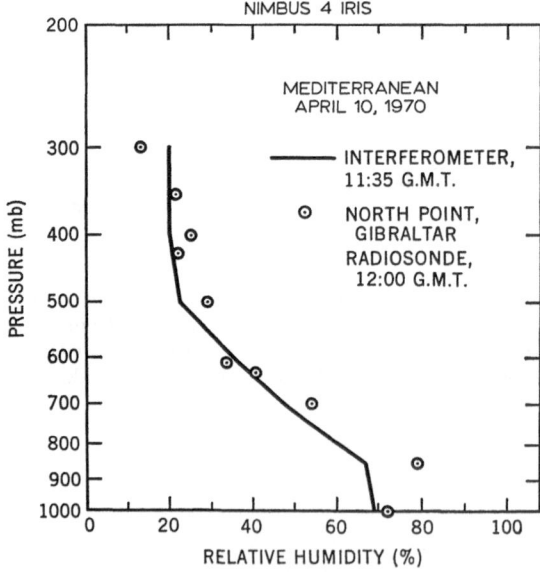

Fig. 11. Humidity profile derived from the same Mediterranean spectrum (Figure 8) is compared to radiosonde data.

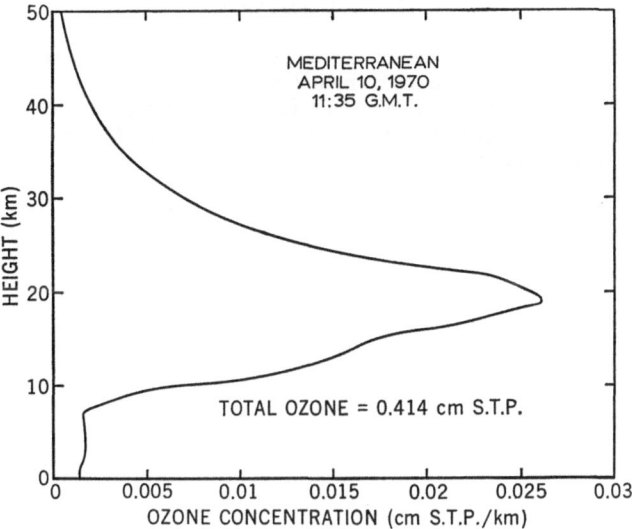

Fig. 12. Ozone profile obtained from the Mediterranean spectrum of Figure 8. The details of the structure come from the use of empirical orthogonal functions derived from radiosondes. For details on the method, see Prabhakara *et al.* [5].

The paper presented at the conference on 'Infrared Detection Techniques for Space Research' in Noordwijk, June 1971, summarizes the results from the IRIS experiments. To save space, only the figures shown in the presentation are reproduced here. The interested reader is referred to the references for further details on the instrument, on the scientific results, and on future plans for planetary exploration.

Acknowledgements

Figures 1, 2, 4, 5, 6, and 7 have first appeared in *Appl. Opt.*, and Figures 8, 9, 10, 11, and 12 in *Nature*. We are grateful to the editors who gave permission to have these figures reproduced.

References

[1] Hanel, R. A., Schlachman, B., Clark, F. D., Prokesh, C. H., Taylor, J. B., Wilson, W. M., and Chaney, L.: 1970, *Appl. Opt.* **9**, 1767.
[2] Hanel, R. A., Schlachman, B., Rodgers, D., and Vanous, D.: 1971, *Appl. Opt.* **10**, 1376.
[3] Hanel, R. A. and Conrath, B. J.: 1969, *Science* **165**, 1258.
[4] Conrath, B. J., Hanel, R. A., Kunde, V. G., and Prabhakara, C.: 1970, *J. Geophys. Res. Oceans Atmospheres* **75**, 5831.
[5] Prabhakara, C., Conrath, B. J., Hanel, R. A., and Williamson, E. J.: 1970, *J. Atmospheric Sci.* **27**, 689.
[6] Randall, C. M. and Rawcliffe, R. D.: 1970, *Proceedings of the Society of Photo-Optical Instrumentation Engineers, 15th Technical Symposium, Sept. 14–17, 1970, Anaheim, Calif.*
[7] Hanel, R. A. and Conrath, B. J.: 1970, *Nature* **228**, 143.
[8] Hanel, R. A., Conrath, B. J., Hovis, W. A., Kunde, V. G., Lowman, P. D., Prabhakara, C., and Schlachman, B.: 1970, *Icarus* **12**, 48.

DISCUSSION

W. Hoffmann: How does your accuracy of the temperature profile compare to what is needed as a boundary condition for global weather computations.

R. A. Hanel: At mid latitudes and higher latitudes it is quite an improvement. In the tropical regions one can predict the temperatures to about 2°, and therefore the remote technique is about equivalent to what one can predict.

THE METEOSAT DUAL-CHANNEL RADIOMETER

A. L. PERALDI

Engins Matra, Vélizy, France

The subject of the present lecture is the description of the dual-channel high resolution radiometer which is the main scientific payload of the French synchronous meteorological satellite Meteosat, which is to be flown in early 1975. This radiometer is now in development phase at Engins Matra, Vélizy, under CNES contract.

The mission is to provide full time cloud coverage of the Earth zone as seen from synchronous orbit. The radiometer makes pictures of the Earth both in the visible and infrared parts of the spectrum. The pictures are transmitted in real time to the ground at a rate of 1 picture every 30 min.

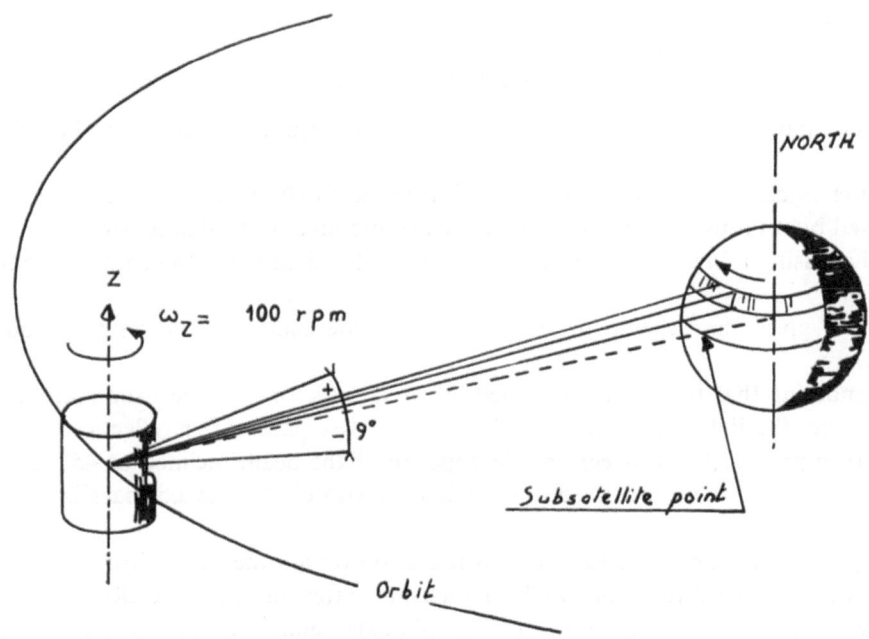

Fig. 1.   Imaging geometry from synchronous orbit.

Figure 1 shows how the system operates: The spacecraft is spin stabilized, the spin axis being normal to orbit (equator) plane. The spin (100 rpm) provides the line East–West scanning, the vertical North–South scanning being provided by a mechanism which is part of the radiometer. After each spacecraft rotation, the optical axis is tilted so that the completed picture has been scanned after about 25 min.

Table I shows the major specifications of the instrument.

Manno and Ring (eds.), Infrared Detection Techniques for Space Research, 93–99. All Rights Reserved
Copyright © 1972 by D. Reidel Publishing Company, Dordrecht-Holland

TABLE I
System specifications

Detection parameters	Visible channel	IR channel
Number of channels	2	1 (+1 redundant)
Wavelength band	0.5–1 μ	10.5–12.5 μ
Field of view	0.65 10^{-4} rd	1.4 10^{-4} rd
Detector	Photodiodes	Hg Cd Te
Electrical bandwidth	58.8 kHz	29.4 kHz

Main system specifications

Spin rate	100 rpm
Picture rate	30 mn
Number of line per picture	2500 (IR channel) 5000 (Visual channel)
Angular scanning range	±9°
Collecting aperture diameter	400 mm
Life time	2 yr
Weight	50 kg

Instrument Description

After a comparative study, the so-called 'Terrestrial mount' configuration has been selected.

Major characteristic of this configuration is that the North–South (vertical) scanning is achieved by rotating the complete telescope around an axis normal to S/C spin axis, rather than using a big flat mirror as usual. Principle advantage is a dramatic saving of weight.

Figure 2 gives the scheme of operation for Siderostat and terrestrial mount respectively.

Remembering that the passively cooled infrared detector must be fixed relative to the structure, the light has to follow a rather complex path. After reflection on the secondary mirror of the Cassegrain telescope, the light beam inclined at 45° and is first reflected by a flat mirror, fixed to the telescope (which puts its axis parallel to the bearing axis).

The light is then submitted to a second reflection on another flat mirror, and falls on a movable dihedral reflector, used for focusing; after another reflection, the light beam, whose axis is now parallel to the S/C spin axis, finally falls on the detector set.

I will now describe the different parts of the radiometer. The instrument can be divided in four major units or sub-systems as we call them:

– the telescope;
– the scanning and focusing mechanism;
– the detection assembly; and
– the structure.

The telescope uses a Ritchey-Chretièn combination where both primary and secondary mirror have hyperbolic surfaces. This optical combination has been selected because it offers good image quality over a rather large field of view, and is

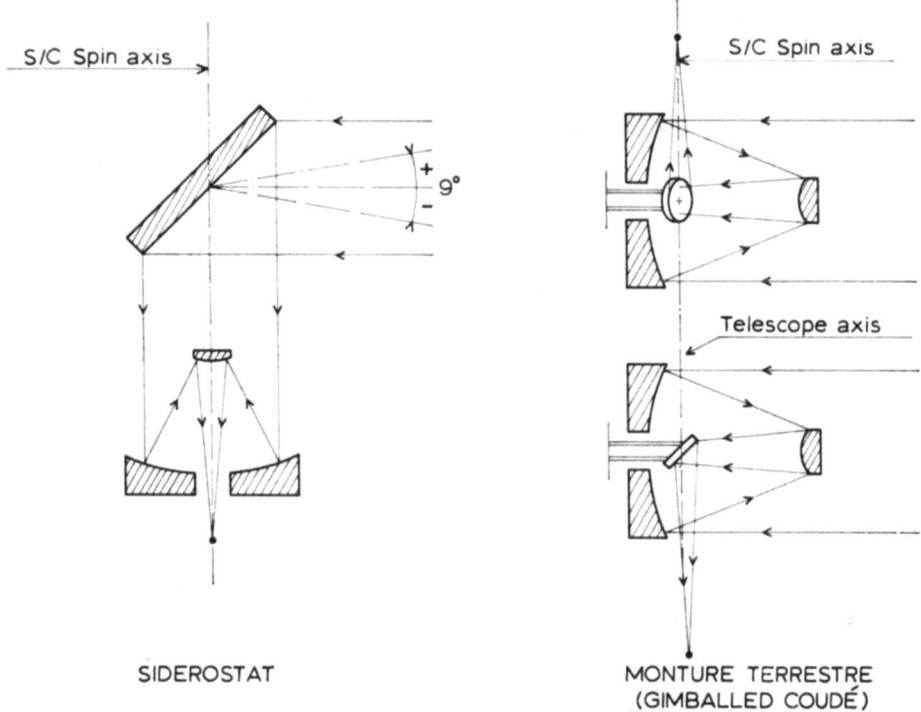

Fig. 2.   Comparison of scanning method.

not very sensitive to mechanical or thermal distortion. Blur circle angular dimensions are 4 arc s over a field of 30′.

The mirrors are made of Cervit; this material has been preferred for its negligible thermal expansion coefficient, and for technological reasons also.

Two different technologies for getting ultra lightweight are presently in competition:

– the first, developed by French Reosc is called 'Tambourin' and used conventional machining techniques.

– the second, developed at Matra used an ultrasonic machining process.

We hope to get a weight reduction of about 70% or better from the solid. The liaison between primary and secondary mirror is a composite structure with a cylindrical part and a tubular (tripole); both are made in Invar alloy to minimize thermal effects. The mirrors are fixed to the telescope structure by three flexible blades and rotule. An isostatic fixation is required to minimize the forces induced on the mirrors, and therefore the surface deformations.

As mentioned previously, the whole telescope is tilted by ±9° for vertical scanning. The telescope is mounted on two flexible pivots. The scanning mechanism consists essentially of a high precision lead screw driven by a stepping motor. This unit is completely sealed by 2 bellows, and lubricated by oil and oil vapor. A set of flexible blades (called 'cabstan' system) transforms the translation of the screw into rotational motion of the telescope.

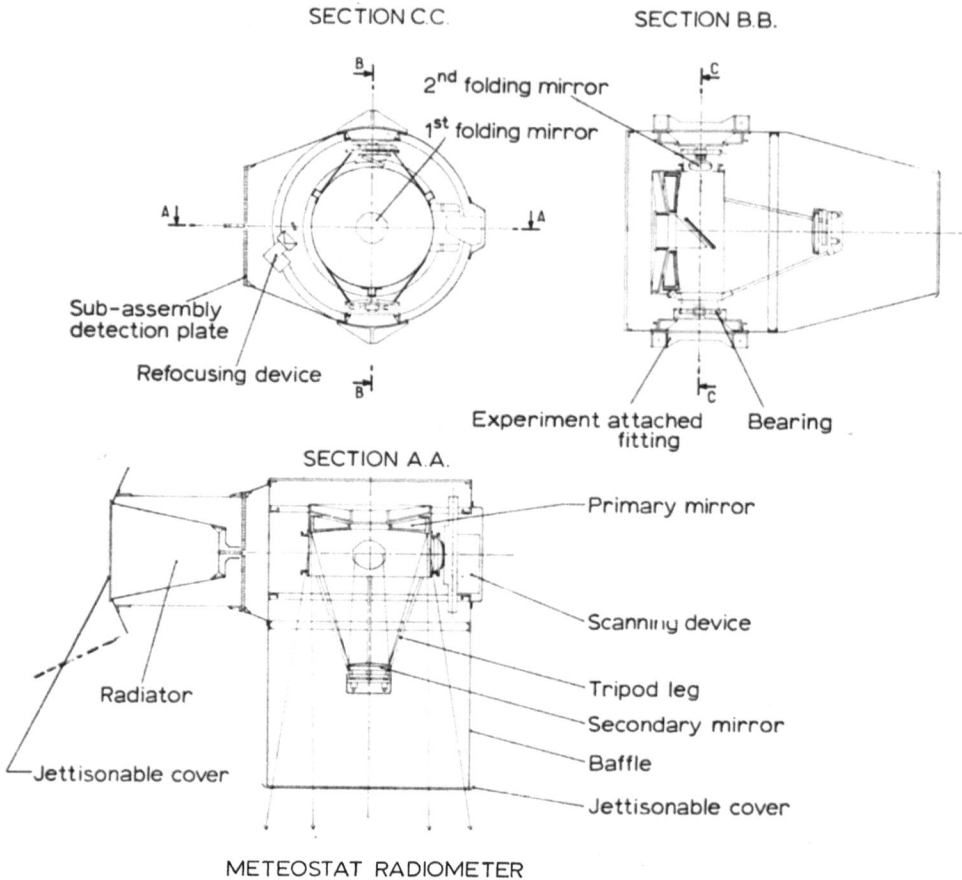

SECTION C.C.   SECTION B.B.

2nd folding mirror

1st folding mirror

Sub-assembly
detection plate

Refocusing device

Experiment attached   Bearing
fitting

SECTION A.A.

Primary mirror

Scanning device

Tripod leg

Secondary mirror

Radiator

Baffle

Jettisonable cover

Jettisonable cover

METEOSTAT RADIOMETER

Fig. 3.   Radiometer overall configuration.

Another device has the purpose of adjusting the focus position, in order to get good image quality. This system consists of two orthogonal flat mirrors fixed at the end of a lead screw driven by a stepping motor. The motion is controlled from the ground by TC.

We will now discuss the detection sub-assembly.

Figure 4 shows the implementation of the detectors in the telescope focal plane. Two infrared detectors are mounted, one as a back-up (stand-by redundancy); on the same line there are two visible detectors, their FOV being half the FOV of the infrared detector, which gives a visible resolution twice as good as the infrared one.

In fact, the detectors are not located at the telescope focus. An auxiliary optic is used to match the effective focal length and the infrared detector dimensions. This optic consists of a doublet of germanium lenses and a field lens. This unit is fixed to the radiator, and operates at very low temperature (about 77 K). The detectors will be either photo-voltaic or photoconductive. The choice will occur at the end of this

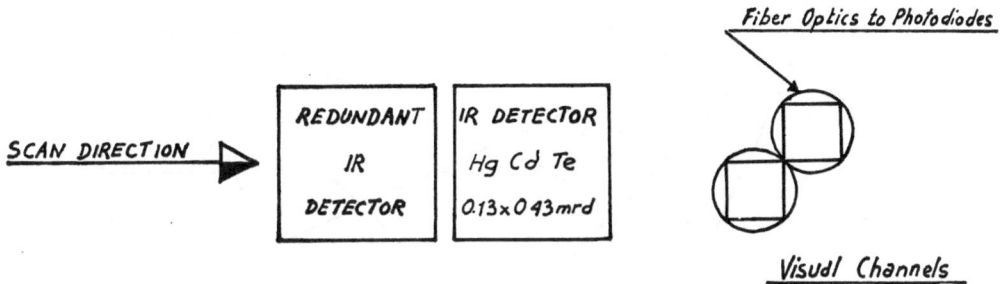

Fig. 4. Detector arrangement at telescope focal plane.

year after comparative tests which are to be conducted by CNES and LMD. Dimensions of the infrared detectors will be between 80 and 100 μ. A field lens is used to conjugate the telescope exit pupil and the aperture stop at the level of the infrared detector.

For visual channels, two fiber optics carry the light to the photodiodes. Possible configuration is shown on Figure 4. In this solution, the fibers are fixed on the 'cold unit', in order to have a fixed geometry of each channel relative to the other, whatever the mechanical distortion. After passing through an afocal system (2 lenses) the light finally falls on the diodes, which are fixed to the structure at ambient temperature. Other solutions are still being considered at this moment.

Detection sub-assembly includes an on-board calibration system on the infrared channel. Two references sources can be used: (1) a black body whose temperature is accurately measured and transmitted to the ground by TM, (2) the Sun when it lies in the angular scanning range ($\pm 9°$), that is to say 1.5 months around equinoxes. As the direct solar flux in 10.5–12.5 μ interval is much higher than the Earth average flux (by about two orders of magnitude) and therefore out of the channel dynamic range, the useful aperture is reduced during sun calibration operation, only the light passing through a pin-hole being transmitted to the detector. A set of two spherical (elliptical) mirrors driven by a small torque motor are used for source commutation. The device is designed in such a way that the field remains free in case of command failure.

An important unit and a critical part of the radiometer is the passive radiator which cools the infrared detector. Remember that normal temperature operation is 77°; at this temperature, a black body radiates only 2 W m^{-2}. As usual on spacecraft, the available volume is strictly limited. The problem is specially difficult in the present case because of the position of the Sun, which can be 23° above the equatorial plane; therefore, a Sun shield much bigger than the effective radiator itself is mandatory. An interesting solution for overcoming this difficulty could be to rotate the S/C spin axis of π every six months; for the moment this solution (which costs a rather important amount of propellant) is not under consideration.

Figure 5 shows the radiator configuration.

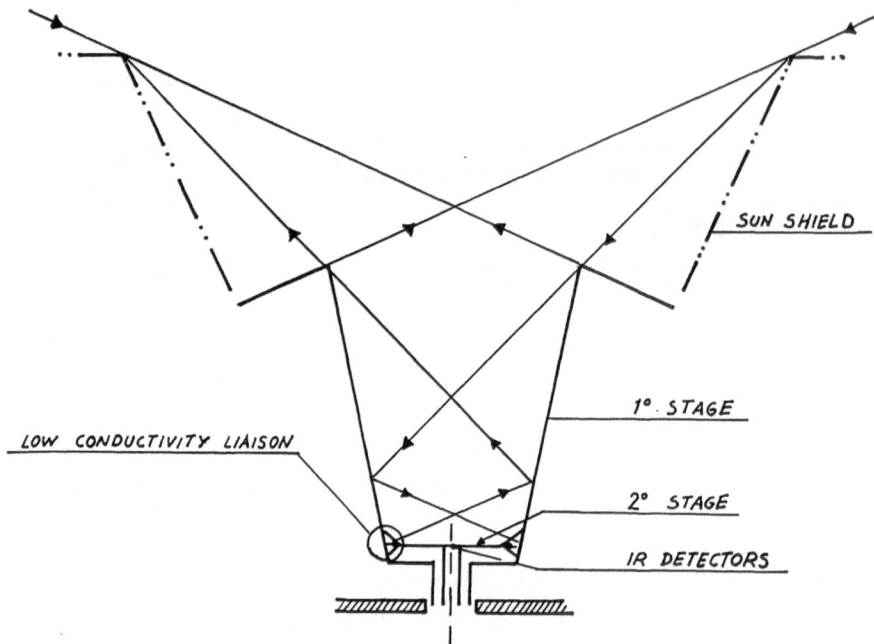

Fig. 5.   Passive radiator geometry.

The upper cone is the Sun shield. Geometry of the system is such that the Sun's rays never reach the internal cone, but are rejected back into space by specular reflection. A very good specularity is even required, which causes some manufacturing problems.

This upper cone (Sun shield) is not part of the radiometer, but is fixed to the S/C structure.

The radiator itself is a two-stage design: First stage is the lower cone, which will operate at about 150 K. Internal coating of this cone is specular too. Second stage is A flat piece of metal of 170 mm diameter.

The most critical part of the radiator assembly is the liaison between second and first stage. A major problem is to have a very low conductivity fixation, able to withstand mechanical constraints (vibration, acceleration) during launch phase.

A solution using a mesh of thin glass wires has been designed by Franch Bertin.

For a good operation of the system any thermal leakage to the radiator second stage should be minimized. Major thermal inputs are due to radiative exchanges between radiometer structure and the cold unit, direct solar flux and detector wires and screening.

Overall radiator design is by CNES.

The various parts of radiometer equipment are held together by a structure. Main structural part is an annular frame supporting the telescope bearings. An upper cylindrical part supports the radiator cone and the detection assembly. Around the telescope is a baffle which thermally insulates the radiometer from the spacecraft. During handling, ground transportation and launch phases, the radiometer is closed and sealed to protect the optics against contamination. Two protective covers, one behind the baffle, the second at the top of the radiator are released on orbit by pyrotechnic devices. Other devices forming part of the structure are:

(1) the telescope bearing locking device (flexure pivots are not able to withstand launch conditions); and

(2) possibly a rubber suspension to damp vibrations.

To complete this brief speech, I will just mention some of the most difficult problems among a lot of others one has to face to develop a space-borne experiment of this type and size. First is to build ultralight mirrors with good optical quality, and to make a mirror fixation able to withstand severe mechanical constraints without inducing deformation of the mirror surface.

A second difficulty is to get a very accurate and reproducible motion of the scanning mechanism.

This involves both the performances of the actuator itself, and the overall mechanical and thermal stability of the radiometer.

Finally, one of the most tricky problems is trying to ensure a good transmission factor of the optics during all the life. This involves contamination of mirrors, and especially of the cold unit which obviously tends to condense outgassing products.

Due to the lack of statistical data in this particular field, maximum care is taken at the design level: First is to avoid, as far as possible, the use of outgassing products, or products which could cause a degradation of the transmission factor. An important test program is under way to clear this problem. Second, is to handle and launch a sealed radiometer, as previously mentioned. Finally, for orbital decontamination of the cold unit, a heater will be mounted on the radiator, in order to stop temporarily the cold operation.

DISCUSSION

R. A. Hanel: What is the lowest temperature you hope to achieve and are you taking steps to prevent any water vapour from forming on your detectors?

A. L. Peraldi: The operational temperature is 77 K. As for the second question, I would like to say that the instrument is completely sealed during launch operations. It will be filled with dry nitrogen so we hope to avoid condensation on ground. In orbit, one of the two jettisonable covers will be ejected first and later the second one, so that the instrument will outgass before being cold.

H.-J. Bolle: What will be the absolute accuracy of the calibration and what does the black body look like.

A. L. Peraldi: The specification is now 2% accuracy of the signal flux for a cloud at 240 K.
The black body is simply a cavity with a hole.

3. DETECTORS

INFRARED DETECTORS. SURVEY OF THE PRESENT STATE OF THE ART

P. LÉNA

Université de Paris et Observatoire de Meudon, 92 Meudon, France

Abstract. In order to make a proper evaluation of the ultimate sensivity the infrared detectors may attain for astronomical and space applications, we survey the present achievements in the field of bolometers, photoconductors and Josephson junctions. We also evaluate potential use of heterodyne detection applied to both spectroscopy and wide-band detection. Finally, some attention is paid to exotic types of detectors which may become of interest in a near future.

Special emphasis is made upon astronomical observations, which is the common goal all detectors are supposed to be fitted with.

1. Introduction

The subject of infrared detection is very broad, and we shall restrict the discussion to special applications of astronomical observations either from the ground or from space-bound platforms (aircraft, balloons, rockets, or satellites).

The spectral range shall be restricted to 10–1000 μ. At shorter wavelengths, classical photoconductors behave properly and nearly reach ultimate sensitivities due to basic limitations of thermodynamics.* Beyond 1000 μ, one may consider that the radio-range begins, and that conventional radio techniques either presently apply or shall be extended in a near future to the millimetric wavelength. Applications of new detectors, such as Josephson junctions, may well beat on their own field the conventional radio detection schemes.

1.1. DEFINITION OF AN OBSERVING PROBLEM

We shall assume some source emitting in the infrared, and restrict the detection to a wavelength range $\lambda_m < \lambda < \lambda_M$. Its power is received by a telescope and/or collecting optics, having a beam étendue $S \cdot \Delta\Omega$, S being the entrance aperture, and $\Delta\Omega$ the subtended solid angle. S defines a diffraction size of full width at half maximum of about $\alpha \approx \lambda/\sqrt{S}$.

We shall define a source as 'punctual' for a given instrument, if its angular size is smaller than α. In the alternate case, it shall be said 'extended'. Let us consider that the *minimum size* of the detecting element placed beyond the optics corresponds to an area $A \geqslant E/2\pi$ to accept the same étendue E: in this case we will say the system is in the diffraction-limited case, which obviously corresponds to the smallest possible detector.

The observation is made through an eventually fluctuating atmosphere and/or scanning system and the response time τ of the detector has to be considered. Although multidetectors arrays are available and may be of considerable help in imaging or improving signal-to-noise ratios, the discussion will be restricted to a single detector.

* Moreover, background radiation, which shall play an important role in the discussion ,is essentially small or negligible below 10 μ, for usual backgrounds at temperature lower than 300 K.

Manno and Ring (eds.), Infrared Detection Techniques for Space Research, 103–113. All Rights Reserved

The instrumental efficiency – i.e. transmission – of the system will be assumed to be 1, excluding the efficiency of the detector to be discussed separately. Typical overall system efficiencies, limited by atmospheric transmission, mirrors, filters, windows ... etc. range from 1% to 50%.

The radiation emitted by the source may have very different spectral characteristics and behavior. It could be a continuum, either thermal (black or grey) or non-thermal (e.g. synchrotron); it could be made up of discrete lines either in emission or in absorption. The relative width of the lines may vary from extremely narrow ($\Delta\lambda/\lambda \lesssim 10^{-6}$ for maser lines [1]), to narrow (turbulent speeds up to 100 km s^{-1}, hence $\Delta\lambda/\lambda \lesssim 5.10^{-4}$ [2]) or broad (typical pressure broadened lines in planetary atmospheres: $\Delta\lambda/\lambda \simeq 5 \times 10^{-3}$ [3, 4].

The goal of the observation is to obtain the best signal-to-noise ratio either on a single line at the time, or on many lines, or on a continuum defined by the band $\{\lambda_m, \lambda_M\}$. Optical techniques most often study a large spectral range by photography, but may also concentrate on a single line (photoelectric scanning, or spectrohelio-grams, or Fabry-Pérot). Radio techniques may also concentrate on lines (heterodyne detection) or on a large spectral range (video detection).

In order to intercompare detectors, let us characterize a given detector by its Noise Equivalent Power or NEP (W Hz$^{-1/2}$), equivalent to the minimum power (in Watts) which can be detected in a 1 s integration time, assuming signal to noise ratio to be unity. This value of P_{min} ultimately puts limits on the observation of a given object in the sky.

A general review of detectors properties may be found in classical books [5, 6, 7, 19].

The following will distinguish between wide-band detection, even used for spectroscopy, and narrow-band detection restricted to very high resolution work.

2. Wide-Band Detection

2.1. THERMAL DETECTORS

The basic thermodynamical fluctuations in such systems as Golay cells, pneumatic detector, pyroelectric detectors, bolometers set an ultimate limit:

$$P_{min} = 4(\sigma k)^{1/2} T^{5/2} A^{1/2} \text{ W Hz}^{-1/2} \tag{1}$$

where σ is Stefan constant, k Boltzmann constant, T temperature of the detector. Gains are obtained by decreasing either T or A.

Two factors limit the decrease of A. First, practical feasibility of detectors and coupling to the radiation field: detectors cannot be very much smaller than the wavelength to measure. Second, the detector has to be matched to the collecting optics, which sets $A \geqslant E/2\pi$ for the diffraction limit. On Figure 1 is shown how to match, with a $f/10$ system, detector size to optics in this diffraction limit. One sees that a $200 \times 200 \mu$ detector is optimum at $\lambda = 10 \mu$ and that the same size would need a $f/1$ optics at $\lambda = 100 \mu$. As soon as the field of view is increased, larger detectors are needed. Hence we shall consider $A_0 = 1$ mm^2 as the standard area being easy to fit with a suitable optics at any infrared wavelength.

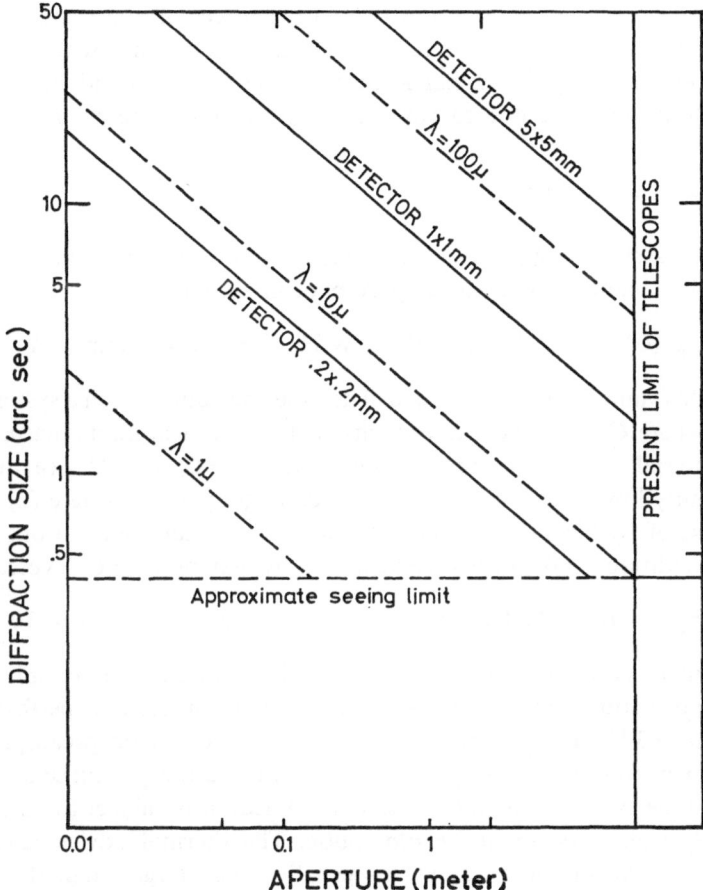

Fig. 1.   Diffraction limit versus telescope aperture, for various wavelengths (– – –). For an $f/10$ system, the corresponding detector size (——) is plotted for different physical sizes. Note that the detector size is approximately equal to the wavelength for a $f/1$ system (field lens).

As far as temperature is concerned, Equation (1) is not valid for $T \ll 10^{-2} - 10^{-3}$ K. Above this value, where most of the cases are, the limits are:

$$
\begin{aligned}
T &= 300\ \text{K} & P_{\min} &= 6 \times 10^{-12}\ \text{W Hz}^{-1/2} \\
T &= 77\ \text{K} & &= 2 \times 10^{-13} & (A &= A_0) \\
T &= 4\ \text{K} & &= 10^{-16}
\end{aligned}
$$

Room temperature thermal detectors (pneumatic receiver) reach 2×10^{-10} for $A = 4\ \text{mm}^2$ and are therefore not too far from theoretical limits. They have intrinsically poor response time $\tau \lesssim 0.1$ s.

Pyroelectric detectors have larger $P_{\min}$ by a factor of ten but fast response ($\tau \approx 10^{-7}$ s).

Germanium bolometers may operate between 4.2 K (He^4) and 0.3 K (He^3). Lower temperatures have not been considered up to now but may be feasible.

This type of bolometer suffers from two basic sources of noise, independent of the radiation field to be measured: fluctuations in photon exchange with the surrounding and fluctuations in phonons exchange through the thermal conductance G with the bath, then Johnson noise due to current [8]. Both limits merge into the following equation:

$$P_{\min} = 4(kG)^{1/2}T \tag{2}$$

assuming the system completely surrounded by a thermostat at temperature T. Lowest possible value for G assumes pure photon exchange

$$P_{\min}(G) > P_{\min}(G_{\text{rad}}) = 10^{-16} \text{ W Hz}^{-1/2}, \quad A = 1 \text{ mm}^2, \quad T = 4 \text{ K}.$$

Practically, this value cannot be reached. The time constant of such a system is given by $\tau \approx 0.7\, C/G$ where C is the thermal capacity. C has practical limits (nature and thickness e of material ...) and corresponding value $\tau(G = C_{\text{rad}}) = 10^3$ s ($e = 10\,\mu$). Beside being inherently low, such a value corresponds to frequencies where (2) is no longer valid, because of $1/f$ flicker noise and $1/f^2$ bath temperature fluctuations noise. Practically, the condition $\tau > 0.1$ s must be achieved, by playing with G. Typical figures are

$$P_{\min} = 10^{-14} \text{ W Hz}^{-1/2} \quad (A = 1 \text{ mm}^2, \quad T = 4 \text{ K}, \quad G = 10^4\, G_{\text{rad}}).$$

Real bolometers are now coming close to this limit, and are reasonably matched by noise figures of good preamplifiers. Any improvement of detector performances (e.g. lowering T to 0.3 K or lower) would need a parallel effort on preamplifiers. Better materials may be found decreasing C, hence τ, but Equation (2) still sets the limit.

As soon as the system is open to observe an object, it is subject to an extra photon noise, usually called *background photon noise*. The thermal power radiated by the optics, windows, atmosphere ... etc. is usually much larger than the signal-to-be detected, except in very special cases where the whole system could be cooled at detector temperature in space.

Background photon noise is given by [5, 21]

$$P_N = 2.8 \times 10^{-18} \varepsilon_b T_b^{5/2} E^{1/2} \{J_4(x_m) - J_4(x_M)\}^{1/2} \tag{3}$$

T_b being background temperature, ε_b background emissivity,

$$J_4(x) = \int_0^x \frac{x^4 e^x}{(e^x - 1)^2}\, dx, \quad x = \frac{h\nu}{kT} = \frac{hc}{kT\lambda}.$$

For large wavelength intervals, this reduces to

$$P_N = 4(\sigma k)^{1/2} T^{5/2} E^{1/2} \tag{3a}$$

which is similar to Equation (1) above.

For

$$\Delta x \ll x, \quad \text{or} \quad \frac{\Delta\lambda}{\lambda} \ll 1,$$

$$P_N = 6 \times 10^{-20} (T^4/\lambda^{3/2}) E^{1/2}. \tag{3b}$$

With the best detectors available at present, this noise becomes far from negligible. Let us take a few typical examples:

(i) Observation of an object from the ground, wide spectral range, diffraction limited, through the 9–13 μ window

$$\varepsilon_b = 0.1, \quad T_b = 300 \text{ K}, \quad \lambda \approx 10 \mu$$

$$P_N \approx 10^{-14} \text{ W Hz}^{-1/2};$$

(ii) Observation at 100 μ, at aircraft altitude, with narrow-band metallic grid filter, diffraction limited.

$$\Delta\lambda/\lambda \approx 0.1, \quad \lambda = 100 \mu, \quad T_b = 200 \text{ K}, \quad \varepsilon_b = 0.5$$

$$P_N \approx 3.10^{-15} \text{ W Hz}^{-1/2};$$

(iii) same case as (i), but wider field of view for survey or extended source ($A: 100 A_{\text{diff}}$):

$$P_N \approx 10^{-13} \text{ W Hz}^{-1/2}.$$

Let us briefly see how this may be improved. E is degrading P_N for wider field of view (case iii). For diffraction-limited system, $E = \lambda^2$, hence P_N varies as $\lambda^{-1/2}$ for narrow-band ($\Delta x \ll x$). Not much is gained by going to longer wavelengths.

ε_b may be decreased by selecting proper coatings for mirrors. Values smaller than 0.03 per mirror have not yet been achieved even in the far infrared. Mylar windows a few microns thick have typical values $\varepsilon_b < 0.03$ beyond $\lambda = 500 \mu$, and $\varepsilon_b \sim 0.05$ around 100 μ. Even placing cold filters in front of the detector does not reduce the emission of hot windows or mirrors, if any, since the cold filters have to be transparent in the very same range where the hot windows have some finite emissivity. The system has to be either completely cooled or not. Nevertheless, it may be, although not proved yet, that thermal fluctuations of some parts of the system introduce extra noise (these fluctuations being due to lack of thermal uniformity). Cooling is then a convenient way to ensure thermal stability (this point was brought up to the author by Dr Beckman).

Even when cooling the system, full gain will not be realized if residual atmosphere is not eliminated. Typical emissivities in the range 10–1000 μ range from $\varepsilon_b = 0.1$ to 0.01 at aircraft altitude ($T_b \simeq 220$ K, 1 to 5 μ precipitable water), to $\varepsilon_b < 0.01$ at balloon altitudes ($T_b \simeq 240$ K, less than 0.5 μ of precipitable water) [20].

Clearly the background photon noise is going to set the limits of low-altitude, non-cooled systems and will eventually become a major reason to place systems in space, without any window.

Other gains in thermal systems may come from some technology improvements, as improving the absorption processes (in other words the 'equivalent quantum efficiency' of thermal detectors), the selectivity which may intrinsically reduce the sensitivity outside the $\{\lambda_m, \lambda_M\}$ interval, lower temperature operation. But the present state of the art seems close to the ultimate limits.

2.2. JOSEPHSON JUNCTIONS (DC MODE)

The Josephson DC effect has been described by various authors [9, 10] and allows detection of wide spectral-band signals. It is difficult at the present time to ascertain fundamental limits of such systems. Fundamental noise is essentially due to lack of coherence of the wave function of the electron pairs.

As far as NEP is concerned, some published values have already been given, which compare very well with the best available bolometers: a wide-band measurement [11] quotes 2×10^{-14} W Hz$^{-1/2}$, other narrow-band measurements (see below 3.2) quote 10^{-14} W Hz$^{-1/2}$ [12].

Response time as fast as $\tau \approx 10^{-10}$ s is obviously an advantage.

Stability of junctions against time, vibrations, etc. is a problem not yet solved. Several types of junctions are of interest: etched, evaporated, point-contact. The problem of optimizing the coupling of the junction to the input radiation field is an open one; great progress may be expected in the future.

An interesting feature of supraconducting systems, especially for space applications, is their possible use at some temperatures between 4 to 20 K. This advantage may become very important for long space missions.

2.3. PHOTOCONDUCTORS

Recent developments of photoconductors have allowed the spectral range 10μ to 1 mm to be covered (Ge:Au, Ge:Hg, Ge:Cu, Ge:Zn, Ge:As, Ge:Sb, InSb) – (see the following paper by Shivanandan).

Fundamental limitation is again background photon noise; other noise sources, especially current noise, may become important if this background is reduced. Another result of such a decrease is to increase detector resistance (as an example, detector impedance used in a 4–10 K background in NRL's rocket experiment goes up to $10^{12} \, \Omega$). Practical quoted values for photoconductors in the submillimetre range are of the order of 10^{-15} W Hz$^{-1/2}$ for $A = 3$ mm^2, which gives a $P_{min} = 6 \times 10^{-16}$ for the normalized value $A_0 = 1$ mm^2.

Clearly photoconductors exhibit advantages: fast response time, intrinsic high quantum efficiency at their peak spectral response. The variation of their sensitivity with wavelength may be a positive or negative factor, depending on the type of experiment. The ease to manufacture arrays is another advantage as soon as multi-detector systems are built for imaging purposes.

In short, it is not clear yet which type of detector: bolometer or photoconductor will first reach the theoretical limits if background is greatly reduced.

3. Narrow-Band Detection

Interesting sources usually discriminate between continuum and discrete emission. Continuum emission is relevant to the detectors discussed above, combined with a suitable analyzing system: filters (interference, Fabry-Pérot, meshes, powder, Rest-strahlen ...), Michelson interferometer or lamellar grating, the resolution being adapted to the type of continuum radiation.

Line emission studies have up-to-now been confined to the same detection scheme, either by will or by force. In the former case, spectral mapping is of interest on a large spectral interval: atlas of planetary atmospheres spectra, star spectra. Fourier spectroscopy is a perfect tool for such purposes, but is, at long wavelength, limited to resolutions of the order of 0.01 cm^{-1} for reasonably sized instruments. Very high resolution spectroscopy (up to 10^6 or 10^7) will likely be essential in the future for the study of one single line position, profile time-evolution. . . . Identification of molecular species radiating in the infrared, measurement of clouds velocities through Doppler shifts call for such a spectroscopic tool.

Several techniques are in sight that shall be reviewed and compared to existing infrared detection methods.

3.1. HETERODYNE SPECTROSCOPY

Although this is a very classical field for radio astronomers it is a new one in infrared, but may prove of great interest [5, 6, 13, 14].

The incoming radiation must be mixed up in a non-linear element (e.g. a quadratic detector) with local oscillator single frequency radiation. Intermediate beat frequency (or 'down-conversion' of the infrared photons) is then detected by conventional RF techniques. Spectral scanning and resolution are achieved by tuning ω_{LO} and/or by adjusting intermediate frequency band pass.

For infrared applications, main problems are:

– have a local oscillator in the spectral range of interest, giving stable and continuum power, at a level high enough to overcome inherent detector noise, and being tunable over some frequency range, in order to scan the spectral range of interest;

– have a suitable detector responding at the beat frequency in a reasonable intermediate frequency (IF) bandwidth.

The first problem is solved in the near infrared (6 to 32 μ) with non-stoichiometric tunable $Pb_{1-x}Sn_xTe$ solid state lasers, giving 1 mW of CW power [14]. Possible extension to longer wavelengths would have to be scrutinized in the future. Such systems seem low-weight and reliable. Much heavier and cumbersome systems are available at longer wavelengths: gas lasers, carcinotrons (travelling-wave tube). Harmonic generation may then fill the gap with the solid state lasers. Tunability for long wavelength lasers is not yet achieved, but the very large number of available lines may overcome this point.

Finally, the Josephson junctions emit an AC electric field at a frequency $\nu = 2 \text{ eV}/h$ when biased with a voltage V. Emitted power could reach 1 μW [9], and high degree of coherence is achieved (measured upper limit for $\Delta\lambda/\lambda$ is 10^{-5}). The wavelength range seems limited above 80 μ.

The second problem is basically solved with the existing photoconductors (see above), with response in the 0–10 MHz IF range. Pyroelectric detectors respond to the same frequency range, and Josephson junctions, used in the DC mode (see above) respond at frequencies up to 100 MHz. If very low IF is used, as may be the case in beating a molecular maser local oscillator against a radiation field due to an emission by the same molecule, then even a slower detector may be convenient.

Let us briefly study the basic limitations of such systems.

Noise behavior differs according to the value of $h\nu/kT$ compared to unity, T being the temperature of the mixer stage and ν the frequency. (In wavelength units, this comes to $\lambda T = 15\,000$ Kelvin-microns.)

If $\lambda T \gg 15\,000$ (e.g. $\lambda = 1000\,\mu$, $T = 300\,K$), then

$$P_{\min} = kT B_{\mathrm{IF}}^{1/2} \, \mathrm{WHz}^{-1/2},$$

where B_{IF} is the IF stage bandwidth [5]. Practical cases have achieved

$$P_{\min} = 7 \times 10^{-15}\,\mathrm{W\,Hz}^{-1/2} \quad \text{for} \quad B_{\mathrm{IF}} = 30\,\mathrm{MHz}.$$

If $\lambda T \ll 15\,000$, which is the most usual case as soon as cooled detectors are used, the noise is given by [15]:

$$P_{\min} = \frac{h\nu}{\eta}\frac{\Delta\nu}{B_{\mathrm{IF}}^{1/2}} = \frac{hc^2}{\eta}\frac{\Delta\lambda}{\lambda}\lambda^{-2}B_{\mathrm{IF}}^{1/2} \tag{3}$$

where $\Delta\nu$ is the spectral bandwidth which illuminates the detector, η the quantum efficiency of the detector.

In the case of wide-band detection of continuum, or quasi-continuum, signals, this may be written:

$$\frac{\text{Signal}}{\text{Noise}} = \frac{1}{h\nu}\frac{\partial P}{\partial\nu}B_{\mathrm{IF}}^{1/2}, \tag{3a}$$

where $\partial P/\partial\nu$ is the received power per unit frequency.

In the special case where an optically thick source of Planckian radiation at temperature T is observed through a diffraction-limited optics, (3a) becomes

$$\frac{\text{Signal}}{\text{Noise}} = \frac{1}{\exp(h\nu/kT) - 1}B^{1/2} \tag{3b}$$

which is a convenient formula.

Consider now two extreme, but interesting practical cases: very narrow-band spectroscopy ($\Delta\lambda/\lambda < 10^{-4}$) and average or large band detection ($\Delta\lambda/\lambda \gg 10^{-4}$).

Assume a spectral line of width $\Delta\lambda$, with $\Delta\lambda/\lambda \sim 10^{-6}$ (typical Doppler broadened interstellar line or coronal line), and assume this line to be the predominant source of radiation to fall on the detector, in order for $\Delta\nu$ to be essentially the spectral range where the line emits.

Consider the Si II line at $\lambda = 34.9\,\mu$ [2], $B_{\mathrm{IF}} = \Delta\nu = 10\,\mathrm{MHz}$

$$P_{\min} = 2 \times 10^{-16}\,\mathrm{W\,Hz}^{-1/2}.$$

For a molecular maser radiating at longer wavelength ($\lambda = 300\,\mu$), and having an assumed equal relative width, $\Delta\lambda/\lambda$, the minimum power becomes:

$$P_{\min} = 2 \times 10^{-18}\,\mathrm{W\,Hz}^{-1/2}.$$

Measurements [16] at $10.6\,\mu$ give practical limits only a factor 5 above theoretical values.

One therefore sees that $P_{\min}$ decreases as λ^{-2} with increasing wavelength, and that very high sensitivities may be in sight, comparable to what has been achieved in the radio range.

The spectral resolution is essentially limited by IF bandpass. Numbers as high as $R = 3 \times 10^8$ have been quoted at 10.6 μ [14]. Ultimately phase fluctuations in the LO power may limit the resolution.

The other extreme case is to observe a continuum radiation field. Assume a 100 K source, observed at 10 μ. To compare with wide-band detectors we may assume $B_{IF} = 1 \mu$, although the latter is not physically feasible. Equation (3b) gives:

$$(S/N) = 5 \times 10^{-3} \text{ for 1 s integration time.}$$

A similarly diffraction limited bolometer with a NEP $= 10^{-14}$ would give for the same integration time:

$$S/N = 30.$$

This gives an 'equivalent NEP' for the heterodyne:

$$\langle NEP \rangle_{equiv} = 6 \times 10^{-11} \text{ W Hz}^{-1/2}.$$

It is clear that this type of detection is very poor in terms of noise figure for wide-band signals, compared to the existing wide-band systems. Nevertheless, its intrinsically high spectral selectivity may be useful to extract information on a continuum being perturbed by numerous stray lines (as atmospheric absorption).

In conclusion, technological developments seem necessary before this technique can be applied to astronomy, especially in space. But, they may be very fast, and an entirely new tool in infrared range would drastically enhance the development of high resolution spectroscopy.

The other known advantage of this down-conversion of photons is the open field of long base-line interferometry, as has been suggested by Townes [17], but this is beyond the scope of the present review.

3.2. JOSEPHSON AC EFFECT

Since Josephson junction exhibit the dual property to emit monochromatic coherent radiation and to detect it, the junction could be used simultaneously as local oscillator and detector to generate the IF. When an infrared field at frequency ν is applied together with a bias voltage V, an intermediate frequency signal may be observed at the frequency $\nu_{IF} = |2 eV/h|$.

One of the problems is to give to the junction the proper geometrical shape, or to place it in a suitably resonant cavity, in order to enhance proper modes and coupling between the field and the junction.

This device, although technologically difficult, seems interesting because it may be intrinsically compact.

3.3. OTHER SCHEMES

Other infrared detection schemes have been proposed and are reviewed elsewhere [13], as are basic noise sources for these methods.

The general idea is to convert one infrared quantum into a quantum more easily detectable, e.g. a visible quantum.

– *Molecular beam spectrometer* is aimed to detect the specific emission of a given molecule. The radiation field excites the same transition in a molecular beam, creating a change in the rotational state J of the molecule. Magnetic field then allows discrimination of this quantum state.

– *Up conversion spectroscopy* observes the photon at frequency $\omega = (\omega_{signal} + \omega_{LO})$, the frequency addition being achieved in a non-linear crystal such as barium niobiate. Proper matching of polarization and direction of propagation ensures the condition

$$\mathbf{k} = \mathbf{k}_{signal} + \mathbf{k}_{LO} \quad (\mathbf{k} \text{ wave vector})$$

to be fulfilled for a small frequency range around ω_{LO}. Up-conversion has been achieved between 1.7 μ and visible but unexpected noise sources have been found, and more work has to be done before this technique could be applied to infrared spectroscopy.

4. Conclusion

At the present time, it is possible to say that the whole infrared range is now covered by sensitive detectors, some of them being close to limits imposed by background radiation.

Lowering their temperature may still improve detectors performances, then the background radiation will have to be eliminated. This will be achieved, either by narrowing the spectral range with cooled filters, and then losing some of the multiplex advantage of Fourier spectroscopy, or by removing the hot surrounding, either cooling the whole instrument if the Earth's atmosphere does not contribute to background, or going out into space.

Narrow-band spectroscopy may gradually apply to specific astrophysical problems, and balance its virtues against the wide-band systems which have, up to now, led to the most important discoveries in the field.

References

[1] Knowles, S. H., Mayer, C. H., Cheung, A. C., Rank, D. M., and Townes, C. H.: 1969, *Science* **163**, 1055.
[2] Pottash, S. R.: 1968, *Bull. Astron. Inst. Neth.* **19**, 469.
[3] Eddy, J. A., Léna, P. J., and MacQueen, R. M.: 1969, *J. Atmospheric Sci.* **26**, 1318.
[4] Encrenaz, T., Gautier, D., Vapillon, L., and Verdet, J. P.: 1971, *Astron. Astrophys.* **11**, 431.
[5] Martin, D. H. (ed.): 1967, *Spectroscopic Techniques for far infrared, Submillimeter and Millimeter Waves*, North-Holland, Amsterdam.
[6] Smith, R. A., Jones, F. E., and Chasmar, R. P. (eds.): 1968, *The Detection and Measurement of Infrared Radiation*, Clarendon Press, Oxford.
[7] Mitra, S. and Nudelman, S. (eds.): 1970, *Far Infrared Properties of Solids*, Plenum Press.
[8] Low, F. J.: 1961, *J. Opt. Soc. Am.* **5**, 1300.
[9] Langenberg, D. N., Scalapino, D. J., and Taylor, B. N.: 1966, *Proc. IEEE* **54**, 560.
[10] Mercereau, J. E.: 1969, in R. D. Parks (ed.), *Superconductivity*, Vol. 1.
[11] Ulrich, B. T.: 1970, *Proc. of the 12th Conference on Low Temperature Physics*, Kyoto.
[12] Richards, T. L.: 1969, *Appl. Phys. Letters* **14**, 394.
[13] Smith, H. A. and Townes, C. H.: 1969, in *Polarisation, Matière et Rayonnement*, Presses Univ. de France.
[14] Hinkley, E. D. and Kelley, P. L.: 1971, *Science* **171**, 635.
[15] Cummins, H. Z. and Swinney, H. L.: 1971 in Wolf (ed.), *Progress in Optics*, North-Holland, Amsterdam.
[16] Teich, M. C.: 1968, *Proc. IEEE* **56**, 37.
[17] Townes, C. H.: 1970, *Proc. of the Conference on Quantum Electronics*, Kyoto.

[18] Midwinter, J. E. and Warner, J.: 1967, *J. Appl. Phys.* **38**, 519.
[19] Wolfe, W. L. (ed.): 1965, *Handbook of Military Infrared Technology*, U.S. Government Printing Office.
[20] Zander, R.: 1971, Private communication.
[21] Jacobs, S. F. and Sargent, M.: 1970, *Infrared Phys.* **10**, 233.
[22] Gay, J.: 1970, *Astron. Astrophys.* **6**, 327.

DISCUSSION

P. E. Clegg: I would like to comment on the cooling of a Michelson interferometer before a detector. First of all the temperature stability of the instrument is important and the easiest way to keep something at constant temperature is to put it in a liquid bath at the boiling point of the liquid.

A second point concerns the position of the chopper. If one has this after the Michelson system and before the detector, then the noise one is looking at with the detector, although with a narrow band, is a low frequency noise, very low temperature drifts in the apparatus. On the other hand, if one puts the chopper before the apparatus then one has a hot chopper and therefore very large signals to deal with. It is very much simpler if one cools the whole lot down.

F. Kneubühl: I have a comment on the probable extension to long wavelengths of the Lead Tin Telluride laser.

If one wishes to extend this laser to longer wavelengths one runs into the same problem as Putley did when trying to extend his detector to long wavelength. This is a solid state device, the band edges of which are not well defined and for this reason Putley had to introduce magnetic fields. There are still discussions going on on the validity of this.

A further point concerns the travelling wave tubes which are heavy and unstable.

The stability depends on the stability of the high voltage. The same problem applies to the Josephson A.C. junctions. In this case the frequency is just defined by the voltage, and again one needs a very stable voltage.

LIQUID HELIUM-COOLED BOLOMETERS

G. CHANIN

Service d'Aeronomie de CNRS, Verrières-le-Buisson, France

Abstract. High responsivity, large surface infrared bolometers have been constructed using three types of thermo-sensitive elements: carbon resistors, germanium, and silicon. All three types have been successfully operated in the liquid helium range (1.5 to 4.2 K) as part of ground-based or balloon-borne experiments using the cryostats described in a separate paper.

1. Carbon Resistor Bolometers

Allen-Bradley $\frac{1}{4}$ W resistors of 8000, 10 000, and 12 000 Ω were ground down to thin wafers of 80 to 200 μ thickness. A 2 mm length of lead wire was retained, but turned down to 0.3 mm diameter before the grinding operation. The resulting metal lead protruding from the wafer was found to be in good mechanical, thermal, and electrical contact with the carbon wafer. Measurements of resistance versus temperature below 4.2 K yielded the temperature coefficients of resistance, $\alpha = 1/R \, dR/dT$, for these elements: for the 8200 Ω resistor $\alpha = -0.40$ deg^{-1}; for the 10 000 Ω resistor $\alpha = -0.64$ deg^{-1}; and for the 12 000 Ω resistor $\alpha = -0.22$ deg^{-1}. The 10 000 Ω resistor was therefore selected for mounting as the thermo-sensitive element in the three detector support shown in Figure 1. The resulting bolometer system was then mounted on the baseplate of the small cryostat. The optical system consisted of an entrance slit, a diffraction grating, and a collimating mirror and is similar to the fixed wavelength monochromator described in the paper of M. Herse. This system permitted simultaneous detection at 3 wavelengths of the light imaged on the entrance slit. Cooled filters and baffles incorporated in the bolometer support ensure isolation of a single

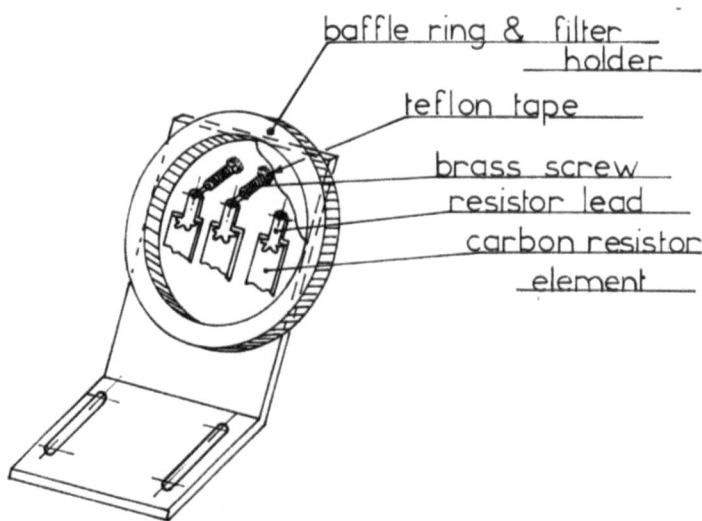

Fig. 1.   Mounting system for a three detector array.

TABLE I

Typical operating characteristics of infrared bolometers

	Carbon-resistors		Ga-doped germanium 0.120 Ω cm		Silicon
	Allen-Bradley 10 000 Ω	Texas instruments			Molectron Si-1.4 R
	Rectangle	Rectangle	Octagon		Square
Size (mm)	$2 \times 5 \times 0.08$	$2 \times 5 \times 0.100$	$5 \times 5 \times 0.100$		$5 \times 5 \times 0.4$
Operating temp (deg K)	4.2 K	4.2 K	4.2 K	1.85 K	1.80 K
Impedance (Ω)	3.05 MΩ	3.42 MΩ	153 kΩ	5.22 MΩ	725 kΩ
e/k (deg K)	–	17.6	12.4	12.4	8.1
$\alpha = 1/R(dR/dT)$ (deg K)$^{-1}$	0.64	1.0	0.7	3.8	2.5
Responsivity S (kV/W)	195 kV/W	213 kV/W	23	444	284
Conductance Ge (μW/deg)	10	33	9.4	22	2.5
Time Constant τ (ms)	30	40	100	6	10
Noise at 10 Hz (nV Hz$^{-1/2}$)	$\simeq 200$	67	<10	200	17
Noise Equiv. Power NEP (W Hz$^{-1/2}$)	10^{-12}	3×10^{-13}	$<3 \times 10^{-13}$	5×10^{-13}	6×10^{-14}
Specific detectivity D (cm Hz$^{1/2}$ W^{-1})	10^{11}	3×10^{11}	$>1.3 \times 10^{12}$	8×10^{11}	8.4×10^{12}

spectral order. The results for one detector mounted in this way are summarized in Table I. The device was employed to scan the lunar surface with a 6 arcsec spatial resolution and spectral resolving power of about 30. One visible light wavelength and two near infrared wavelengths between 1.5 and 3.5 μ were employed. Using a ground-based 152 cm telescope, signal-to-noise ratios of up to 50 were observed.

2. Semi-Conductor Bolometers

Semi-conductor single crystals have often been employed as helium-cooled bolometer elements because of their high responsivity and low noise. The most frequently used material has been gallium-doped germanium. The resistance vs temperature curves for several different concentrations of gallium in germanium were obtained in order to select material likely to be highly responsive. To avoid self-heating effects the measur-

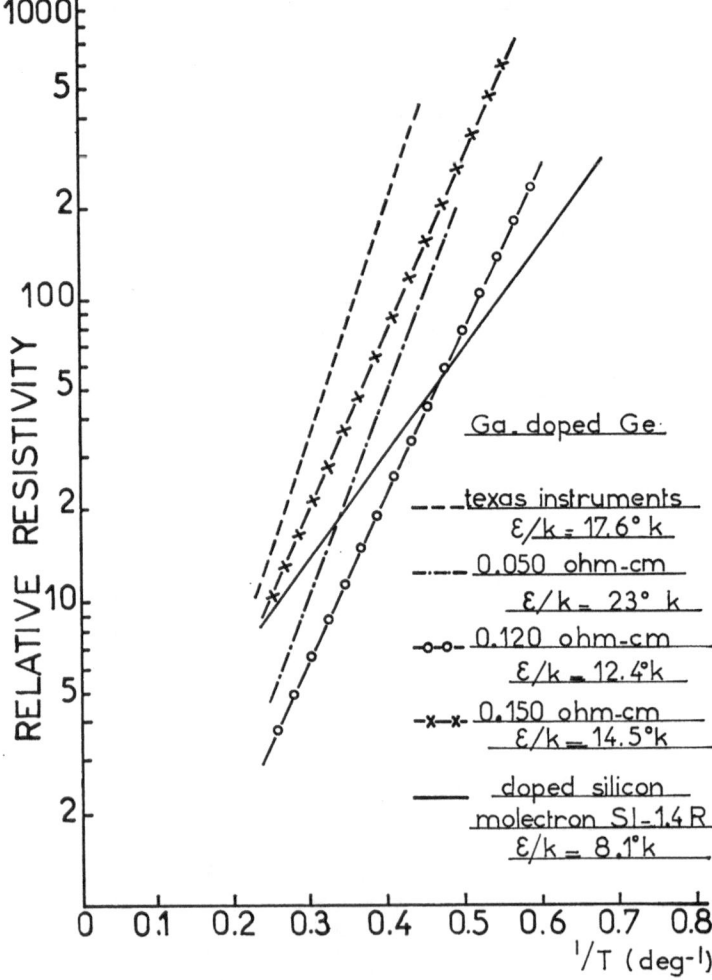

Fig. 2.   Temperature vs resistance curves of germanium and silicon bolometer elements.

ing current was kept small enough to limit dissipation to 10^{-8} W and the samples were immersed in liquid helium. The results for these measurements are shown in Figure 2 where, for comparison, two commercially obtainable detectors are also shown: gallium-doped germanium and doped silicon. The value of the temperature coefficient $\alpha = -[(\varepsilon/k)/T^2]$, may be easily obtained for a given operating temperature from these curves.

The mounting of the selected thermo-sensitive element depended upon the application. In one case, three rectangular elements of the 0.120 Ω cm material, $2 \times 5 \times 0.08$ mm, were mounted as were the carbon resistor elements of Figure 1. The lead wires were, in this case, two platinum wires 3 mm in length and 60 μ in diameter. Responsivities of 35 kV/W were obtained for a detector impedance of 1.2 MΩ at 4.2 K and with a time constant of 2 ms. No noise measurements were made of these detectors.

Other mountings involving octagonal elements $5 \times 5 \times 0.100$ mm and rectangular elements $2 \times 5 \times 0.100$ mm gave the results summarized in Table I. Note that although very high responsivities are possible, the noise level limits the improvement in detectivity over carbon resistor bolometers to about one order of magnitude. This is largely excess noise, the noise level in the absence of current being an order of magnitude lower.

The use of silicon as the thermo-sensitive element permits the quasi-total elimination of the excess noise because the current passes from the wire leads to degenerate semiconductor material (gold diffused) before entering the thermo-sensitive region. In addition, a shorter time constant and/or higher responsivity may be obtained because its Debye temperature is 75% greater than that of germanium, leading to a more than factor of five decrease in thermal heat capacity. Finally, far infrared radiation is more completely absorbed because the absorptivity is higher and the index of refraction is lower.

The curve marked Si-1.4 R in Figure 2 was obtained for a square element $5 \times 5 \times 0.4$ mm supplied by Dr McCaul of Molectron Inc. Its mounting was somewhat complicated because of the relatively high thermal conductivity of the leads (doped gold wire). The mounting system adopted is shown in Figure 3. Results with the entrance to the light cone blanked off are shown in Figure 4 and summarized in

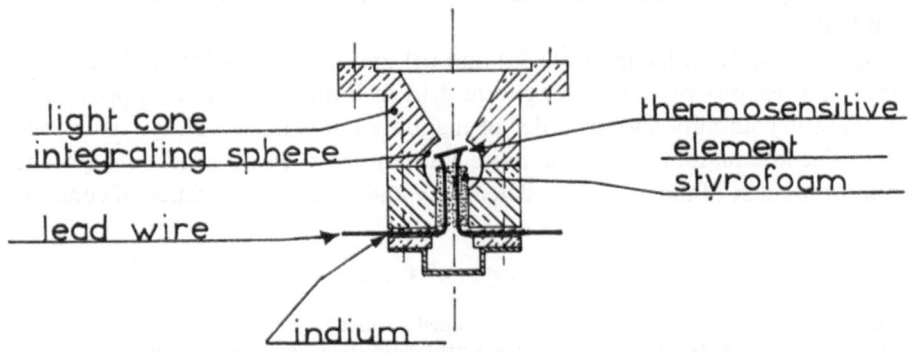

Fig. 3.   Silicon bolometer mounting in integrating sphere.

Table I. Note that the specific detectivity obtained for this detector classes it among the best yet obtained in the infrared region. Verifications of this value using submillimetre radiation are now in progress.

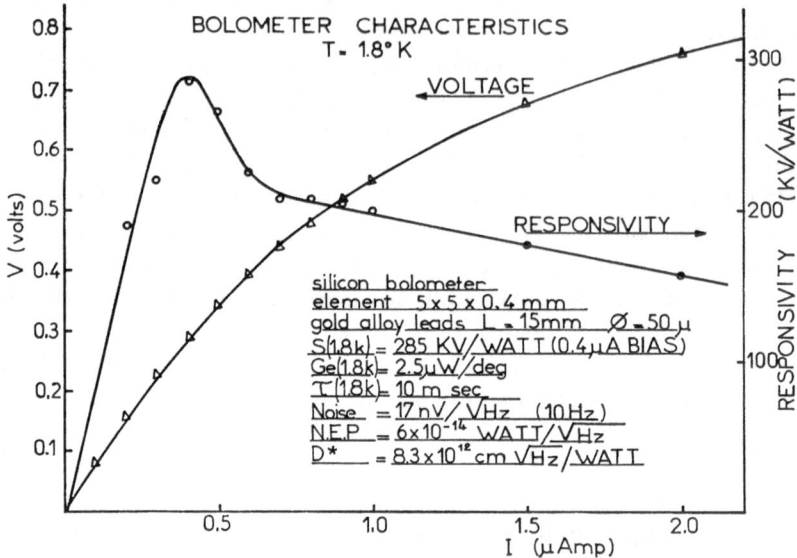

Fig. 4.   Electrical response of silicon detector.

The mounting and operating conditions necessary to obtain optimum performance from these (or any other) high-responsivity bolometers are as follows:

(a) cooled filters and baffles to avoid radiation heating of the detector and to limit background photon noise;

(b) adequate heat sinking of detector leads to the liquid helium bath (passing through the bath if possible);

(c) constant current bias source with cooled bias resistances;

(d) magnetic shielding of detector (if possible with superconductive shields);

(e) damping of mechanical and acoustic vibrations of the element and of the helium bath.

Failure to comply with conditions (a) and (b) may result in a lowered detector impedance and responsivity and a long term drift in both. Condition (c) provides optimum operating stability and an order of magnitude less noise voltage from the bias resistor. The attenuation of outside noise sources provided by conditions (d) and (e) permits attainment of the maximum detectivity under actual experimental conditions.

DISCUSSION

M. Renard: What is the efficiency of your blackened surface?

G. Chanin: None of the surfaces have been artificially blackened, except the Germanium detector for use in the middle infrared range.

Fig. 1.

The silicon detector is not blackened at all. The absorption is very high in the middle and far infrared. What one can do is to apply a coating to the back surface of the silicon detector in order to match its impedance to that of the free space, 377 Ω. This technique is feasible and would result in a detector absorbing a minimum of 50% of the incident radiation.

F. Melchiorri: I would like to report briefly (also on behalf of G. Dall'Oglio and G. Fantoni) on some of our results (already presented at the Meeting of the Italian Physical Society, October 1970, Venezia).

We realized high responsivity, low noise carbon bolometers cutting slabs from Allen Bradley 150 Ω resistors. The slab thickness is around 30 μ in the case of our most sensitive detector. The bolometer system is shown in Figure 1. We found a responsivity of about 10^6 V/W and a NEP lower than 10^{-12} W Hz$^{-1/2}$ at 1.2 K. The opacity of the bolometer in the far infrared has been tested using a Hitachi FIR spectrophotometer between 20 and 330 μ.

The transmission curve is shown in Figure 2.

The frequency response is shown in Figure 3. We found that the response time is due to the thermal capacity and conductivity of the bulk of the sample rather than of the leads; in fact the bolometer immersed in liquid helium has the same time constant, although the signal becomes a hundred times lower.

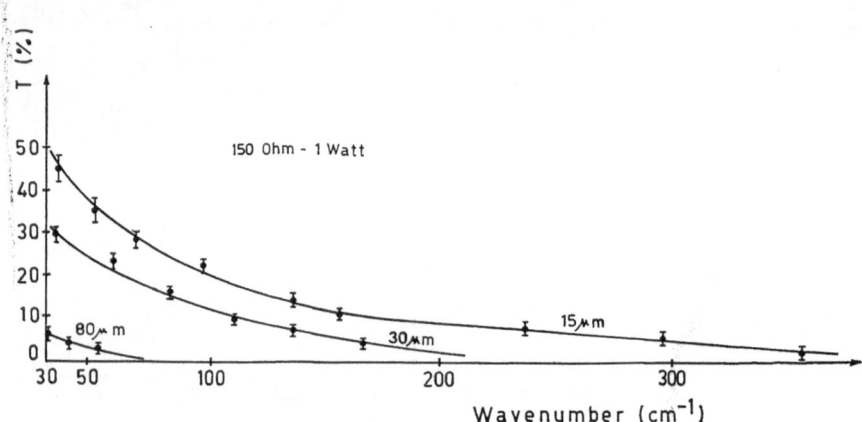

Fig. 2.

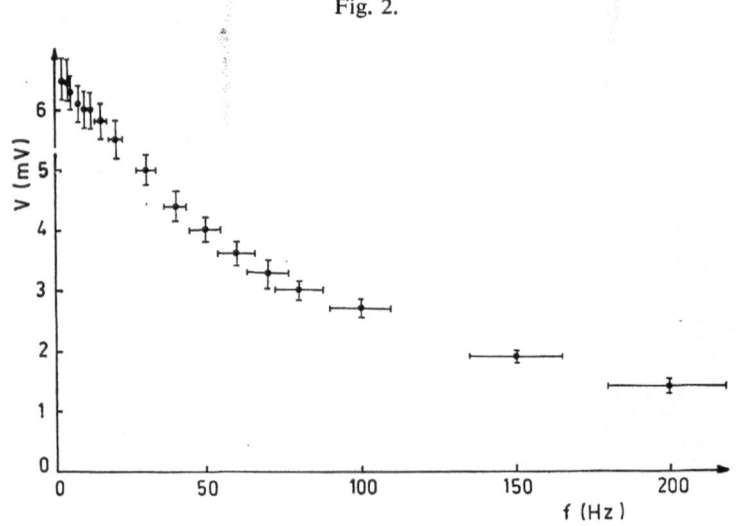

Fig. 3.

A NEW TYPE OF HELIUM-COOLED BOLOMETER

N. CORON, G. DAMBIER, and J. LEBLANC

Laboratoire de Physique Stellaire et Planétaire et Groupe Infra-Rouge Spatial de Meudon,
Meudon, France*

1. Introduction

Astrophysical far infrared experiments are still often limited by the detectors available, though high performances have been obtained, in particular with cooled bolometers [1, 2].

Detector noise becomes very limiting if cooled optics are used in experiments at sufficiently high altitude or with high frequency of modulation, so that photon and sky noise are low.

Is it possible to improve performances of cooled bolometers? How?

Let us see what are the theoretical and real limits before describing the new type of cooled bolometer we developed with improved performances.

2. Theoretical Limits

If detection of the temperature of the bolometer is made by a crystal having a resistance:

$$R(T) = R(T_0)\left(\frac{T_0}{T}\right)^A \tag{1}$$

the maximal responsivity is [1]:

$$\Phi_0 = h(A) \times \left(\frac{R_0}{T_0 G}\right)^{1/2}; \tag{2}$$

the time constant is

$$\tau = K(A)\frac{C}{G}, \tag{3}$$

where G is the heat conductivity between the bolometer and the thermostat due to phonon and photon exchange; and C is the heat capacity of the bolometer; $h(A)$ varies from 0.3 for $A=1$ to 1.0 for $A=9$; and $K(A)$ varies slightly from 0.80 for $A=1$ to 0.7 for $A=9$;

It is then possible to put under similar form the Johnson and the thermodynamic noises [1] so that the theoretical noise equivalent power is:

$$\text{NEP} \geqslant \frac{[8kT^2Gg(A)]^{1/2}}{\varepsilon_\lambda}, \tag{4}$$

where

$$g(A) = \tfrac{1}{2}\left(1 + \frac{1}{h(A)^2}\right); \quad \begin{array}{l} g(1) \sim 6.0 \\ g(4) \sim 1.5 \\ g(9) \sim 1 \end{array}$$

ε_λ is the effective emissivity of the bolometer at the wavelength of use.

* B.P. No. 10 – 91 – Verrières-le-Buisson – France.

Manno and Ring (eds.), Infrared Detection Techniques for Space Research, 121-131. All Rights Reserved
Copyright © 1972 by D. Reidel Publishing Company, Dordrecht-Holland

The ultimate limit is for the bolometer exchanging only photons with the thermostat:

$$\text{NEP} \geqslant \frac{[32k\sigma\varepsilon g(A)ST_0^5]^{1/2}}{\varepsilon_\lambda}, \tag{5}$$

where ε is the average emissivity from $\lambda=0$ to $\lambda=\infty$; S is the whole area of the bolometer; and T_0 is the background temperature.

Fig. 1.   Photograph of far infrared bolometers: on the left, bolometers with integration sphere; on the right, bolometers with light-cone, one having additional cooled filter.

At 4 K for 1 mm² detector, it gives limit NEP of 1×10^{-16} W/$\sqrt{(\text{Hz})}$

But, to eliminate parasitic noise and slow derivation, the bolometer is used at a frequency superior to f_0 (from 1 Hz to 10 kHz); so we must have:

$$2\Pi f_0 \tau \leqslant \tfrac{1}{2}. \tag{6}$$

From Equation (4) we have:

$$\text{NEP} \geqslant [32\Pi K\rho(A)f_0]^{1/2} \times \frac{\sqrt{C}}{\varepsilon_\lambda} \times T \tag{7}$$

where

$$\rho(A) = K(A)g(A); \quad \rho(1) = 4.8$$
$$\rho(4) = 1.2$$
$$\rho(0) \sim 0.7$$

We see the importance of having a very low value for $\sqrt{C}/\varepsilon_\lambda$.

For most materials and for a bolometer with metallic and dielectric parts:

$$C \sim C_0 T + C_1 T^3.$$

We have the two limit cases for a given type of bolometer with fixed geometry and technology:

$$T > T_x \quad \text{NEP} \sim T^{5/2}$$

$$T < T_x \quad \text{NEP} \sim T^{3/2}.$$

For doped silicon: $T_x \sim 1.5$ K.
For doped germanium: $T_x \sim 0.6$ K   (Ref. [3] and Figure 2).

So under T_x there is less to gain with lower temperature and we must try to diminish $\sqrt{C}/\varepsilon_\lambda$ by optimal choice of materials.

Till now, far infrared cooled bolometers used absorption in the crystal itself which is blackened (very thin black) only for the near infrared [2, 4, 5, 6].

We have then:

$$C = [\tfrac{1}{3}C \text{ (wires)} + C \text{ (black for near infrared)} + C \text{ (solders)}]$$
$$+ C \text{ crystal)} = C_p + C_a.$$

With good technology, we have generally for the classical bolometer:

$$C_p < C \text{ (crystal)} = C_a$$

$$C_a = S \times e \times C_0(T), \tag{8}$$

where S is the area of the bolometer; e the thickness; and C_0 the intrinsic heat capacity of the crystal.

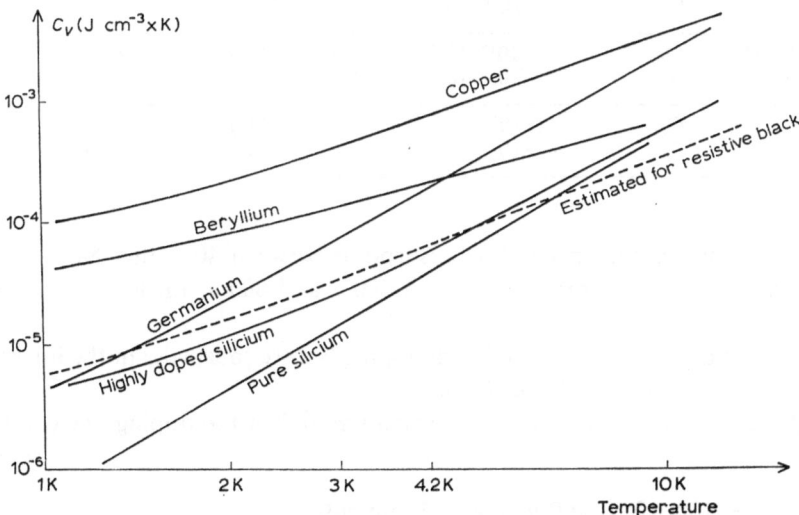

Fig. 2. Variation of volume heat capacity with temperature for typical materials used in classical bolometer or in the three parts bolometer.

We want $\varepsilon_\lambda \sim 1$ and we choose e so that:

$$\alpha_\lambda e \sim 1 \quad \text{and then} \quad \varepsilon_\lambda \sim 1 - R \sim 0.7 \text{ for Ge and Si;} \tag{9}$$

α_λ is the absorption coefficient of the crystal.

We then have from Equation (7):

$$\text{NEP} \geqslant \sqrt{32 \Pi K \rho(A)} \times \sqrt{f_0 C_0(T)} \times \sqrt{\frac{1}{\alpha}} \times T \times \sqrt{S} \tag{10}$$

or

$$D^* = \frac{\sqrt{S}}{\text{NEP}} \sim \frac{1}{T} \left(\frac{\alpha_\lambda}{f_0 \times C_0(T)} \right)^{1/2}. \tag{11}$$

The importance of α_λ comes from the several orders of magnitude by which α_λ can change from one material to another or between different wavelengths.

In the far infrared, we can have high absorption in silicium or germanium, either by impurity ionization [7, 8] or by the photon induced hopping mechanism [9] or in the very far infrared by the hot electron effect [6].

In Table I, we compare absorption coefficients and corresponding part of the spectrum for these mechanisms.

TABLE I

Far infrared absorption coefficients in Ge and Si and wavelength ranges of different mechanisms

Absorption mechanism	Material	Doping (at cm^{-3})	Wavelength range	Peak absorption coefficient	Ref.
Impurity ionization	Ge	10^{15}–10^{17}	$10\,\mu \rightarrow 150\,\mu$	$\leqslant 500$ cm^{-1}	[7]
	Si	10^{16}–10^{18}	$3\,\mu \rightarrow 40\,\mu$	$\leqslant 1000$ cm^{-1}	[8]
Photon induced hopping mechanism	Ge	10^{15}–10^{17}	$500\,\mu \rightarrow 1$ cm	~ 100 cm^{-1}	[9]
	Si	10^{16}–10^{18}	$100\,\mu \rightarrow 1$ mm	~ 500 cm^{-1}	[9]
Hot electron effect	Ge	$\geq 10^{16}$	$\lambda \geq 500\,\mu$	1–10 cm^{-1}	[6]
	Si	$\geq 10^{17}$	$\lambda \geq 500\,\mu$	1–10 cm^{-1}	–

We see that doped silicon has low absorption between $40\,\mu$ and $200\,\mu$ while germanium has minimum absorption between $200\,\mu$ and $600\,\mu$. In these parts, we have $\alpha \sim 1$ cm^{-1}.

To have high absorption, we need high doping but we must stay in the intermediate conduction zone in order to have $A > 3$.

By appropriate choice of silicon or germanium and of the doping, we can hope to reach:

$$\alpha \sim 100 \quad \text{for} \quad 200\,\mu < \lambda < 3 \text{ mm with Si}$$

$$\alpha \sim 100 \quad \text{for} \quad 50\,\mu < \lambda < 200\,\mu \text{ with Ge.}$$

To have a good response in the whole far infrared for the same bolometer we must use a thickness e so that:

$$\alpha e \sim 1$$

So, to have a bolometer in the far infrared using semiconductor absorption we need thickness superior to $100\ \mu$ in all cases.

3. Principle of the Three-Parts Bolometer

We have developed a new type of cooled bolometer in which the area absorbing the infrared is distinct of the crystal used to detect temperature variations of the system. In most cases, we have in fact three parts: the absorber very thin which is deposit on a material of high heat conductivity (substrate) and the crystal soldered on this (cf. Figure 3).

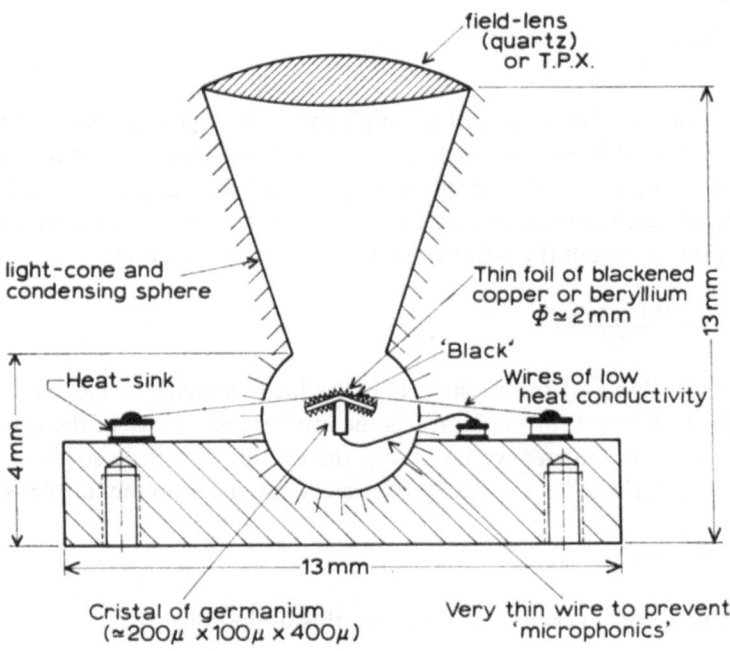

Fig. 3.   Schematic of the three parts bolometer with optics for the far infrared and antimicrophonics suspension.

A similar idea was proposed by F. Low in 1969 with the composite Si-Ge bolometer but we don't know what performances he obtained [10].

This permits to optimize the factor $\sqrt{(C)}/\varepsilon_\lambda$ for the absorbing area independently of the crystal.

We must have for the absorbing area, at the same time $\varepsilon_\lambda \sim 1$ and a high heat conductivity so that there is no temperature gradient in the bolometer at the frequency of modulation. This gives the condition with (6):

$$S < \frac{1}{\omega} \frac{G_s}{C_s}, \tag{12}$$

where G_s is the heat conductivity of the composite area; C_s is the heat capacity of the composite area; and $\omega = 2\pi f_0$.

At very low temperatures, mean free path λ of electrons or phonons can be greater than the thickness e of the substrate of the area [11].

Then:

$$G_s = G_{s0} \times \frac{e}{\lambda} \log \frac{\lambda}{e}, \tag{13}$$

where G_{s0} is the heat conductivity of a large piece.

For a given material at low temperature, this limit is nearly independent of T because:

$$\frac{G_{s0}}{C_s} \times \frac{1}{\lambda} \sim C_{s:e}.$$

This limitation may be important for high speed or large area bolometers.

Generally for $S > 0.3$ mm^2 the heat capacity of S is superior to that of the crystal which can be cut very little though technological problems may arise.... But there is the limitation of having the electrical field lower than the impact ionization value E_I at the polarization current of the bolometer. This gives the condition:

$$V \gg \frac{0.1 T_0 \rho \omega C}{E_I^2}, \tag{14}$$

where V is the volume of the crystal; C is the heat capacity of the bolometer; $\omega = 2\Pi f_0$; $E_I \sim 3 - 10$ V cm^{-1} for Ge; and ρ is the intrinsic resistivity of the crystal.

To optimise the bolometer we try to cut the crystal very little so that: $V \times C_0 \ll C$ where C_0 is the intrinsic heat capacity of the crystal. This is compatible with (14) if:

$$\frac{0.1 T_0 \rho \omega C_0}{E_I^2} \ll 1. \tag{15}$$

Adequate value of ρ permits to overcome this limitation.

A. EXPERIMENTAL

We have used for the high diffusivity substrate thin foils of copper or beryllium with thickness between 3 μ and 20 μ. To obtain high absorption and flat response, we have blackened these foils with resistive absorbers chosen so that between λ_{min} and λ_{max} they are a 'dilute resistive medium'.

$$\lambda_{min} = 2\pi c \tau < \lambda < \lambda_{max} = \frac{2\pi c \varepsilon_0}{\sigma}, \tag{16}$$

where τ is the relaxation time of electrons; and σ the conductivity.

The impedance is matched to the vacuum impedance of 377 Ω and the thickness is made superior to the skin depth at λ_{max}:

$$e > \frac{\lambda_{max}}{\pi\sqrt{2}\sqrt{\mu_r}}. \tag{17}$$

We have used ferrites to optimize absorption because they have high plasma frequency ($\tau \sim 2 \times 10^{-15}$ s, [12]) and high permeability.

We have, for example, measured reflectivity of a 10 μ thickness deposit of ferrous ferrite at 45° incidence. It was less than 25% in the whole far infrared. This is less than reflectivity of Ge or Si.

There is yet much work to do to know far infrared properties and conduction mechanism of ferrites but they are very promising as absorbers for infrared spectrometers and other experiments.

The crystal of selected doped germanium is soldered at the perpendicular of the metallic foil. Sizes are of the order of 100 $\mu \times$ 100 $\mu \times$ 400 μ for low G value bolometers.

In Figure 4, we have plotted variations of resistivity with temperatures of characteristic crystals. By choice of the appropriate crystal, we can match the impedance of the bolometer to the minimum noise impedance of the preamplifier at the wanted temperature. Some crystals have a value of A superior to 8.

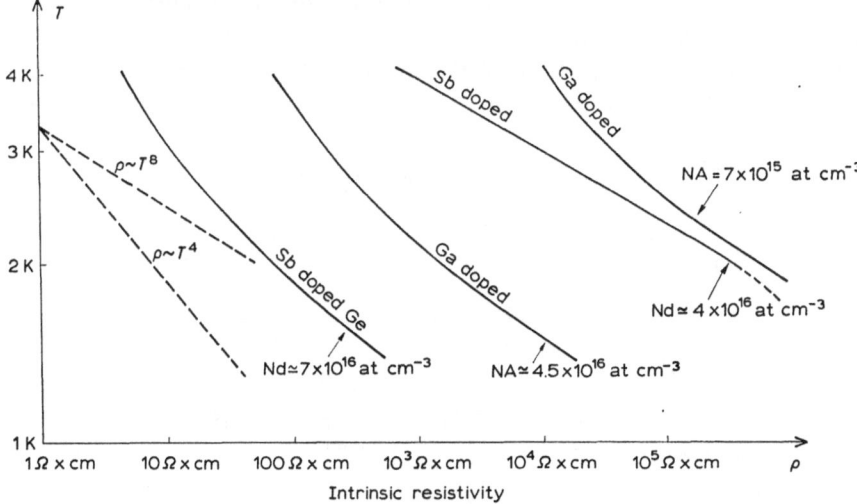

Fig. 4. Intrinsic resistivity versus T (with logarithmic scales) for selected doped crystals of germanium with low compensation ratio.

Suspension wires are soldered on one side at the substrate and on the other at an insulated heat-sink. The other extremity of the crystal is connected to the electronics by a very thin wire so that piezoresistance in Ge is eliminated (see Figure 3).

For the far infrared, the bolometer is mounted inside an integration sphere and in many cases with a light-cone and a field-lens at the entrance. So we can use the maximal throughput of the bolometer, $\Pi \cdot S$ with any aperture of the optics.

In Figure 2, we compare volumic heat capacities of the different components. It can be seen that between 2 K and 5 K, beryllium or copper substrates are more interesting than silicium because they can be obtained technologically in very thin foils (3 μ). Besides, the heat capacity of the absorber for the very far infrared is superior to this of the substrate. We have plotted an estimate of the heat capacity of the 'black' in agreement with our measurements and with [13].

In the case of fast bolometers, or of near infrared ones, we are looking for the possibility of having better substrates but technological problems have to be solved.

4. Calibration – Noise – Performances

We have tested bolometers with the standard following procedure:

– measurement of the *V-I* curve of the crystal at different temperatures in liquid helium (this permits to test hot electron effects).

– the bolometer is mounted on the cold plate of the cryostat, in the vacuum and we measure *V-I* characteristics and the electrical noise with 1 Hz bandwidth wave analyzer in function of *T*, *I*, and *f*.

– optical calibration is made in the near infrared (2–14 μ) with narrow bandwidth interferential filters and a reference black body.

In the far infrared, we make broad band calibration using cooled glass and Irtran 6 filters giving a peak of energy at 400 μ on a 300 K black body. This is correlated with the study of the resistive coating in a far infrared monochromator.

With good precision ($\pm 5\%$) we have agreement in the near infrared between optical and electrical responsivity. But in the far infrared calibrations are very imprecise

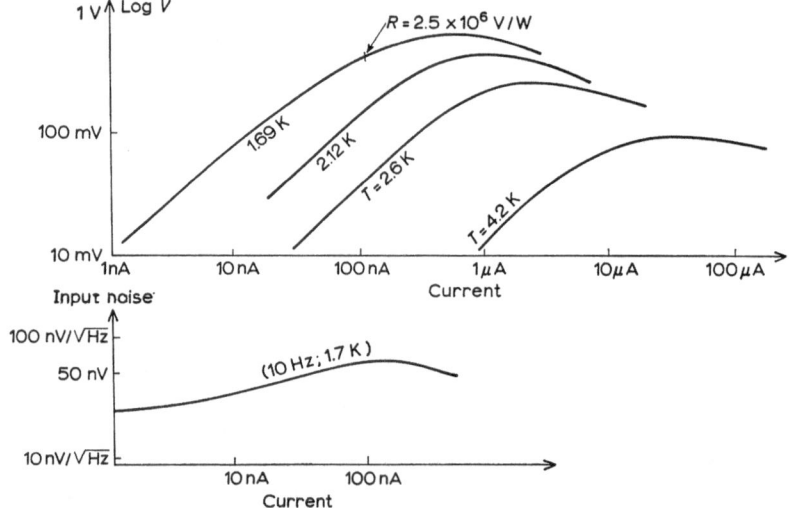

Fig. 5. Typical voltage-current characteristics for three parts bolometer G21 and input equivalent noise at 10 Hz (with 1 Hz bandwidth versus *I*. It has an effective diameter of 2 mm and the time constant is $\tau(4.2 \text{ K}) \simeq 20$ ms, $\tau(1.8 \text{ K}) \simeq 10$ ms. At 1.7 K: responsivity $= 2.5 \times 10^6$ V/W, $G = 7 \times 10^{-7}$ W/K. Far infrared NEP is better than 3×10^{-14} W/$\sqrt{\text{Hz}}$ at 1.8 K.

because of diffraction and of uncertainties in the black body emissivity. We can just conclude that the efficiency of our bolometers is superior to 40% but independent study of the absorber permits to affirm efficiency superior to 70% for the bolometer itself till 1 mm wavelength.

Extraneous sources of noise will be discussed in another paper. With good ohmic contacts and careful mounting of the detector, we have as a rule, not more than twice the theoretical noise at 10 Hz at the point of maximum responsivity (Figure 5).

We have compared the theoretical (Equation (7)) and real performances obtained by the three parts bolometer with these reported in literature by Low *et al.* (refs. [2, 4, 10, 14]).

Comparison appears in Figure 6 with NEP$=f(S)$. Theoretical limits have been calculated using heat capacity of Figure 2 [3, 13, 15] and assuming:

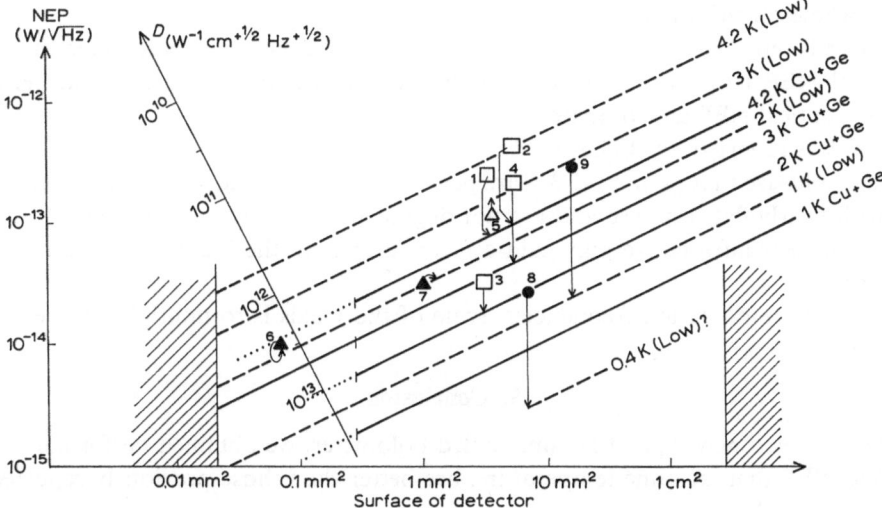

Fig. 6. Comparison of performances of classical and three parts bolometers at modulation frequency of 10 Hz for different bath temperatures. Straight lines are limits in case of no extraneous noise and perfect absorption. Points 1-2-3-4 are real far infrared performances of three parts bolometers. Point 5 is a classical type bolometer we built for near infrared. Points 6-7-8-9 are performances reported in literature (see text and refs. [2, 4, 10, 14]). Arrows indicate at which bath temperature and theoretical limit it corresponds.

– for Low's type:
$e \sim 250\ \mu$ (corresponding to $\alpha \sim 40\ \mathrm{cm}^{-1}$)
$\varepsilon_\lambda \sim 1 - R \sim 0.7$
$A = 4 \rightarrow \rho(A) \sim 1.2$
– for the three parts bolometer:
$e\ (\mathrm{Cu}) \sim 3\ \mu$
$e\ (\mathrm{ferrite}) \sim 10\ \mu$ with density ~ 1
minimal $\varepsilon_\lambda \sim 0.7;\ A = 6$
$V\ (\mathrm{crystal}) \sim 100\ \mu \times 100\ \mu \times 400\ \mu$

The comparison is made at low frequency (10 Hz) for detectors built with minimal G value compatible with:

$$2\pi \times 10 \times \frac{C}{G} < \frac{1}{2}.$$

Because of logarithmic scales, we can plot NEP and D^* on the same graphic. Points 1-2-3-4 correspond to reproductible performances we obtained with our three parts bolometer. The NEP is the far infrared NEP and the arrow indicates at which temperature and theoretical limit it corresponds.

Performances of 4 have been obtained in airborne experiment without degradation at 3 K.

Point 5 is a classical type bolometer we built for the near infrared with $A = 8$ and $e \sim 200\ \mu$.

Points 6 and 7 are reported by Low [2, 10] but it is not clear if the NEP is the same in the whole far infrared.

Point 8 is reported by Drew and Sievers for a 0.4 K Low's type bolometer and is an electrical NEP [16]. So at 2 K we obtained equivalent performances, and perhaps better if optical NEP are compared.

Point 9 is reported by Richards [4].

We emphasize that points 1-2-3-4 are performances obtained in the whole far infrared with nearly flat response and are not degraded in airborne experiments.

For the near infrared we may gain a factor 2 or 3 on the NEP; in the same other conditions.

For higher modulation frequencies, scale of the y axis is to multiply by the factor $\sqrt{(f)}/10$.

5. Conclusion

By developing a new type of helium cooled bolometer, we obtained performances on D^* and NEP that, at same temperature, are better than those previously reported in the far infrared.

Analysis of real limitations shows that improvements are still possible, in particular by careful optimization of parameters for each experimental conditions. A very promising one is the construction of selective bolometers specially adapted to narrow-band photometry or high resolution spectroscopy. The development of the three parts bolometer makes it possible by using selective absorbers instead of a resistive 'black'.

Acknowledgements

This research was mainly supported by the Centre National d'Études Spatiales. Additional support came from the CNRS through general laboratory equipment.

We wish to thank specially Professor P. Léna for encouragement and helpful discussions.

References

[1] Low, F. J.: 1961, *J. Opt. Soc. Am.* **51**, 1300.
[2] Low, F. J.: 1965, *Proc. I.E.E.E.* **5**, 516.

[3] Keesom, P. H. and Seidel, G.: 1959, *Phys. Rev.* **113**, 33.
[4] Richards, P. L.: 1965, *J. Pure Appl. Chem.* **11**, 535.
[5] Zwerdling, S., Theriault, J. P., and Reichard, H. S.: 1968, *Infrared Phys.* **8**, 271.
[6] Oka, Y., Nagasaka, K., and Narita, S.: 1968, *Japanese J. Appl. Phys.* **7**, 611.
[7] Putley, E. M.: 1964, *Phys. Stat. Solidi* **6**, 571.
[8] Soret, R. A.: 1967, *J. Appl. Phys.* **38**, 5201.
[9] Blinowski, J. and Mycielski, J.: 1964, *Phys. Rev.* **136**, 266.
[10] Low, F. J.: 1970, *Cryogenics and IR Detection*, Boston Technical Publishers, p. 21.
[11] Carruthers, J. A., Cochran, J. F., and Mendelssohn, K.: 1962, *Cryogenics*, p. 160, March.
[12] Samokhvalov, A. A. and Afanas'ev, A. Ya.: 1969, *Soviet Phys. Sol. St.* **10**, 2172.
[13] Kowel, J. S.: 1956, *Phys. Rev.* **102**, 1489.
[14] Zwerdling, S.: 1970, *Cryogenics and IR Detection*, Boston Technical Publishers, p. 38.
[15] Johnson, J.: 1960, Compendium of Properties of Materials at Low Temperature N.B.S.
[16] Drew, H. D. and Sievers, A.: 1969, *Appl. Opt.* **8**, 2067.

PERFORMANCE TESTS OF InSb AND Ge BOLOMETERS

P. E. CLEGG and J. S. HUIZINGA

Physics Dept., Queen Mary College, London, U.K.

1. Introduction

The theory of the InSb hot electron bolometer was first considered by Rollin [1] and a working detector was constructed by Kinch and Rollin [2]. We have investigated carefully the optimization of this type of detector and have made absolute, wideband, calibration measurements under operating conditions. Comparison measurements have also been made with a gallium-doped germanium bolometer under a restricted range of operating conditions; it is found that under these conditions the performance of the InSb detector is better by a factor of ~25. The indium antimonide detector has been used in the laboratory for high resolution spectroscopy using the polarizing interferometer described by Martin and Puplett [3].

2. Detector Circuitry

The detector circuit is shown schematically in Figure 1. This circuit is the same as that used by Kinch and Rollin except for the use of a polycarbonate blocking capacitor C_B in place of a backing-off circuit. L_p and L_s are the primary and secondary inductances of the transformer used to match the low impedence crystal to the noise characteristics of the amplifier. The effective turns ratio, n, of the transformer is defined by

$$n^2 = L_s/L_p$$

R_s represents both ohmic resistance and losses in the transformer. Both the blocking capacitor and the transformer are in the cryogenic enclosure at the same temperature as the crystal. C is an ambient temperature tuning capacitor, adjusted to make the source presented to the amplifier real.

Study of the circuit equations indicates that, for optimum performance, C_B should be tuned with L_p at the operating frequency; however, the space available in the cryostat limits the size of C_B to about 4.7 μF. The transformer is wound on a Standard Telephones and Cables permalloy core, code number WP3090/27. In practice about 10 000 turns can be wound on this core giving, a secondary inductance L_s of 9 H. Measurements gave a value for R_s of 700 Ω, practically independent of turns ratio and of temperature.

A denotes a low noise FET amplifier, described by Cantarano and Pallottino [4], which has an optimum noise factor, referred to a source at 1.8 K, of 1.94 for a source resistance of 260 kΩ.

A computer programme was written to calculate the overall noise factor for the

Manno and Ring (eds.), Infrared Detection Techniques for Space Research, 132–140. All Rights Reserved
Copyright © 1972 by D. Reidel Publishing Company, Dordrecht-Holland

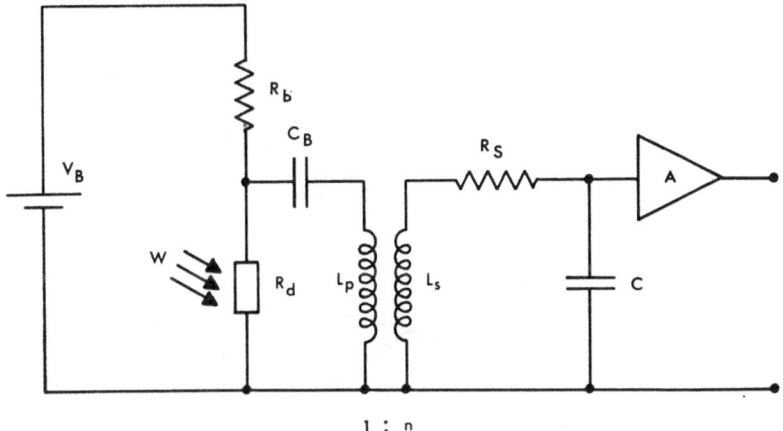

Fig. 1. InSb detector circuitry.

transformer-amplifier system for various values of n and for various crystal resistances. For a crystal resistance of 70 Ω at a temperature of 1.8 K operating at 1 kHz, the optimum noise factor, *referred to a source at 1.8 K*, was 2.44 at a turns ratio of 47.

3. d.c. Measurements

For a detector element of d.c. resistance R_d, biassed with a current I through a bias resistor R_B and loaded with an impedence Z, it can be shown that the a.c. responsivity $\mathfrak{R}$ is given by

$$\mathfrak{R} = \frac{\partial v}{\partial W} = \frac{1}{I}\left|\frac{r\alpha}{1 + 2r\alpha}\right|, \tag{1}$$

where

v = a.c. output voltage
W = a.c. (modulated) incident power
$r = (Z')/(R_d + Z')$
$Z' = R_B \| Z$

and

$$\alpha = -\tfrac{1}{2}\frac{d(\ln R_d)}{d(\ln V)},$$

where

V = d.c. voltage across the crystal.

Figure 2 shows measured values of V plotted as a function of I. From this curve, R_d and hence α have been calculated; R_d is shown. On the assumption that $r = 1$, which is justified for this case, $\mathfrak{R}$ has been calculated from Equation (1). From the plot of $\mathfrak{R}$ as a function of I it can be seen that the maximum value is 320 V W^{-1}.

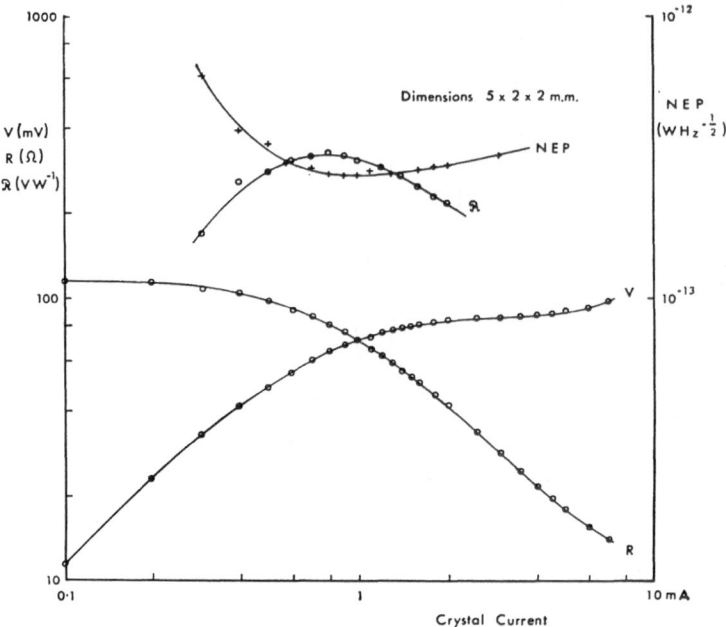

Fig. 2.   Parameters of InSb crystal obtained from d.c. measurements at 1.8 K.

If the detector were Johnson noise limited, the noise equivalent power (NEP) would be given by

$$\text{NEP} = (1/\Re)(4kTR_d)^{1/2} \tag{2}$$

The NEP has also been plotted in Figure 2 and it can be seen that the minimum value is 2.7×10^{-13} W $\text{Hz}^{-1/2}$. This (shallow) minimum occurs at a current of 1 mA. Since the noise factor of the transformer-amplifier system is 2.44 we can expect an overall NEP of $\sqrt{(2.44)} \times 2.7 \times 10^{-13} = 4.2 \times 10^{-13}$ W $\text{Hz}^{-1/2}$.

So far we have considered the responsivity and NEP of a specific detector element. Both $\Re$ and NEP depend upon the dimensions of the crystal. For comparison with other elements it is convenient to define quantities which depend on the crystal material alone. We therefore define a specific responsivity, $\mathfrak{r}$, and a specific noise equivalent power, nep, in the following way

$$\mathfrak{r} = (\partial E)/(\partial w)$$

where E is the a.c. electric field strength in the material and w is the power incident per unit volume of the material. Clearly

$$\mathfrak{r} = A\Re \tag{3}$$

where A is cross section of the crystal, transverse to the current flow.

From Equation (1) we have

$$\mathfrak{r} = \frac{1}{J}\frac{\alpha}{1 + 2\alpha}$$

where J is the current density. (N.B. α is independent of crystal dimensions. We have put $r=1$ since we can always arrange for r to have the same value for two crystals.) The specific noise equivalent power is defined by

$$\text{nep} = \frac{1}{\mathfrak{r}} (4kT\rho)^{1/2} \tag{4}$$

where ρ is the resistivity of the crystal. From Equations (2), (3), and (4) it can be seen that

$$\text{nep} = \text{NEP}/\sqrt{v} \tag{5}$$

where v is the volume of the crystal.

For the present crystal of dimensions $5 \times 2 \times 1\frac{1}{2}$ mm, we have from the d.c. measurements, at an optimum bias current density of 330 A m^{-2},

$$\mathfrak{r} = 9.6 \times 10^{-4} \text{ V m}^2\text{W}^{-1}$$

$$\text{nep} = 2.2 \times 10^{-9} \text{ W m}^{-3/2} \text{ Hz}^{-1/2}$$

and for the overall nep we get

$$\text{nep}_0 = 3.4 \times 10^{-9} \text{ W Hz}^{-1/2}\text{m}^{-3/2}.$$

4. Responsivity Measurements

Figure 3 shows the apparatus used for measuring the responsivity of the detector. In order to obtain an absolute callibration of the system, it was decided to attempt to produce black-body radiation in the wavelength region longer than 300 μm. Various shaped bodies, marked 'body under consideration', were tested as black-body radiators using the method suggested by Shivanandan et al. [5]. The body under consideration is covered successively by a series of baffles, each successive baffle in the series having a greater part of its area cut away. We assume that, for the baffle with the smallest area cut away, the radiation emitted is black-body, since it is true cavity radiation. A plot is made of the total radiation from the baffled body versus the area of opening in the baffle.

If this plot is a straight line it indicates that the radiation from body is itself black-body, since otherwise there would be a falling off in energy emitted with increasing exposed area. It was found that a simple cone, of apex semi-angle 30°, and painted internally with Parsons Optical Black, approximated well to a black body.

In the actual callibration the cone was heated by stirred, electrically heated, water, and the detector measured directly the difference between the radiation emitted by the cone and the (constant) radiation, reflected in the angled chopper, emitted by the cooled walls of the lower part of the apparatus. For mechanical reasons this measurement was made at a chopping frequency of 150 Hz. A plot was made of a.c. detector output voltage v versus cone temperature T and hence the slope, $(\partial v)/(\partial T)$, was obtained. Such a plot is shown in Figure 4.

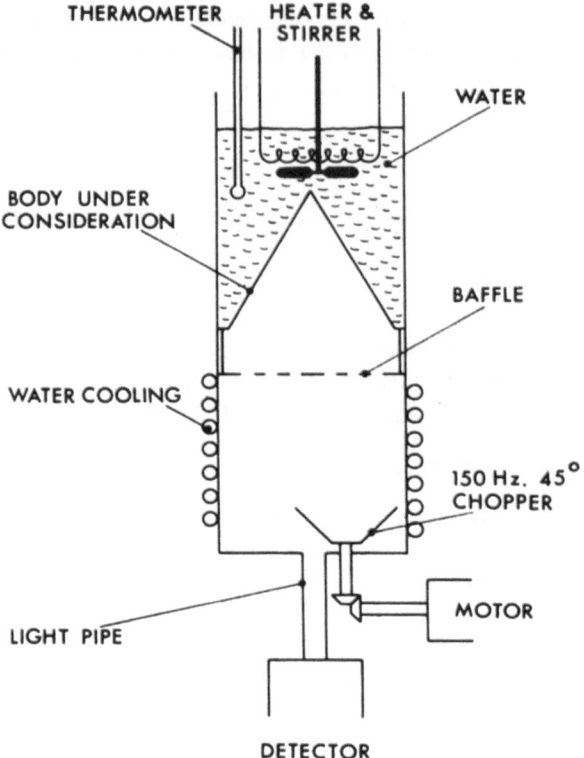

Fig. 3.   Schematic diagram of apparatus used in measuring detector responsivity.

Now the detector responsivity is defined as

$$\Re = \frac{\partial v}{\partial W} = \frac{\partial v}{\partial T}\frac{\partial T}{\partial W} \tag{6}$$

and for black-body radiation in the Rayleigh-Jeans region we have

$$W = 2A\Omega ckT \int_0^\infty f(\tilde{v})\tilde{v}^2 \, d\tilde{v} \tag{7}$$

where A is the detector aperture, Ω is the solid angle of acceptance of the detector and $f(\tilde{v})$ is the filter characteristic as a function of wavenumber $\tilde{v}$. For this experiment we used sooted polythene for which [6]

$$f(\tilde{v}) = e^{-0.21\tilde{v}t} \tag{8}$$

where t is the thickness of the sooted polythene filter. For the thickness used of 4 mm this gives a sharp high frequency cut-off and, when combined with the Rayleigh-Jeans spectrum, gives a maximum intensity at 400 μm, well above the detector cut-off. Insertion of Equation (8) in Equation (7) gives

$$\frac{\partial T}{\partial W} = \frac{(0.21t)^3}{4A\Omega ck} \tag{9}$$

and from Equations (6) and (9), the responsivity can be found.

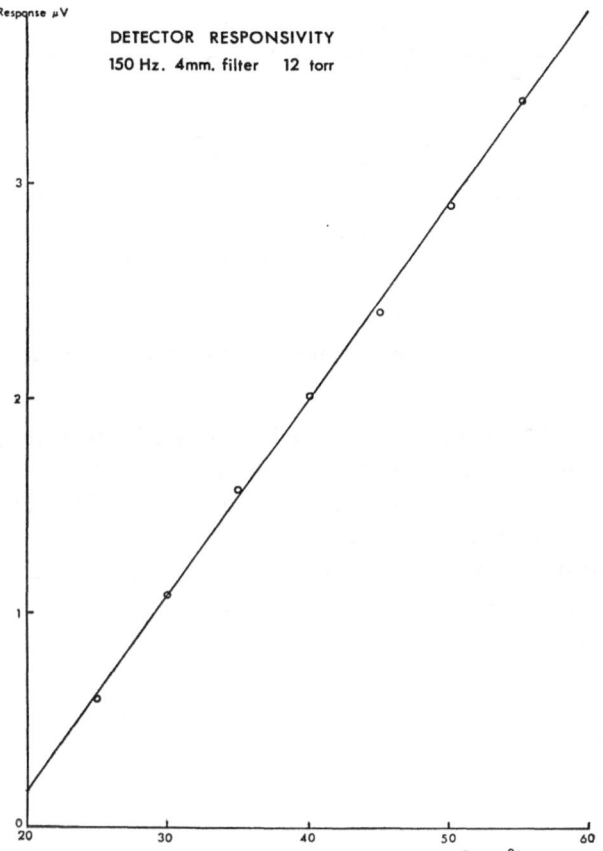

Fig. 4. InSb detector responsivity for 300 K black-body radiation modulated at 150 Hz and filtered through 4 mm sooted polythene; crystal temperature 1.8 K. Detector output voltage is plotted against temperature of black body.

The detector element is at the end of 66 cm of 2 cm diameter light-pipe, and its response falls sharply off-axis. The effective angle of acceptance was determined separately and was found to be 6.15×10^{-2} sterad, corresponding to an f-number of 3.5.

From the measurements the values of $\Re$ and τ obtained were 168 V W^{-1} and 5.05×10^{-4} V m^2 W^{-1} respectively. The responsivity for incident radiation is therefore down by a factor of 1.8 on that deduced from d.c. measurements.

A similar calibration was carried out on a gallium doped germanium detector of dimensions $3 \times 3 \times 0.4$ mm, giving values for $\Re$ and τ of 236 V W^{-1} and 2.83×10^{-4} V m^2 W^{-1}.

5. Noise Measurements

Shown schematically in Figure 5 is the arrangement used for measuring the noise of the detector. Narrow-band noise is measured for various resistive input terminations

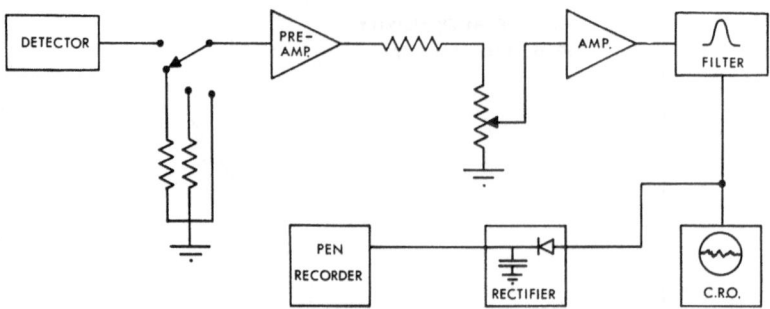

Fig. 5.   Schematic diagram of apparatus used for noise measurements.

of the preamplifier. The potentiometer is used to bring all signals to the same level on the pen recorder, thus avoiding the effects of any non-linearity in the system. The use of a rectifier rather than a true rms meter is justified provided the statistics of all noises involved are the same.

For source impedances which are low compared with ~1 MΩ the narrow-band output noise from the pre-amplifier may be written

$$v_0^2 \propto (R_{\text{source}} + R_A)$$

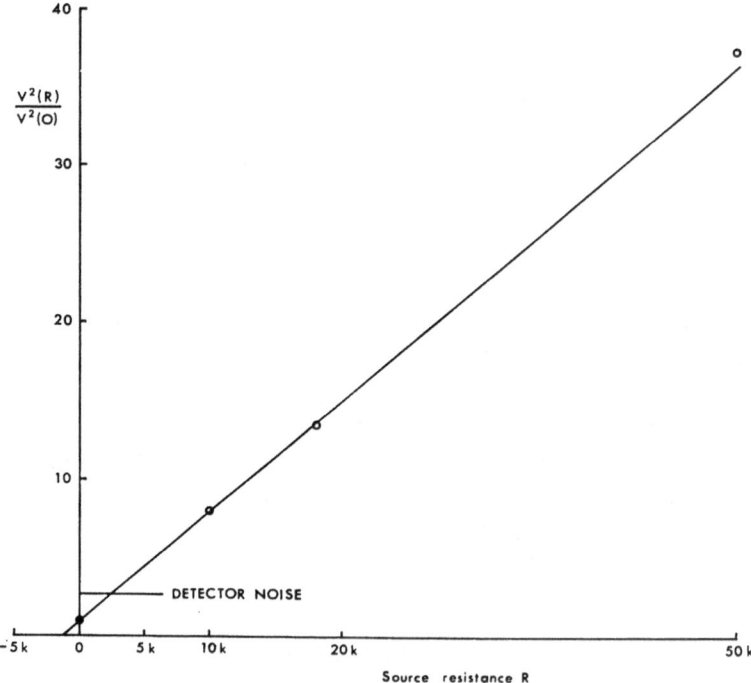

Fig. 6.   InSb detector and amplifier noise. Narrow band noise at 1 k Hz output from the amplifier is plotted against source resistance. Also shown is noise level with detector as source.

where R_A is a constant of the amplifier. Plotted in Figure 6 is v_0^2 versus R_{source} at 1 k Hz; the intercept gives the value of R_A as 800 Ω.

Also shown is the value of v_0^2 when the amplifier input is terminated with the detector circuit. The value of R_{source} corresponding to this noise gives the equivalent *room temperature* noise resistance of the detector and transformer as 4.4 kΩ. The equivalent noise resistance of the whole detector-transformer-amplifier system is therefore 5.2 kΩ. It is convenient to refer this to the crystal by dividing by the measured voltage gain of the transformer squared, and to multiply by 293/1.8 to obtain an equivalent low temperature noise resistance. This gives a value of 1.6 kΩ. The actual crystal resistance is 70 Ω so that the measured noise factor of the system is 15.1 rather than the calculated value of 2.44. It should be pointed out here that during the measurements the detector was exposed to the same area solid angle product at ambient radiation as under operating conditions; in obtaining the noise factor only Johnson noise has been assumed.

Substitution in Equations (2) of the measured responsivity and equivalent noise resistance gives for the NEP and nep the values 1.93×10^{-12} W Hz$^{-1/2}$ and 1.55×10^{-8} W m$^{-3/2}$ Hz$^{-1/2}$ respectively.

Similar measurements on the Ge detector give values for NEP and nep of 3.8×10^{-11} W Hz$^{-1/2}$ and 6.4×10^{-7} W m$^{-3/2}$ Hz$^{-1/2}$.

The results are summarized in Table I.

TABLE I

Detector performance figures

Detector	$\Re$ V W^{-1}	$\mathfrak{r}$ V m^2 W^{-1}	NEP W Hz$^{-1/2}$	nep Wm$^{-3/2}$ Hz$^{-1/2}$
Ge	236	2.83×10^{-4}	3.8×10^{-11}	6.4×10^{-7}
InSb d.c.	320	9.6×10^{-4}	4.2×10^{-13}	3.4×10^{-9}
InSb radiation	168	5.1×10^{-4}	1.9×10^{-12}	1.6×10^{-8}

Finally we should like to compare our results for the InSb detector with those obtained by Kinch [7]. Kinch used a $2 \times 1 \times 0.25$ mm crystal mounted in 4 mm waveguide with a tuning plunger. He obtained a responsivity for 4 mm radiation of 1.7 kV W^{-1} and an NEP of 1.3×10^{-13} W Hz$^{-1/2}$ for a temperature of 1.2 K. These figures give a specific responsivity, $\mathfrak{r}$, of 4.25×10^{-4} V m^2 W^{-1} and an nep of 5.8×10^{-9} W m$^{-3/2}$ Hz$^{-1/2}$. These figures are comparable with those we obtain for wideband radiation.

6. Conclusions

For wavelengths greater than 300 μm it is clear that, of the two detectors tested, the InSb is very much more satisfactory. Furthermore it is reproducible and simple to construct. In particular there is no problem connected with the thermal conductance of the leads; the relevant conductance is that between the electrons and the lattice which is not under the control of the experimenter. Lastly the detector has a fast response, limited by the transformer, which is an advantage in many astrophysical applications. In this connection it should be pointed out that the responsivity of the Ge bolometer was measured at 150 Hz. A somewhat better figure would have been obtained at lower frequencies. However, this would probably have been more than compensated by an increase in $1/f$ noise.

Acknowledgements

We should like to thank D. G. Vickers for invaluable technical advice and assistance, and E. I. Robson for writing the computer programme and carrying out tests on the black bodies.

References

[1] Rollin, B. V.: 1961, *Proc. Phys. Soc.* **77**, 1102.
[2] Kinch, M. A. and Rollin, B. V.: 1963, *Brit. J. Appl. Phys.* **14**, 672.
[3] Martin, D. H. and Puplett, E.: 1969, *Infrared Phys.* **10**, 105.
[4] Cantarano, S. and Pallottino, G. V.: 1970, *Electronic Eng.* **42**, 57.
[5] Shivanandan, K. Houck, J. R., and Harwit, M.: 1968, *Phys. Rev. Letters* **21**, 1460.
[6] Ade, P. A. R.: 1971, Private communication.
[7] Kinch, M. A.: 1968, *Appl. Phys. Letters* **12**, 78.

DISCUSSION

P. L. Marsden: Do you have any idea where the excess noise is coming from?

P. E. Clegg: It is not very clear. In the case of the InSb we have to account for a factor three. As for the crystal itself, there is a slight increase in noise when the bias current is on, but very little.

The excess noise may come from the transformer though we have not yet been able to test it alone.

G. Chanin: It should be emphasized that the Ge detectors which you use are very different from the commercial ones with high NEP. Your detector has very large volume in order to get sufficient absorption in the far infrared and you also require to be very rapid. Because of that the responsivity of your detector is low.

N. Coron: We have made a comparison of an InSb detector without transformer with one of our bolometers.

With a high impedance crystal we get an NEP of 3×10^{-12}, as compared to the bolometer NEP of less than 10^{-13}. This for a peak transmission at 400 μ. We seem to have conflicting results.

K. Shivanandan: This derives from the different physics involved. One has first to understand the noise before discussing the detectors.

AN INFRARED TRANSDUCER FOR SPACE APPLICATIONS

M. CHATANIER and G. GAUFFRE

ONERA, Chatillon, France

Abstract. A new model of ONERA capacitance-varying pneumatic infrared detector withstands the stresses of space environment. The detector is operated with an electronic measuring circuitry, thus forming a transducer unit named TRIAS (TRansducteur Infrarouge pour Applications Spatiales). With a 3 mm² sensitive area, the overall NEP is 2×10^{-10} W Hz$^{-1/2}$ at 30 Hz.

1. Introduction

Infrared detectors used in space have so far been restricted to two types, namely quantic receivers in the near infrared, and bolometers for applications over a wider spectral range. Other thermal detectors such as pyroelectric or pneumatic types are now available; their laboratory test shows performances close to uncooled bolometer ones. Their application to space ends does not reveal insuperable difficulties any longer.

Over the past years, ONERA has been working on a capacitance varying pneumatic detector. The earliest prototype, designed by K. Luft, has now gone into industrial production. Another more recent model has been able to stand up to the stresses of the space environment. Two successful experiments aboard a sounding rocket have demonstrated its resistance to the impact of environmental lift-off conditions. Further new improvements in reliability open the way to its use on satellites.

As the information supplied by a passive sensor is not of direct interest, the infrared detector is coupled to an electronic measuring circuitry, thus forming a transducer unit named TRIAS (TRansducteur Infrarouge pour Applications Spatiales).

2. Operation

A. DETECTOR (Figure 1)

Incident infrared radiation enters the front chamber through a window; the window material determines the useful spectral bandwidth. The rays fall on an infrared-absorbing target, which is warmed, raising the gas temperature. The target has a constant emissivity over the entire spectral range. The resulting overpressure changes the shape of the metallized membrane acting as the movable plate of the measuring capacitor.

The fixed capacitor plate is located in the rear chamber.

The two chambers are interconnected by a calibrated fluid leakage to ensure the long-term maintenance of equal pressure on either side of the capacitor membrane. This arrangement shows that the detector is not intended for the reception of a continuous incident flux. The filling gas is xenon, distinctive for having the lowest thermal conductivity among neutral gases. Filling pressure is set to provide a gas heat

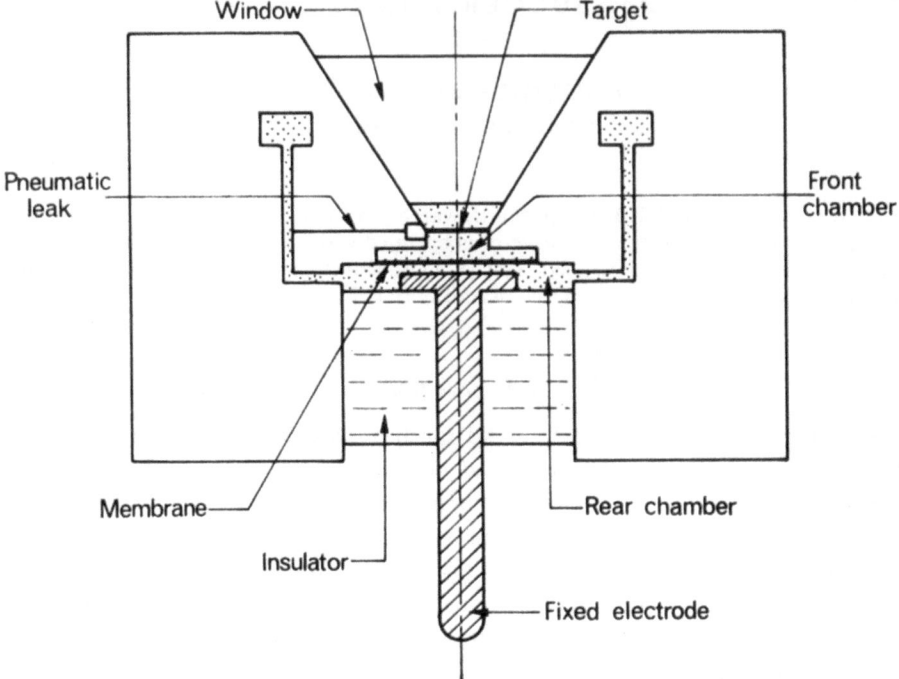

Fig. 1.   Schematic diagram.

capacity in the front chamber ensuring the optimum balance between responsivity and bandwidth.

B. ELECTRONIC MEASURING SYSTEM (Figure 2)

The information is measured by means of an electronic capacitance bridge.

The functional block diagram splits into five subassemblies:

(1) A RF oscillator feeds the bridge and supplies the reference voltage of the synchronous detector stage. In order to obtain an overall detectivity equal to the detector one, this oscillator has a very low amplitude modulating noise.

(2) A bridge whose one arm is formed by the capacitor incorporated in the infrared detector.

(3) A low noise and high input impedance RF amplifier to obtain a high level modulated signal from the bridge output without reduction of the signal to noise ratio.

(4) A synchronous detector stage controlled by the local oscillator, and followed by an equalizing amplifier circuit to ensure constant gain in the transducer passband.

The equalizer circuit also reduces the residual RF carrier level present in the demodulated signal.

(5) A servoamplifier for feedback control of the bridge (in the very low frequency

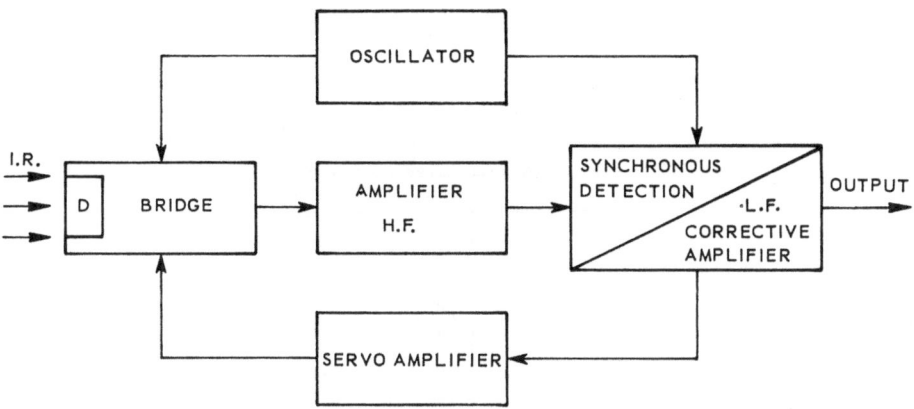

Fig. 2. Block diagram of electronic circuitry.

range) in order to maintain the bridge at null against long-term ambient temperature variations. Action on the bridge balance is obtained by electrostatic attractive force on the membrane, controlled by the DC signal component present at the synchronous detector output.

The control loop maintains at a constant mean value the measuring capacitor.

3. Chosen Methods of Meeting Environmental Conditions

In designing TRIAS, several technical solutions were worked out in order to obtain satisfactory operation in presence of the two kinds of stresses operating on equipments intended for orbiting, namely those inflicted by launching (acceleration, vibrations, shocks) and those attendant on orbital flight (temperature variations, behaviour under vacuum, radiation resistance).

The detector is highly compact. The sensitive target is well protected by a housing of some millimeters wall thickness sealed by indium joint. These features safeguard against outer-space radiation and pressure variation hazards.

To cushion the impact of acceleration upon the equipment, the design has observed Golay's [1] geometry principle for volumes of gas, in that the volumes present on either side of the capacitance membrane have coinciding centres of gravity. In first approximation, the overpressures created in the gas by forces of inertia should cancel at membrane level.

With an eye to a high degree of accuracy and adequate constancy of the electrode spacing (about 4 μm), the membrane holder and the electrode-support plus fixed electrode assembly are tangent to a single sphere (Figure 3). The electrode support is made from two materials of equal heat expansion coefficients over the entire range of working temperatures. This arrangement tends to mitigate the effect of temperature fluctuations and makes the electrode gap easier to match.

Throughout the production process, the transducer undergoes a number of partial

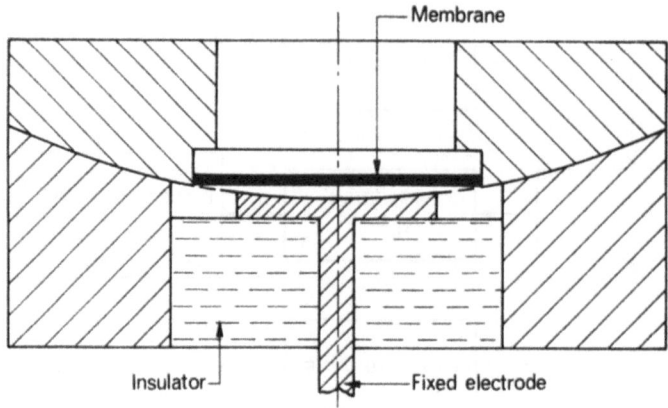

Fig. 3. Detail of capacitance electrodes.

checks for the quality of the detector unit proper and the associated electronic measuring circuitry (active and passive components, subassemblies, interfaces, etc. . . .). Though lightweight, the equipment is very rigidly built and is mechanically so designed as to ensure freedom from internal stresses.

The electronic component parts are immovably embedded in a polymerizing resin: the transformer and inductance coil are moulded in their magnetic circuits.

4. Specifications

Measurements

Space requirement: diameter 70 mm, height 70 mm (270 cm^3)
Weight: 400 g
Target area: 2 mm dia. disc. (3.14 mm^2)
Angle of sight: 60°

Power supply

Supply voltage: ±15 V stabilized
Current: +30 mA, −20 mA
Power consumption: 0.75 W

Response to Environmental Conditions

Acceleration: 20 g rms white noise from 20 to 2000 Hz along 3 axis
Shock: 100 g for 5 ms
Temperature: −20 °C to +40 °C
Operative under vacuum

Measurement Performance Data

Spectrum coverage: 0.5 to 1000 μm wavelength. The spectrum bandwidth depends on the nature of the window, the materials used include KRS 5, ICs, Ge, Crystal quartz.

KRS 5 0.6 to 40 μm

ICs   0.5 to 55 μm

Ge   1.2 to 25 μm

Quartz 50 to 1000 μm

Response 30 kV W^{-1} at $\pm$1 db from $-$20 °C to 40 °C (Figure 4)

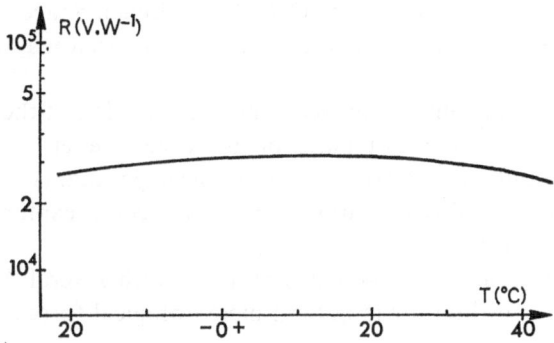

Fig. 4.   Responsivity vs temperature.

Pass band: 2 to 200 Hz at $-$3 db (Figure 5)

NEP: 2.10^{-10} W Hz$^{-1/2}$ at 30 Hz, 4×10^{-10} W Hz$^{-1/2}$ at 10 and 120 Hz (Figure 5)

Measurement range: 1 V, rms (or at input 30 μW rms IR)

Linearity better than 1% over full scale

Output impedance 50 Ω

Min. load impedance 2 kΩ

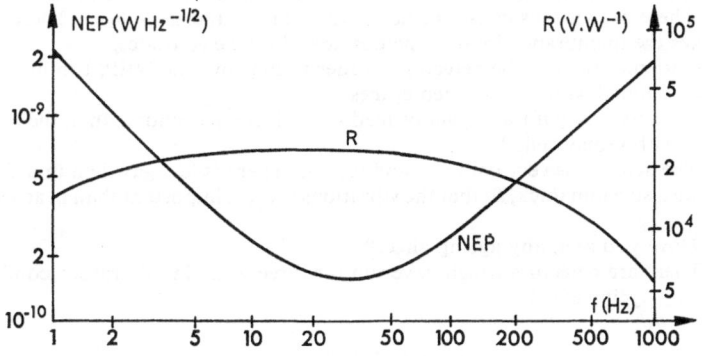

Fig. 5.   NEP responsivity vs Frequency.

5. Conclusion

The distinctive features of TRIAS include a high sensitivity, wide optical acceptance and sturdy construction. Over and above space uses, it lends itself to any ground application requiring a dependable and easy to handle transducer.

The laboratory prototype has already proved its worth in flight tests as part of an airborne equipment [2].

Six units of this kind of transducer had been successfully used for two experiments on sounding rockets performed by ONERA in 1965 and 1968. These experiments were named TACITE: endeavour of analysis of the earth-space infrared contrast [3–4]. Several units of the TRIAS were subjected recently, without failure to the main resistance tests in space environment by CNES, the French Space Agency. This result opens the way to applications on satellites. Among those that may be considered, one may mention:

– stabilization of space vehicles, by detecting the variation of the thermal radiation around the geometrical horizon of Earth or any other planet;

– Earth resource study or detection of atmospheric pollutants;

– determining the vertical temperature profile by spectral exploration of the Earth and atmosphere radiation;

This type of measurement, at present performed with several American satellites, is very important for medium or long term meteorological forecasting.

References

[1] Golay, M. J. E.: 1947, *Rev. Sci. Instr.* **18**, 347.
[2] Eddy, J. A., Lee, R. H., Léna, P. J., and McQueen, R. M.: 1970, *Appl. Opt.* **2**, 439.
[3] Girard, A.: 1967, *Proceedings of the 7th International Space Science Symposium*, North-Holland, Amsterdam, 934.
[4] Girard, A. and Mme Lemaître, M. P.: 1970, *Appl. Opt.* **2**, 903.

DISCUSSION

G. Chanin: What is the limiting noise on the Golay cell due to? Is it due to temperature noise of the gas itself and if so should not the noise itself decrease when you cool the detector slightly?

G. Gauffre: There are various causes of noise which are at the same level. There is noise in the gas which makes the membrane vibrate. There is noise in the electronics.

We made no attempt to cool the detector in order to improve the NEP. I think a small gain on an NEP can be obtained with smaller membranes.

R. D. Joseph: Can you say if the capacitor read-out is less microphonic than the ordinary photo-electric readout with Golay cells?

G. Gauffre: The detector is very compact and rugged. There is less gas than in a Golay detector. The windows are also 5 mm thick, so that the vibrations are less important than in an ordinary Golay cell.

R. Zander: Have you seen any ageing effect?

A. Girard: There are detectors which have run for three years in laboratory conditions and we have not seen any ageing effect.

JOSEPHSON DETECTION

M. RENARD

CNRS – CEDEX 166 – Grenoble – France

1. Superconductivity

In superconductors, the state of the electrons is more ordered than in normal state, as can be shown by specific heat measurements. The kind of order existing in super-conductors is well described by BCS theory:

– At first electrons are paired in states of opposite momentum and antiparallel spins. These Cooper's pairs are particles with 0 spin quantum number and so can be considered as Bosons.

– Then, these Cooper's pairs are described by a coherent wave function. This is needed in order to optimize the attractive interaction existing in sc materials. It can be shown that if a phase factor ϕ can be given to a single pair, its contribution to cohesive interaction is multiplied by $(\cos \phi)$.

– A last feature is of interest here: the existence of an energy gap $2\Delta = 3.5 k_B T_c$. This is the minimum energy for an excitation. If we send a perturbation with quantum energy higher than 2Δ, we can create excitations. So Δ or a few Δ seems to be the higher limit for the use of Josephson detectors. (But see [1].) For Niobium $2\Delta = 80 \ \mu$m in λ.

2. Josephson Effect

So, below T_c, a piece of bulk superconductor exhibits phase coherence in space, every Cooper's pair having the same phase.

If we achieve a tunneling contact between two superconductors, pairs can tunnel by second order terms, involving T^2, T being the transmittance of the barrier.

The wave function modulus for Cooper's pairs coming from (I) is given in Figure 1. (And conversely by the broken line for pairs coming from II.)

So, they spend some time in part II, and contribute to the cohesive energy, by interaction with pairs that are already in II. These interactions are of course multiplied by $(\cos \phi)$. ϕ being the phase difference between I and II.

Integrating over all the pairs gives a contribution to the energy:

$$E = -E_0 \cos \phi.$$

If $E_0 \gg kT$, phases are linked together as in the case of a single bulk superconductor. If E_I and E_{II} are the energies of the two superconducting blocks far apart from each other the total energy can be written as:

$$H = E_I + E_{II} - E_0 \cos \phi.$$

Of course E_I and E_{II} do not depend explicitly on the phase difference between the two

Manno and Ring (eds.), Infrared Detection Techniques for Space Research, 147–153. All Rights Reserved
Copyright © 1972 by D. Reidel Publishing Company, Dordrecht-Holland

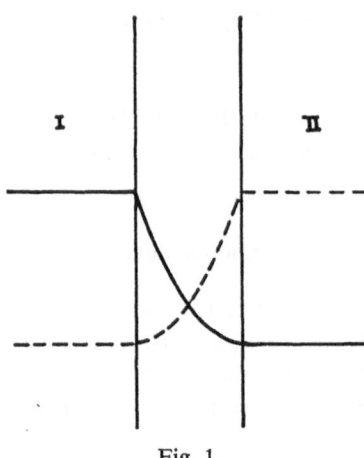

Fig. 1.

blocks. If N_1 is the operator for the number of pairs in excess in block (I), it is well known that a number of particles do not commute with the phase:

$$[\phi, N] = 1.$$

So

$$\hbar(\mathrm{d}N_1)/(\mathrm{d}t) = [H, N_1] = [-E_0 \cos \phi, N_1] = E_0 \sin \phi[\phi, N_1] = E_0 \sin \phi.$$

But $\mathrm{d}N_1/\mathrm{d}t$ is proportional to the current flowing in the junction. We can write:

$$j = j_c \sin \phi. \tag{1}$$

If we calculate:

$$\hbar(\mathrm{d}\phi)/(\mathrm{d}t) = [H, \phi] = [E_\mathrm{I} + E_\mathrm{II}, \phi].$$

But $E_\mathrm{I} + E_\mathrm{II}$ depends on the excess number of electrons N_1, only if there is a difference in chemical potential for pairs, arising from an electrical potential:

$$E_\mathrm{I} + E_\mathrm{II} = 2\,\mathrm{eV} \times N_1.$$

So we have the second Josephson relation:

$$\hbar(\mathrm{d}\phi)/(\mathrm{d}t) = -2\,\mathrm{eV}. \tag{2}$$

So we have shown that:

(a) A current can flow without an external applied voltage, if $i < i_c$.

(b) If a dc voltage is assumed, an alternating current flows through the junction.

3. Josephson Detection (Video mode)

If we polarize with a constant direct current (i), energy given by the source to the junction is:

$$W = \int iV\,\mathrm{d}t = -\frac{\hbar}{2e} \int i \frac{\mathrm{d}\phi}{\mathrm{d}t}\,\mathrm{d}t = -\frac{\hbar}{2e}\, i\phi.$$

Total energy is then:

$$W = -\frac{\hbar}{2e} i\phi - E_1 \cos \phi \propto (-i\phi - j_0 \cos \phi).$$

If $|i| < j_c$ a minimum exists for ϕ_0 such that:

$$i = j_c \sin \phi_0. \quad \text{(Figure 2.)}$$

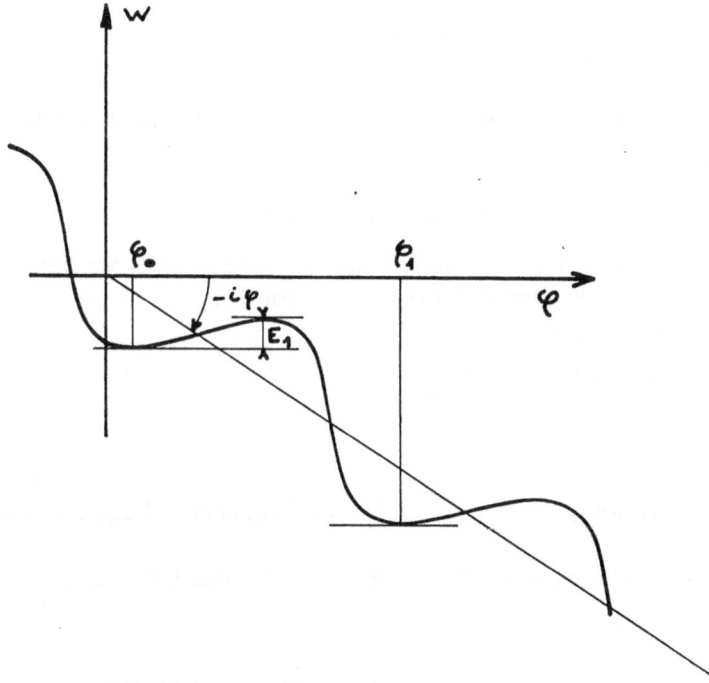

Fig. 2.

Just the relation giving dc current in junction equal to applied one. But you can see from Figure 2 that the energy needed to go from ϕ_0 to the next equilibrium point $\phi_1 = \phi_0 + 2\pi \cdot$ is E_1, much less than E_0 the coupling energy of the junction. And for $i=j_c$, $E_1=0$.

So, we can have $E_1 \sim kT$, even if $E_0 \gg kT$. We get here a thermally, incoherent, activated process giving phase slippage, and so associated to a continuous potential given by a second Josephson equation.

If some excess electromagnetic radiation hits the junction, we have more probability of jumping across the activation energy and so voltage is increased.

By principle this is an incoherent broad band detection. NEP of about 10^{-14} W Hz$^{1/2}$ have been reported by [2, 3, and 4]. Time constant is certainly less than 10^{-6} s. Ultimate theoretical sensitivity has not yet been established (Figure 3).

M. RENARD

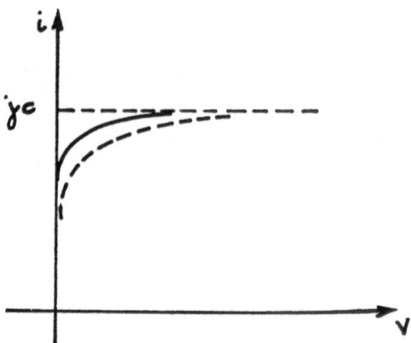

Fig. 3. Full line: thermally activated voltage. Broken line: plus incident radiation.

4. Heterodyne Mode

If $i_{dc} > j_c$, there is apparently no solution. But what happens if we have a regulated dc voltage V_0 plus a little alternating potential $v \cos \omega t$?

Josephson's relations gives:

$$\phi = \phi_0 + \frac{2eV_0}{\hbar} t + \frac{v}{\omega} \frac{2e}{\hbar} \sin \omega t \qquad (3)$$

$$j = j_c \sin \phi$$

$(2eV_0)/\hbar = \Omega_0$, pulsation proportional to V_0, the frequency-voltage equivalence being 583 Mc μV.

It is easy to see that if $\Omega_0 = n\omega$, j is a purely periodic function of time.

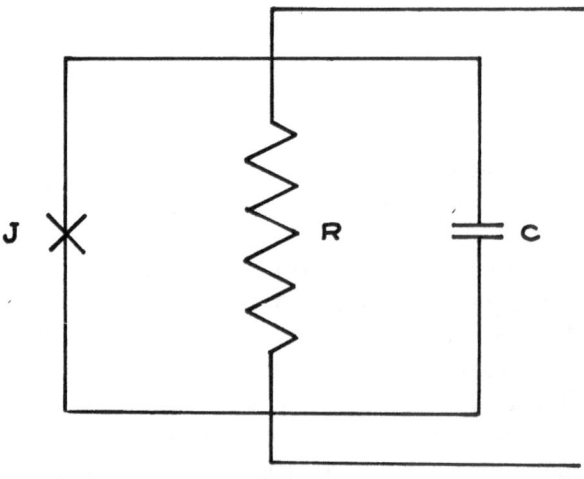

Fig. 4.

It can be expanded in Fourier series, and especially contain a zero frequency harmonic which amplitude is given by:

$$j_{dc} = j_c \left[J_n \left(\frac{2\,eV}{\hbar\omega} \right) \right] \sin \phi_0, \tag{4}$$

where J_n is the Bessel function of nth order. As $|J_n|$ is less than unity, it does not appear that $j > j_c$ can be achieved.

But a Josephson junction always carries resistive current when dc voltage is applied. The equivalent electrical circuit is given (Figure 4). This resistive part arises for example from excitations classical tunneling. So if we polarize with a regulated current source, with also monochromatic alternating power supply, we can get some steps to voltages multiples of the frequency of the alternating perturbation (Figure 5).

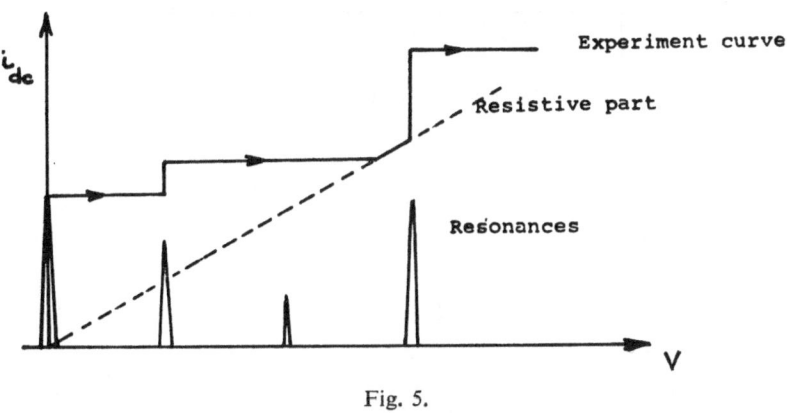

Fig. 5.

But, any junction of finite dimensions is more or less like a wave guide. It has dimensional dependent resonance frequencies. When voltage reaches the equivalent frequency, alternating Josephson current gives excitation of these modes. And we can have a beating between this self-induced mode with itself by the non-linear sinus term in j. The formula is the same as in (4) but with $n = 1$.

These autosteps are very easy to obtain. Their slope (di/dV) is much lower than in the case of monochromatic irradiation-induced one. It depends on the Q of the junction. The potential V of an autostep can range between one tenth to one thousandth of V_0 (Figure 6).

Ulrich [4] has performed the following experiment: Taking a junction with an autostep, he irradiates it with a monochromatic klystron source, the frequency of which falls in the limits (V_0, $V_0 + \Delta V$).

He has given the curve of the excess voltage (δV) when irradiated, versus V in the preceding limits. The curve looks like a resonance curve (Figure 7).

This is, of course, a monochromatic heterodyne type of experiment. But the shape of the curve by itself is rather discouraging in order to extract a monochromatic radiation from a continuous spectra. NEP of 5×10^{-14} W Hz$^{-1/2}$ is quoted in (4), this seems very far above theoretical predictions.

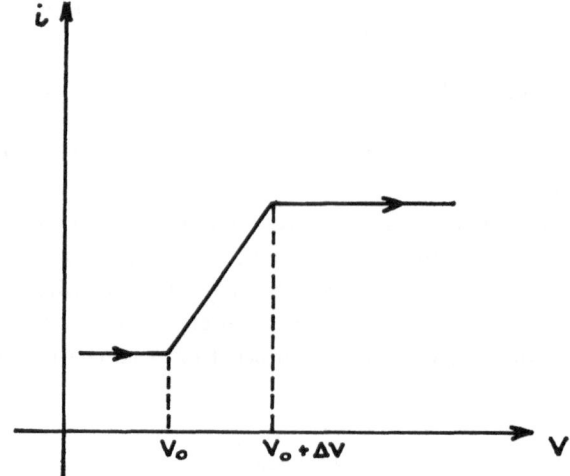

Fig. 6. Step magnified in V.

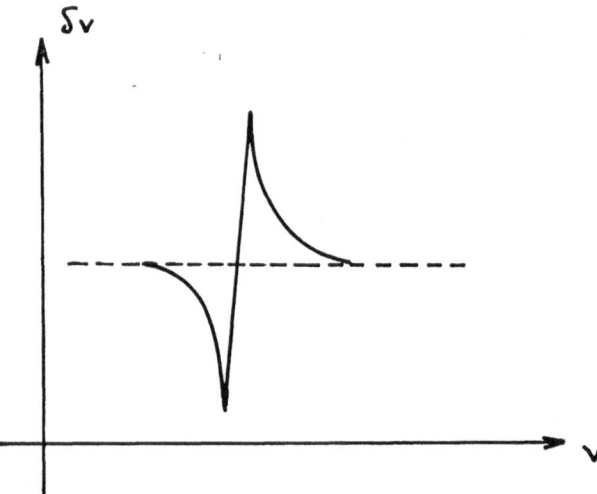

Fig. 7. Excess voltage.

5. Prospects

Many problems are not yet solved. On a theoretical point of view perhaps the coupling between the junction and the electromagnetic field is one of major importance. My feeling is that coupling in certain conditions and for evaporated junctions depends highly of the polarization. Solutions for the inner electromagnetic fields have been given, [5], if we know $\phi(r, t)$. The non-linear Ferrell and Prange equation has to be solved to know ϕ. And it appears to be a very difficult problem.

On the experimental aspect, usually people use point-contact junctions. I think it would be very difficult to avoid problems arising from vibrations in launching conditions.

But, in conclusion, we must say that Josephson effects are investigated since no more than 7 yr, and that carbon resistors sometimes used as bolometers are known in thermometry since about 40 yr. The fact that in video detection Josephson devices have reached in every case NEP factors better than those of the majority of bolometers, with time constants at least 1000 times shorter, seems very promising.

References

[1] Werthamer, N. R.: 1966, *Phys. Rev.* **147**, 255.
[2] Grimes, C. C., Richards, P. L., and Shapiro, S.: 1968, *J.A.P.* **39**, 3905.
[3] Grimes, C. C., Richards, P. L., and Shapiro, S.: 1966, *Phys. Rev. Letters* **17**, 431.
[4] Ulrich, B. T.: 1968, *Phys. Rev. Letters* **20**, 381; W. Hogan and T. Mass (eds.), *Cryogenics and Infrared Detection*, T. Mass – Boston Technic. Publ. p. 120.
[5] Renard, M.: 'Complete Electromagnetic Solution for a Josephson Junction in London approximation', Low Temperature Physics Conference – Kyoto (1970). (To be published.)

DISCUSSION

F. Kneubühl: Would you prefer a layer structure or a point contact? You talked of the resonances of the layer structure and I know the point contact is a problem too.

M. Renard: The problem is not even theoretically solved, because it is much more complex when one has to account for the spatial variation of the phase and one gets large non-linear differential equations.

ASTRONOMICAL POLARIMETRY AT 5 AND 10μ

J. C. D. MARSH

The Hatfield Polytechnic Observatory, Bayfordbury, Herts., U.K.

and

R. F. JAMESON

Dept. of Astronomy, University of Leicester, Leicester, U.K.

1. Introduction

During the past few years many celestial objects have been located which emit a considerable amount of radiation at 5 and 10 μ. Gross spectral measurements on some of these objects indicate a departure from a theoretical Planck curve for a specified temperature, and the high intensities measured lead to the possibility of some mechanism other than thermal emission, producing the flux. One such possibility is synchrotron emission, and should this be so the received flux would contain a component of linear polarization. It is the magnitude and direction of this flux which the polarimeters have been designed to measure. Additionally, polarized infrared emission from the Moon and planets could yield important information about the nature of the surfaces.

2. The Analyzers

It has not been possible to obtain an analyzer having a high efficiency at both 5 and 10 μ, and in any case since the system calls for relatively high speed chopping techniques, a wide band bolometer detector could not be used owing to its comparatively slow response. At 5 μ an Indium Antimonide (InSb) detector cooled to 77 K is preceded by a Rochon Prism made from the material Proustite [1, 2, 3] (Ag_3AsS_3) and an Indium Arsenide (InAs) filter. At 10 μ a Lead-Tin-Telluride (PbSnTe) detector, cooled to 77 K is preceded by a wire grid analyzer [4].

2.1. THE 5 μ ANALYZER

Calculations on the Proustite prism show that the refractive index for the O-ray is 2.6321 and for the E-ray is 2.4779. The beam separation is computed to be 20° for an interface angle of 27.51°. The aperture is 1 cm square and the prism is bloomed for a nominal 5 μ operation. The prism is shown in Figure 1.

2.2. THE 10 μ ANALYZER

It was not found possible to use a crystalline material for a 10 μ analyzer which satisfied the rather stringent requirements:
  (a) An undeviated output ray.
  (b) A wide acceptance angle.

Manno and Ring (eds.), Infrared Detection Techniques for Space Research, 154–159. All Rights Reserved
Copyright © 1972 by D. Reidel Publishing Company, Dordrecht-Holland

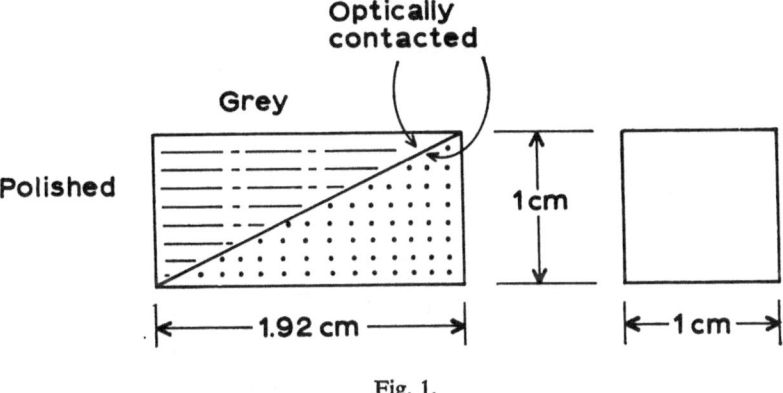

Fig. 1.

(c) High transmission and polarization efficiency.

(d) Easy rotation.

(e) Good mechanical strength coupled with small size.

Consequently a wire grid device (a special form of dichroic analyzer) was used, transmitting without deviation one form of polarization whilst absorbing or reflecting the orthogonal form. The commercial analyzer employed comprised a metallic grid deposited photographically on a mylar substrate. The details are:

Thickness of grid   0.2 μ
Period of grid   4.0 μ
Clear aperture   25.4 mm

The efficiency, defined as $[(k_1 - k_2)/(k_1 k_2)]$ % is 93% at 20 μ and approximately 28% at 10 μ. The analyzer is shown in Figure 2.

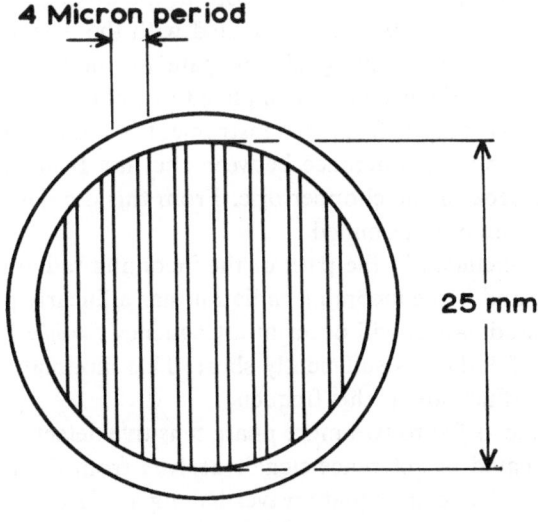

Fig. 2.

3. The Polarimeters

Figure 3 shows the schematic diagram.

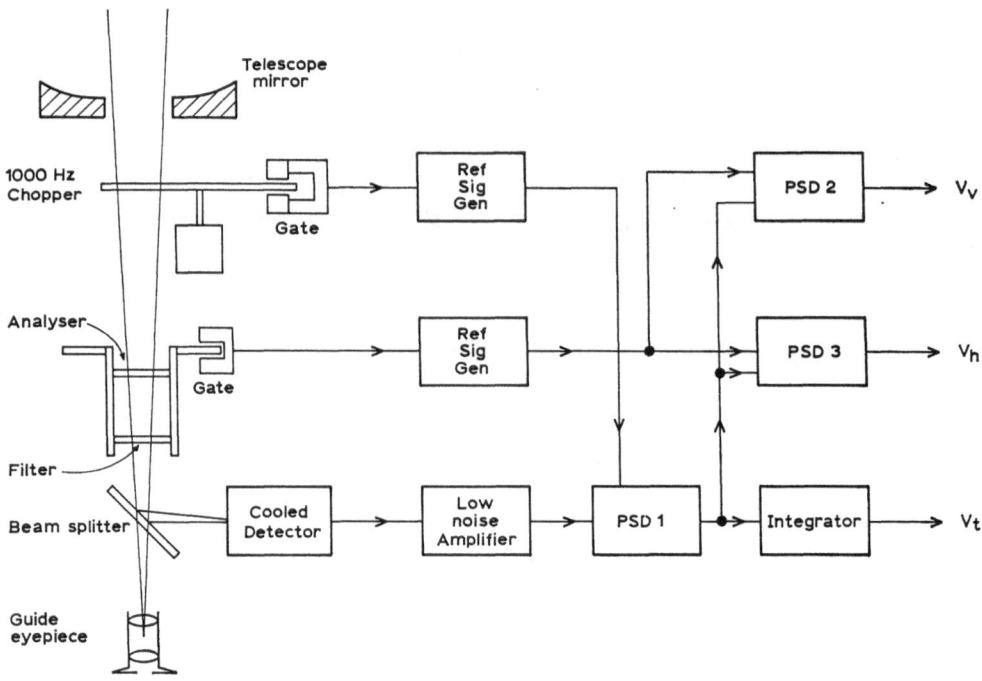

Fig. 3.

The incoming flux from the infrared telescope at the Cassegrain focus is chopped at 1000 Hz; this minimizes the $1/f$ noise associated with the detector. Reference signals are also derived from the chopping disc to gate the first phase sensitive detector PSD 1. The output from the detector is amplified in a commercial low noise amplifier and fed to PSD 1. The output from this instrument is a d.c. level, the magnitude of which is determined by the difference between the flux from the telescope and the radiation from the back of the chopper disc. From this d.c. level the intensity of the astronomical signal can be determined.

If the analyzer is included in the path of the incoming radiation and rotated at an angular velocity ω, and if the incoming radiation has a linearly polarized component, simple amplitude modulation will occur at a frequency $f = \omega/\pi$ (16 Hz) provided that the time constant of PSD 1 is sufficiently short. This modulation will cause the d.c. output of PSD 1 to fluctuate at this frequency.

The fluctuating d.c. is fed to two more phase sensitive detectors, PSD 2, and PSD 3, which are in turn gated by reference signals derived from the rotating cylinder containing the analyzer. These are square waves having a relative phase difference of 90° and mechanically aligned so that a positive going crossover coincides exactly with a chosen physical orientation of the analyzer; this orientation corresponds to the acceptance of a vertically polarized signal.

The result is to produce three outputs corresponding to:

(1) the total intensity of the incoming radiation V_t;

(2) the intensity of the vertical component of the incoming radiation V_V;

(3) the intensity of the horizontal component of the incoming radiation V_H.

The total intensity of the polarized component can be checked from

$$|V_t = (V_V^2 + V_H^2)^{1/2}|$$

and the angle of polarization from

$$\theta = \tan^{-1} V_V/V_H.$$

4. Calibration and Testing

Whilst some infrared signals could possibly contain a high degree of polarization, it is likely that the figure will, in many cases, be less than 10%. Consequently it is important that the instrumentation shall not produce internal polarization, or effects which might be construed as polarization.

4.1. POLARIZATION SENSITIVITY OF THE DETECTOR

It is well known that some optical photo-electric detectors e.g., photomultipliers, are sensitive to the direction of the plane of polarization of the incident radiation. It was not known whether infrared detectors, in particular those employed in the polarimeter, were susceptible. Measurements were made on two PbSnTe cells, and within an upper limit of 0.2% no polarization effects were discernible.

4.2. EFFICIENCY OF THE 10 μ ANALYZER

Measurements were made in the wavelength interval 8.5 to 12.0 μ using the response of the PbSnTe cell as the definitive filter. As there was no source of perfectly linearly polarized radiation available the method of Rupprecht [5] was used. Three analyzers were used in pairs, the following measurements being made:

E_0 total power transmitted by the beam without an analyzer in the path;

E_h power transmitted by a single analyzer with grid bars horizontal;

E_v power transmitted by a single analyzer in the beam with the grid lines vertical;

$E_\perp$ power transmitted by a pair of analyzers in the beam with the grid bars crossed.

From these measurements k_1 and k_2 can be calculated [4].

$$k_1 = \frac{E_h + E_v}{2E_0}\left\{1 + \left[1 - \frac{4E_\perp E_0}{(E_h + E_v)^2}\right]^{1/2}\right\}$$

$$k_2 = \frac{E_h + E_v}{2E_0}\left\{1 - \left[1 - \frac{4E_\perp E_0}{(E_h + E_v)^2}\right]^{1/2}\right\}$$

Percentage polarization is given by:

$$P = \frac{k_1 - k_2}{k_1 + k_2}\%$$

The average value for the three analyzers was 27.0%.

4.3. THE EFFECT OF TELESCOPE OPTICS

Where reflecting surfaces are employed some degree of instrumentally induced polarization is almost inevitable. However it was felt that by using the Cassegrain focus of the telescope, these effects would be minimized, and in the light of investigations at optical wavelengths (Clarke [6]), could be ignored.

A second possible cause of instrumental polarization could be due to imperfect or tarnished mirror surfaces. Again, at optical wavelengths Clarke has indicated the order of magnitude, and at infrared wavelengths the effect may be considered negligible.

5. Conclusions

It is felt that infrared techniques are sufficiently advanced for polarization measurements at 5 and 10 μ to be attempted. The design investigations showed that it was practicable to build a polarimeter, and the laboratory performance at 10 μ has so far proved comparable with the theoretical considerations. First measurements (in a photometric mode) have been made using a 16" (40 cm) telescope and it is expected that a full system calibration will be carried out within the next six months.

Acknowledgement

The authors would like to acknowledge the material aid given by the Plessey Co. Ltd., and by Cambridge Consultants Ltd. They like to thank Mr D. A. Robinson for making the calculations on the Proustite prism and they like to acknowledge the permission from the Institution of Electronic and Radio Engineers for publishing this article.

References

[1] Hobden, M. V.: 1969, 'The dispersion of the Refractive Indices of Proustite Ag_3AsS_3', *J. Opto-Electronics*, Vol. 1, No. 3, p. 159.
[2] Hulme, K. F. *et al.*: 1967, *Appl. Phys. Letters* 10, 135.
[3] Bardsley, W. *et al.*: 1969, *J. Opto-Electronics* 1, 29.
[4] Auton, J. P.: 1967, *Appl. Opt.* 6, 1023.
[5] Rupprecht, G. *et al.*: 1962, *J. Opt. Soc. Am.* 52, 665.
[6] Clarke, D.: 1965, *Monthly Notices Roy. Astron. Soc.* 130, 75 and 82.

DISCUSSION

F. Melchiorri: In your instrumentation there is a 45° beam-splitter used as a mirror in infrared. A spurious polarization should be introduced by this mirror and modulated by the rotating polarizer. Have you observed this fact?

J. C. D. Marsh: We do not think this is important in our case. We have asked ourselves two things: firstly, is a photoconductive infrared cell sensitive to polarized infrared radiation. We have tested a cell and found that it is not sensitive to within 0.2%. Secondly, if one has radiation being reflected from a mirror, will that itself give instrumentally induced polarization? We decided that if there was no 45° mirror then it did not matter at all, and this was shown to be true in the optical range. We feel that by rotating the analyzer right at the front end of the input then it does not matter if you put a 45° mirror right at the backend.

D. Lemke: What objects are you planning to observe?

J. C. D. Marsh: The edge of the Moon to start with, where we expect some degree of polarization. This will give some information on the surface. We would then look at the atmosphere of Venus which seems to show a high percentage of polarization as the wavelength increases.

We would like to look then at galactic sources if we had sufficient sensitivity or a big telescope.

W. Hoffmann: Do I understand from your description that you are chopping the whole sky?

J. C. D. Marsh: We are chopping the sky against the back of the chopper disc, so that we will have a very large offset.

PROBLEMS AND DESIGN OF BLACK-BODY
REFERENCES

H. P. BALTES and P. STETTLER

Infrared Physics Group, Solid State Physics Laboratory,
Swiss Federal Institute of Technology, Zürich, Switzerland

Abstract. This contribution covers the following problems involved in the calibration of far-infrared telescopes:

(i) thermal radiation in finite resonators and the long-wave corrections of the Planck and Stefan-Boltzmann radiation formulae;

(ii) the search for high-temperature black-body wall materials among the refractory oxides; and

(iii) a black body for the absolute calibration of a balloon-borne far-infrared lamellar grating interferometer for solar measurements.

1. Introduction

Far-infrared and submillimetre black-body cavity radiation standards are indispensable reference sources for the absolute calibration of telescopes and radiometers used in astrophysical [1–13] as well as space research [14–17].

Far-infrared black-body cavities of usual laboratory size are characterized by the relations

$$\lambda \approx d/50, \qquad \Delta(1/\lambda) \approx 1/d, \tag{1}$$

where λ represents the wavelength, $\Delta(1/\lambda)$ the spectral resolution, and d a typical linear dimension of the cavity, and where $1/d$ is the approximate separation between two adjacent cavity modes. These relations (1) indicate that the far-infrared black bodies can be approximated neither by microwave thermal noise standards nor by optical cavity black bodies.

The *microwave thermal noise standards* fulfil the conditions

$$\lambda \approx d, \qquad \Delta(1/\lambda) \ll 1/d. \tag{2}$$

If ideal impedence matching is provided, the emission is described by the Nyquist formula

$$I(\nu_0) = kT\Delta\nu_0, \tag{3}$$

where $\Delta\nu_0$ is the bandwidth of the *single* emitted mode with the frequenty ν_0. Near-to-ideal narrow-band thermal noise cavity sources have been realized and successfully used in the microwave astronomy [18].

Optical cavity black bodies, on the other hand obey the relations

$$\lambda \ll d, \qquad \Delta(1/\lambda) \gg 1/d, \tag{4}$$

which are the prerequisites of Planck's radiation formula,

$$u(\nu, T) \, d\nu = \frac{h\nu}{e^{h\nu/kT} - 1} \, D_0(\nu) \, d\nu, \tag{5}$$

Manno and Ring (eds.), Infrared Detection Techniques for Space Research, 160–167. All Rights Reserved
Copyright © 1972 by D. Reidel Publishing Company, Dordrecht-Holland

with the asymptotic spectral density

$$D_0(\nu)\,d\nu = (8\pi/c^3)V\nu^2\,d\nu = D_p(\nu)\,d\nu \tag{6}$$

of the electromagnetic modes in a *lossless closed* cavity of the volume V.

The Equation (5) combined with the mode density (6) is strictly valid only in the limits $\nu \to \infty$ or $V \to \infty$. According to the assumptions (4), optical and near-infrared cavity black bodies with an exit aperture are calculated with the aid of geometric optics. Broad-band optical standards of high quality have been constructed [19].

In the case of poor spectral resolution, Equation (5) remains valid for *closed* cavities, provided that a refined spectral mode density

$$D(\nu) = D_0(\nu) + D_1(\nu) + D_2(\nu) + \cdots \tag{7}$$

is introduced. This is the subject of Section 2. $D_1(\nu)$ and $D_2(\nu)$ are expected to allow for the surface and the linear corrections, respectively. In Section 2, we report on the search for far-infrared high-temperature black wall materials. In the far infrared, the enhancement of the blackness of a given material by constructing astute cavity geometries is limited by diffraction losses. Thus the quality of the wall material is of fundamental interest. In Section 4, we present a preliminary calibration lay-out for our balloon-borne far-infrared interferometer measuring the solar spectral intensity between 15 μm and 1 mm wavelength.

2. Refinements of the Planck and Stefan-Boltzmann Radiation Formulae for Finite Cavities

The frequency density $D(\nu)$ of the electromagnetic resonances in a *lossless closed* cavity is an element of Planck's formula (5). Let us denote by $N(\nu)$ the number of modes with eigenfrequencies ν_n not exceeding ν. Then, by definition, $D(\nu) = d/d\nu\,N(\nu)$. $D(\nu)$ is a sum of delta distributions unless some averaging procedure is introduced, which, e.g., allows for the finite bandwidth of the spectrometer.

For a number of cavity geometries we computed the first 10^6 modes and calculated the average function $\overline{D(\nu)}$ in the spectral region intriguing the submillimetre spectro-copist. We found *second* order corrections of the kind

$$\frac{\overline{D(\nu)} - D_0(\nu)}{D_0(\nu)} = A(c/\nu)^2, \tag{8}$$

where the constant A depends on the cavity geometry. For instance for the *parallel-epiped* with edge lengths L_1, L_2, L_3 we obtained

$$A = -\frac{1}{8\pi}\frac{L_1 + L_2 + L_3}{L_1 L_2 L_3}, \tag{9}$$

and for the *circular cylinder* with the radius R and the length L we found

$$A = -\left(\frac{1}{16\pi}\right)\frac{2R + L}{R^2 L}. \tag{10}$$

Similar refinements were obtained for sectors of circular cylinders [20] corresponding to a *closed* wedge-shaped black body and for spherical sectors [21] corresponding to *closed* cone-shaped resonators. In particular, there is no *first* order correction $D_1(\nu)$ proportional to the cavity's surface area S. The vanishing of such a surface term in the Equation (7) can be shown in general [22, 23]. Some typical averaged computer results valid for poor resolution are shown in Figure 1.

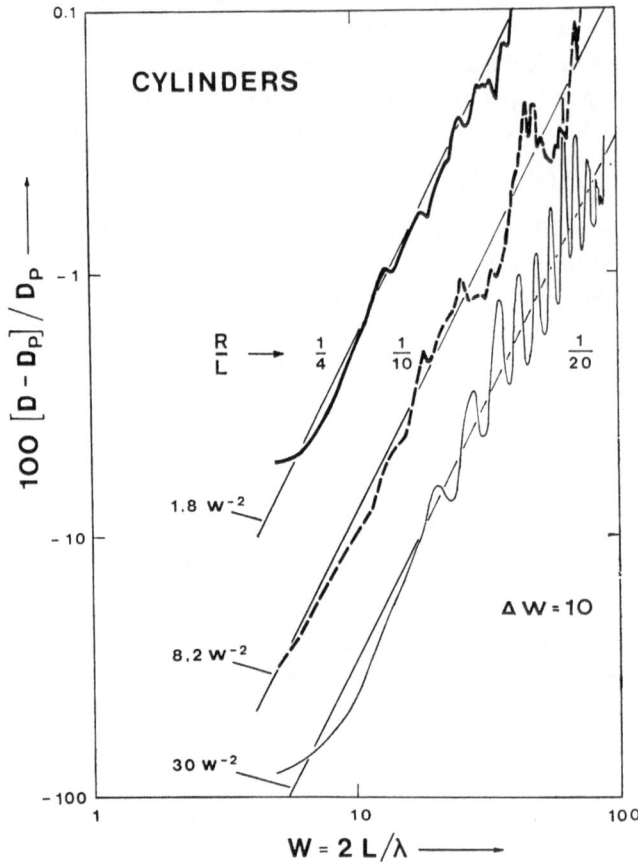

Fig. 1.   Computer results for the mode density of the electromagnetic radiation in cylinder-shaped lossless cavities. – R = radius, L = length of the cylinder. The relative deviation from the Planckian density D_p is plotted as a function of the parameter $W = 2L/\lambda$. The straight lines indicate the average deviations.

For narrow-bandwidth $\Delta\nu$ the fluctuation part of the mode density is relevant. We computed absolute mean derivations of the kind

$$\frac{\overline{|D(\nu) - \overline{D(\nu)}|}}{D_0(\nu)} = B\left(\frac{c^2}{\nu \cdot \Delta\nu}\right), \tag{11}$$

where B depends on the geometry. For instance for the cube with edge length L we found

$$B = 0.1L^{-2}. \tag{12}$$

So, the fluctuations provide a correction showing a first order *amplitude*.

The corresponding expansion for the total radiation energy inside the cavity reads

$$E(T): = \int_0^\infty u(\nu, T) \, d\nu = E_0(T) + E_1(T) + E_2(T) + \cdots, \tag{13}$$

where

$$E_0(T) = \sigma V T^4 \tag{14}$$

represents the well-known Stefan-Boltzmann law. There is *no* first order or surface term $E_1 \sim S T^3$. The relative second order correction reads

$$\frac{E_2(T)}{E_0(T)} = 10A \left(\frac{\hbar c}{kT} \right)^2, \tag{15}$$

where A is defined in Equation (8). For the *parallelepiped* we still know

$$E_3(T) = \tfrac{1}{2}kT. \tag{16}$$

The oscillation term leads to a correction of the type $E_4 = C_1 + C_2 T^3 \exp(-C_3 T)$ with geometry dependent constants C_1, C_2, C_3.

3. The Search for High Temperature Black Wall Material

The rigorous electromagnetic wave theory of a thermal submillimetre cavity source including the *aperture* is a hopeless task. We therefore introduced a one-dimensional model [21] for cavity standard sources which demonstrates the requirement of *impedance matching* between the hole and the black-body wall. This condition affects the construction of cavity black bodies in an essential manner: Baffles, small holes, sharp edges, etc., though useful in the optical and near-infrared regions, have to be abandoned in the long wave range. A *sufficiently large* flat black wall with uniform temperature would be the less intricate reference from this point of view. Thus the quality of the wall material is a much more fundamental problem than it is in the optical case.

As we measure the spectral intensity of the Sun, we are mainly interested in a *high temperature* reference source. We therefore examined refractory oxides, e.g. alumina, as possible black wall material. Actually, many materials of this type are non-transparent and show a spectral emissivity $E \gtrsim 0.8$ in a large part of the spectral region we are interested in. However, marked reststrahlen bands due to optical phonons appear between 20μ and 100μ with a maximum spectral reflectivity $R \approx 0.8$. As a consequence, the spectral emissivity $E = 1 - R$ is as low as 0.2 for the corresponding wavelength. Unfortunately, these reflection bands are particularly strong for the high purity ceramics showing excellent high temperature mechanical strength. As examples we mention the aluminas named Degussit, Giralumine, and Sikor, which have a fairly high emissivity only for $\lambda \gtrsim 50 \mu$.

It is well known that the reststrahlen bands are due to the long-range Coulomb-interaction [24]. This interaction can be screened by bringing imperfections into the crystal, e.g. impurities, or vacancies, or stoichiometric imperfections. So, some natural steatite compounds, e.g. Pyrostea, are nearly non-reflectant in the far infrared. Too much imperfections, however, may lead to a conducting ceramic material which becomes a good reflector when heated. An example is titania.

Ceramic layers with low spectral reflectivity in the optical phonon region can be produced by means of the *plasma spray* procedure described in [25]. We measured the spectral reflectivity of a number of these materials and found that the alumina spray Metco 101 is a promising black-body wall material.

Between 10μ and 50μ, its room temperature reflectivity at $40° \pm 5°$ incidence is $R \lesssim 0.25$. Between 50μ and 100μ we measured the spectral emissivity directly and found $E \approx 0.8$ for temperatures $600 \text{ K} > T > 300 \text{ K}$ and $45°$ emission angle. In Figure 2,

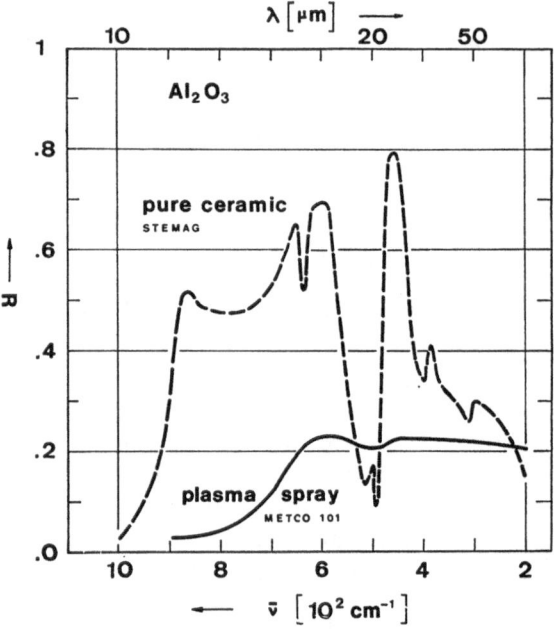

Fig. 2. The spectral reflectivity of alumina in the optical phonon region measured at room temperature and under $40° \pm 5°$ incidence angle. A high purity bulk ceramic material and the polished plasma spray layer Metco 101 are compared.

we compare this Al_2O_3 spray to a high grade bulk Al_2O_3 material produced by the STEMAG, Laufen, Germany. The reflection measurements were made at room temperature. One might hope that the reststrahlen peaks flatten with increasing temperature. Unfortunately, this is not a large effect as long as the temperature of the material is well below the melting point. This fact is well known from the temperature dependent reflection measurement of LiF and MgO due to Jasperse *et al.* [26].

We observe that the plasma spray layers can be made fairly thin (e.g. $\frac{1}{10}$ mm). They are directly applied to a metallic substrate, e.g. inconnel or tungsten. So, the large temperature gradient across the wall of a thick-walled ceramic tube can considerably be reduced.

The surface structure of a *polished* Metco 101 alumina layer is shown in Figure 3. The caves in the surface are due to the porosity of the material. Their diameter varies between 1 and 10 μ, approximately.

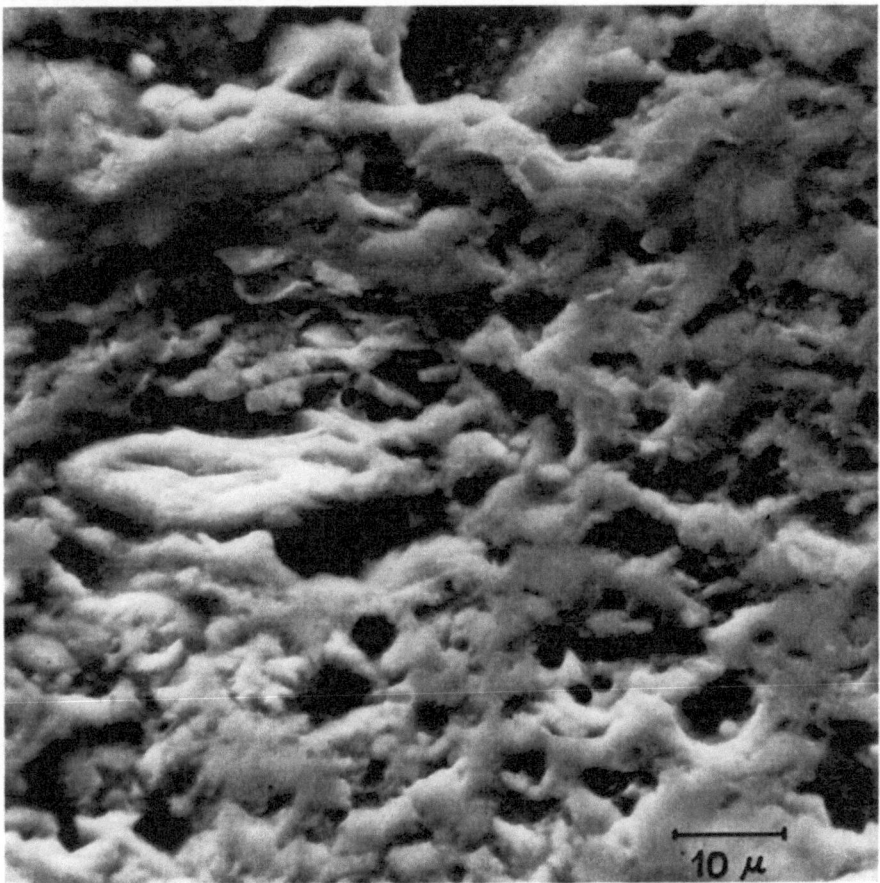

Fig. 3.   Scanning electron micrograph of a polished Metco 101 alumina plasma spray layer. The total edge length of the micrograph corresponds to 75 μm distance on the sample.

4. A Black Body for the Calibration of a Balloon-Borne Lamellar-Grating Interferometer

In 1968–1970 we constructed a balloon-borne lamellar-grating interferometer for the observation of the solar emission at wavelengths between 15 μ and 1 mm [8, 27]. The gondola was provided by the Observatory of Geneva, and the stabilization was designed by the 'Compagnie des Compteurs', Paris. The first successful flight started from the balloon station of the CNES in Aire sur l'Adour, France, in Spring 1971.

The absolute calibration of the interferometer is performed on ground before the flights. Various cones made of stainless steel and covered with the alumina plasma spray Metco 101 are used as rear wall.

The temperature profile along the cylinder axis is measured by three thermocouples. The range of operating temperature is between 800 K and 1500 K. The linear dimensions of the black body are sufficiently large in order to make sure that the correction terms (8) and (11) are below 1% for submillimetre waves. Thus we are entitled to calculate its effective spectral emittance by developing the ray tracing procedure due to Sparrow *et al.* [28]. We improved Sparrow's method by introducing a parabolic temperature profile along the cylinder axis instead of a uniform temperature [29].

The radiation is emitted through a cooled aperture. The image of this black-body exit is transferred to the entrance aperture of the interferometer by a collimator identical with the telescope (Figure 4). Therefore the black-body intensity is only about

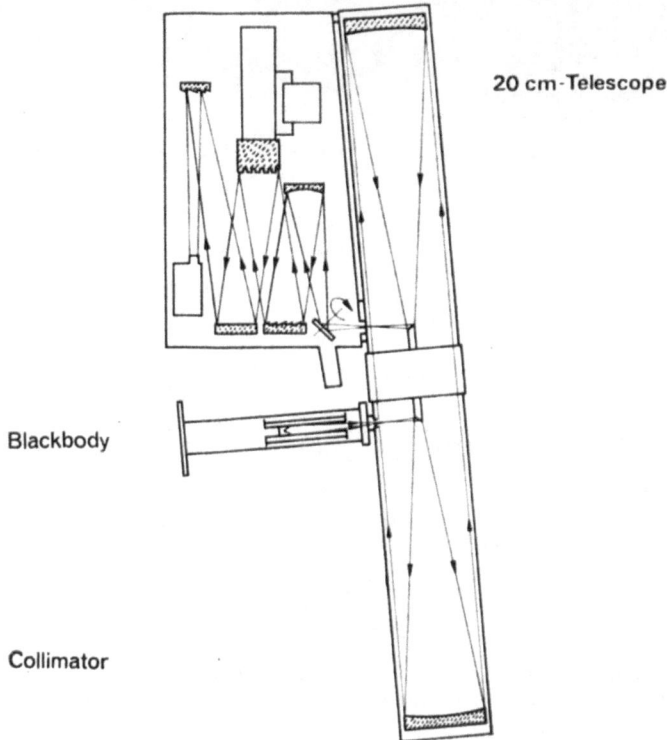

20 cm-Telescope

Blackbody

Collimator

Fig. 4.   Optical lay-out for the absolute calibration of the solar-interferometer. For solar measurements the collimatot-system will be replaced by a stabilized heliostat mirror.

five times weaker than the received solar intensity. Consequently the radiation intensities have the same order of magnitude in the calibration and the measurement arrangement. This seems to be advantageous when compared to calibration procedures using cold references.

Acknowledgement

This work was supported by the Schweizerische Nationalfonds.

References

[1] Shivanandan, K., Houck, J. R., and Harwit, M. O.: 1968, *Phys. Rev. Letters* **21**, 1460.
[2] McNutt, D. P., Shivanandan, K., and Feldmann, P. D.: 1969, *Appl. Opt.* **8**, 2199.
[3] Muehlner, D. and Weiss, R.: 1970, *Phys. Rev. Letters* **24**, 742.
[4] Park, W. M., Vickers, D. G., and Clegg, P. E.: 1970, *Astron. Astrophys.* **5**, 325.
[5] Low, F. J.: 1970, *Astrophys. J.* **159**, L173.
[6] Ammann, H. H. and Low, F. J.: 1970, *Astrophys. J.* **159**, L159.
[7] Noyes, R. W.: 1969, *Phil. Trans. Roy. Soc.* **264A**, 205.
[8] Stettler, P., Kneubühl, F. K., and Müller, E. A.: 1969, *Helv. Phys. Acta* **42**, 630.
[9] Kisljakov, A. G.: 1970, *Adv. Phys. Sci.* **101**, 607.
[10] Gay, J. *et al.*: 1968, *Astrophys. Letters* **2**, 169.
[11] Gay, J.: 1970, *Astron. Astrophys.* **6**, 327.
[12] Eddy, J. A., Léna, P., and MacQueen, R. M.: 1969, *Solar Phys.* **10**, 330.
[13] Eddy, J. A., Lee, R. H., Léna, P. J., and MacQueen, R. M.: 1970, *Appl. Opt.* **9**, 439.
[14] Kendall, Sr., J. M. and Berdahl, C. M.: 1970, *Appl. Opt.* **9**, 1082.
[15] Burroughs, W. J. and Harries, J. E.: 1970, *Nature* **227**, 824.
[16] Kreins, E. R. and Allison, L. J.: 1970, *Appl. Opt.* **9**, 681.
[17] Gurvich, A. S. and Yuferev, V. N.: 1970, *Fiz. Atmosf. Okeana* **6**, 523.
[18] IEEE Trans. *MTT-16*, No. 9, Sept. 1968, Special issue on thermal microwave standards.
[19] Quinn, T. J.: 1967, *Brit. J. Appl. Phys.* **18**, 1105.
[20] Baltes, H. P. and Kneubühl, F. K.: 1971, submitted to *Helv. Phys. Acta*.
[21] Baltes, H. P., Muri, R., and Kneubühl, F. K.: 1971, *Proc. XX Symposium on Submillimeter Waves, New York 1970* (in press).
[22] Balian, R. and Block, C.: 1971, submitted to *Ann. Phys.*
[23] Baltes, H. P. and Kneubühl, F. K.: 1971, submitted to *Opt. Comms.*
[24] Born, M. and Huang, K.: 1954, *Dynamical Theory of Crystal Lattices*, Clarendon Press, Oxford.
[25] (No author given), *Metallurgia* **81**, 94, March 1970.
[26] Jasperse, J. R., Kahan, A., Plendl, J. N., and Mitra, S. S.: 1966, *Phys. Rev.* **146**, 526.
[27] A detailed description of the improved balloon-borne solar lamellar-grating interferometer will be published.
[28] Sparrow, E. W., Albers, L. V., and Eckert, E. R. G.: 1962, *J. Heat Transfer* **84C**, 73.
[29] Jeannet, P.: 1971, Master Thesis, Swiss Federal Inst. of Technology (ETH), Zürich.

DISCUSSION

J. Ring: What are the physical reasons for the oscillations you have shown in the theory of mode density. Did the spectrometer resolve the individual modes?

H. P. Baltes: No spectrometer presently available can resolve these modes. This is the reason why I did the averaging by using realistic bandwidth or resolving powers available in spectrometers. The reason for the oscillations is the following: suppose you have a rectangular cavity where one edge is much larger than the other. If you now consider the eigenvalues of the resonance frequencies in the activated space, you will have dense planes with eigenvalues at a large distance. Then count, by cutting out a sphere, and whenever you touch the next dense plane it will oscillate.

D. H. Martin: I think the cancellation of the first order terms is a beautiful result.

But you considered modes in a closed cavity. Do you think that the introduction of an aperture now, to make it an emitting black body will change your conclusions very much?

H. P. Baltes: I do not think so, as long as the aperture is not too large. But then of course diffraction problems will appear. What one has to do, is to study the complete vector wave electromagnetic theory of a source having that shape, but this is a rather difficult task.

4. CRYOGENICS

COOLING SYSTEMS FOR SPACEBORNE
INFRARED EXPERIMENTS

R. W. BRECKENRIDGE, JR.

Arthur D. Little, Inc., Cambridge, Mass., U.S.A.

1. Introduction

A cryogenic cooler is an important component in many infrared systems, but it is an absolutely critical one on a spaceborne system.

This paper discusses cooling systems for spaceborne infrared experiments. Although the emphasis will be on space applications, it will be obvious that much of the discussion is applicable to airborne and balloon-borne experiments as well. The purpose of the paper is to discuss the characteristics of the various types of systems which can be used for cooling spaceborne infrared experiments. It will present generalized data from which a system designer can identify the type of cooling system most suited for his application. Data will also be presented which will enable him to make a preliminary estimate of the size and weight of a cooling system.

There are three basic cooling techniques which will be discussed:

(a) Passive radiators which cool components to cryogenic temperature levels by radiation to the space environment.

(b) Open-cycle systems which use stored high-pressure gas or a stored cryogenic fluid in either liquid or solid form.

(c) Closed-cycle systems which utilize a mechanical refrigerator to provide cooling at low temperatures.

The temperature level covered will be that below 100 K.

There are several constraints which are common to any cooling system which must operate in the space environment. The first of these are system weight and volume. System weight is of relatively little importance in ground-based installations, is of concern in airborne applications, but is critical in a spacecraft. The system weight includes the weight of all elements, not only the cooler, but also any auxiliaries. In closed-cycle refrigerators, the system weight includes the weight of power supply, control equipment, and radiators for rejecting the cycle heat.

The requirement for system reliability, over extended periods of time in some cases, is also important. Operating lifetimes of years will be required in some instances. Since there is no opportunity for maintenance or repair, the reliability of the cooler must be of the highest order. The simpler the system the better in this respect. But, as will be shown subsequently, the simpler systems, such as radiators and open-cycle coolers, are not universally adaptable to all temperature levels and heat loads.

Vibration affects the system in two ways. First, the vibrations transmitted by the cooler to the detector must be acceptably low. Second, all elements of the refrigeration system must be able to withstand the vibrations and shocks to which it is subjected at launch. These requirements present a major design problem particularly with respect

Manno and Ring (eds.), Infrared Detection Techniques for Space Research, 171–188. All Rights Reserved
Copyright © 1972 by D. Reidel Publishing Company, Dordrecht-Holland

to structural supports because the structural supports for low-temperature elements of the system must have small cross-sectional areas to minimize conductive heat leak. Special attention must be devoted to the dynamic behavior of this support system. Considerable ingenuity is generally required to produce a structurally adequate support system which provides sufficient thermal isolation.

A final consideration is thermal control of the refrigeration system itself, particularly in closed-cycle refrigerators where all cycle heat must be rejected to space via radiators.

2. Passive Radiators

One of the most intriguing methods of developing cryogenic temperatures in space is to utilize the low-temperature sink of space directly by radiating heat to it with a cyrogenic radiator. This concept is particularly attractive since such a system is completely passive, requires no power, and should be capable of high reliability for extended periods. Recently, there has been considerable activity devoted toward the design of passive radiators to maintain the temperature of detectors in electro-optical systems at temperatures in the 70–120 K region. Several passive radiators have been flown by NASA (see [1]).

The effective temperature of the star-speckled sky is less than 10 K. A suitably sized cold plate of high emittance to which one or more detectors are mounted can be made to radiate to this sink. It is, of course, necessary to shield the cold plate against heat inputs from the Sun, the Earth, and the spacecraft. These considerations usually result in a radiator design which is tailored to a particular spacecraft system. The orbit plane, orbit altitude, and the location of the radiator on the spacecraft all significantly influence the design of the radiator.

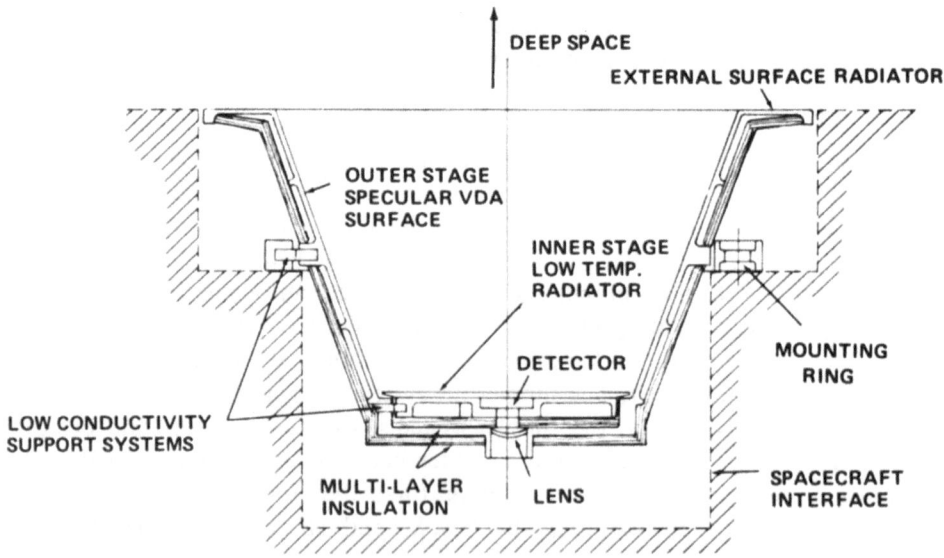

Fig. 1.   Schematic diagram of a passive radiator.

A typical passive radiator design is shown schematically in Figure 1. The assembly consists of an outer stage mounted to the spacecraft with low-conductance supports. The outer stage supports the cold stage also by low-conductance supports. Housed within the low-temperature stage is the detector. The internal surfaces of the outer stage are highly reflecting specular surfaces.

The top surface of the cold stage is of high emittance to dissipate the heat load on the stage by radiation to space. The bottom and side surfaces of the cold stage have low emittance to minimize the radiant heat input from the warmer outer stage. Multilayer insulation is also used between the two stages to further decrease the energy transfer.

The outer stage is designed to operate at a temperature lower than the spacecraft temperature to decrease the conductive and radiant energy transfer to the cold stage. Multilayer insulation is used to insulate the outer stage from the relatively warm spacecraft. The internal surfaces of the outer stage are of low emittance and absorption in both the solar and infrared spectral regions. It should be noted that *both* infrared emittance and solar absorption must be as low as possible – not the usual α/ε ratio.

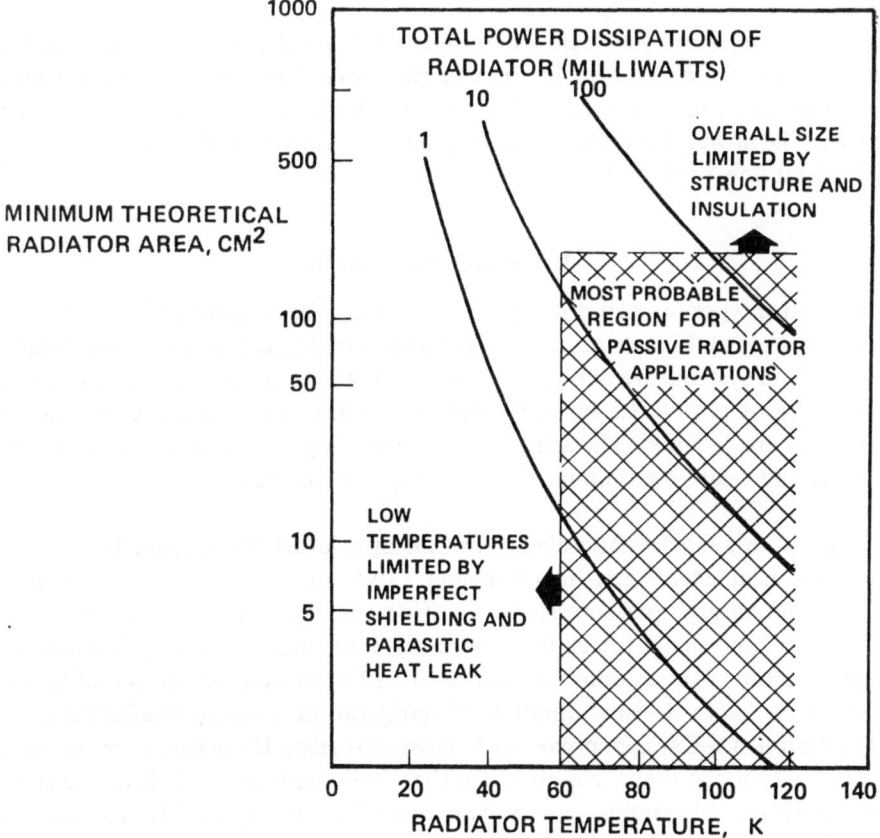

Fig. 2. Minimum theoretical radiator area as a function of radiator temperature and heat load.

Radiative cooling of the outer stage is provided mainly by the external surface radiator and to a lesser extent by the internal surface radiator. On the external surface radiator, the ratio, α/ε, must be small.

Important considerations in the design of low-temperature radiators are: the amount of detector heat load that must be radiatively dissipated and the maximum allowable operating temperature of the detector. The way in which these parameters affect the size of the radiator are shown in Figure 2. This graph shows the required area of a radiator of unity emittance as a function of the total power radiated and the radiator temperature level. The heat leak shown here is the sum of the heat load imposed by the detector plus all parasitic heat leaks. In a typical design the parasitic heat leaks will be 90% of the total heat load. Thus, for the 100 mW curve, only 10 mW are from detectors, and the balance is from conductive and radiative heat leak in the cooler itself.

The area indicated in Figure 2 is the area of the inner stage shown in Figure 1, assuming that this stage has an emittance equal to unity. The area of the complete radiator assembly will be larger than shown because the emittance of the low temperature stage is not unity and because the thermal shielding system and structure increase the overall size of the assembly. The relative dimensions shown in Figure 1 are typical dimensions.

Figure 2 also shows the region of power dissipation and operating temperature which appear to be best suited for passive radiators. The maximum area is the area above which the size and weight of a structurally adequate radiator system become excessive. The lower temperature limit (approximately 60–70 K) is determined by parasitic heat flow to the cold plate.

3. Open-Cycle Systems

There are three types of open-cycle systems which can be considered for space cooling: Joule-Thompson (J-T) systems using stored gas, stored liquid systems, and stored solid systems. These systems are applicable only where the total amount of refrigeration is low since the weight of this type of system precludes its use where large amounts of refrigeration are required. The attractive features of open cycle systems are simplicity, reliability, and relative economy. Further, they can be built within the present state-of-the-art.

Figure 3 shows a schematic diagram of an open-cycle J-T system. It consists of a counterflow heat exchanger, an expansion valve (or throttling orifice), and a gas bottle containing high pressure gaseous refrigerant. It is one of the simplest systems which can be considered for cooling detectors, and therein lies its prime virtue. The refrigerant can be stored at ambient temperature for an extended period of time at no loss. Where a relatively small amount of refrigeration is required after an extended inactive period, a J-T system proves to be most attractive. If the gas supply contains no impurities which can freeze out and plug the small passages in the heat exchanger or throttling valve, this system can be very reliable. The optical device can also be mounted at a distance from the high-pressure gas-storage bottle and articulated with respect to this bottle.

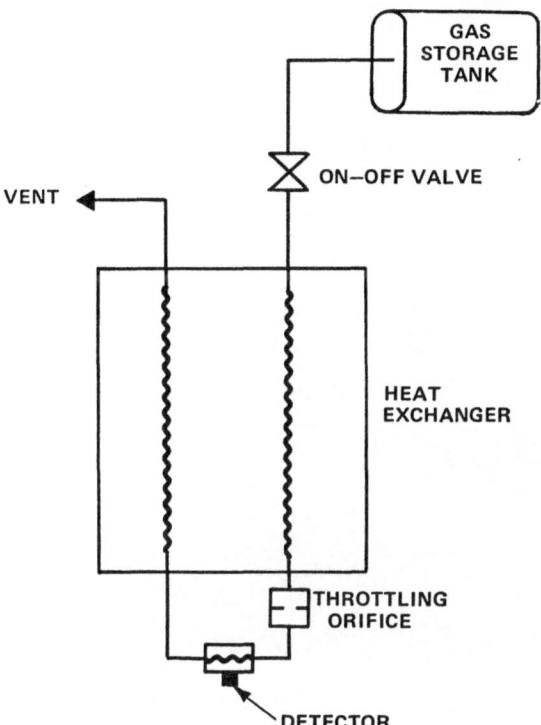

Fig. 3. Open cycle Joule-Thompson system.

Figure 4 is a photograph of the heat exchanger assembly of a representative J-T system. It consists of a capillary tube wrapped around and bonded to a hollow mandrel which is packed with a multiplicity of perforated plates. The assembly is inserted into a miniature dewar in which the detector is mounted. In operation, high pressure gas flows from the gas storage bottle through the capillary tube, and is cooled and throttled in the process. The cold gas from the open end of the capillary tube flows across the detector mounting plate, and back up the central mandrel, where it exchanges heat with the incoming gas stream.

Figure 5 shows a detail of the cold end of the device in which the opening in the capillary tube and the perforated plates may be seen. The tube has an inside diameter of 0.13 mm, and the holes in the plate are 0.20 mm in diameter.

A fundamental drawback to the J-T system is the high-pressure gas supply, which is too large and heavy for long-term operations. Consequently, in spite of its simplicity, the J-T system is applicable in spaceborne systems only where the duty cycle is very short.

Open-cycle systems using stored-liquid cryogens have been used to cool optical components. Figure 6 shows the liquid helium dewar which was used to cool an infrared detector and optical system in a sky-scanning astronomical telescope carried by a scientific sounding rocket [2]. This package was launched in a 38 cm diameter

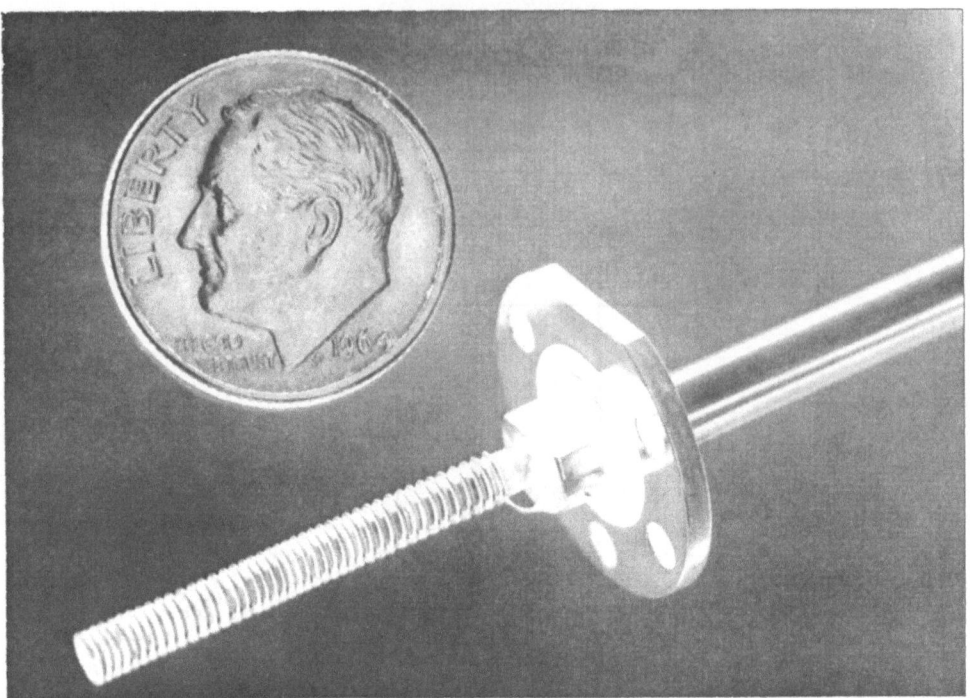

Fig. 4.   Joule-Thompson cooler.

Fig. 5.   Detail of a Joule-Thompson cooler.

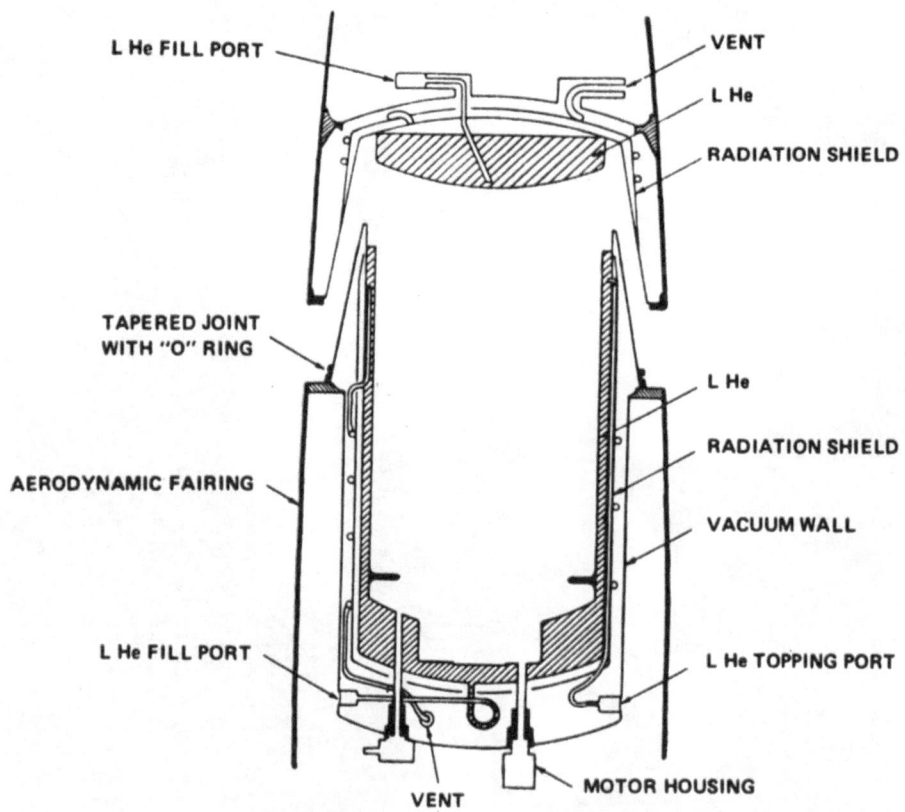

Fig. 6.   Rocket-borne liquid helium dewar.

liquid-fueled rocket. The entire telescope is surrounded by a liquid-helium dewar which maintains it at 4.2 K. There are two liquid storage volumes, one for the telescope and one for the cover. After the rocket reaches altitude, the cover is ejected and observations are made for 50 s.

Figure 7 shows a liquid helium dewar which is designed to maintain a superconducting magnet at liquid helium temperature for a six hour pre-launch hold period and a seven minute flight in a Trailblazer sounding rocket. The magnet is housed inside the vacuum shell of the dewar. The dewar, which has a liquid capacity of approximately one litre, weighs approximately 5 kg and is 19.0 cm in diameter by 25.0 cm long.

The use of liquid cryogens in space appears to be confined to cases of the sort for which the dewars of Figures 6 and 7 are used – i.e. cases in which refrigeration is required for a relatively short period immediately following launch.

In a solid cryogen dewar the item to be cooled is usually mounted on a pedestal which is thermally connected to a vessel containing a solidified cryogen. The ullage space above the stored solid is evacuated to maintain the cryogen in its solid state. Heat entering from the cooled object and the surroundings causes the cryogen to

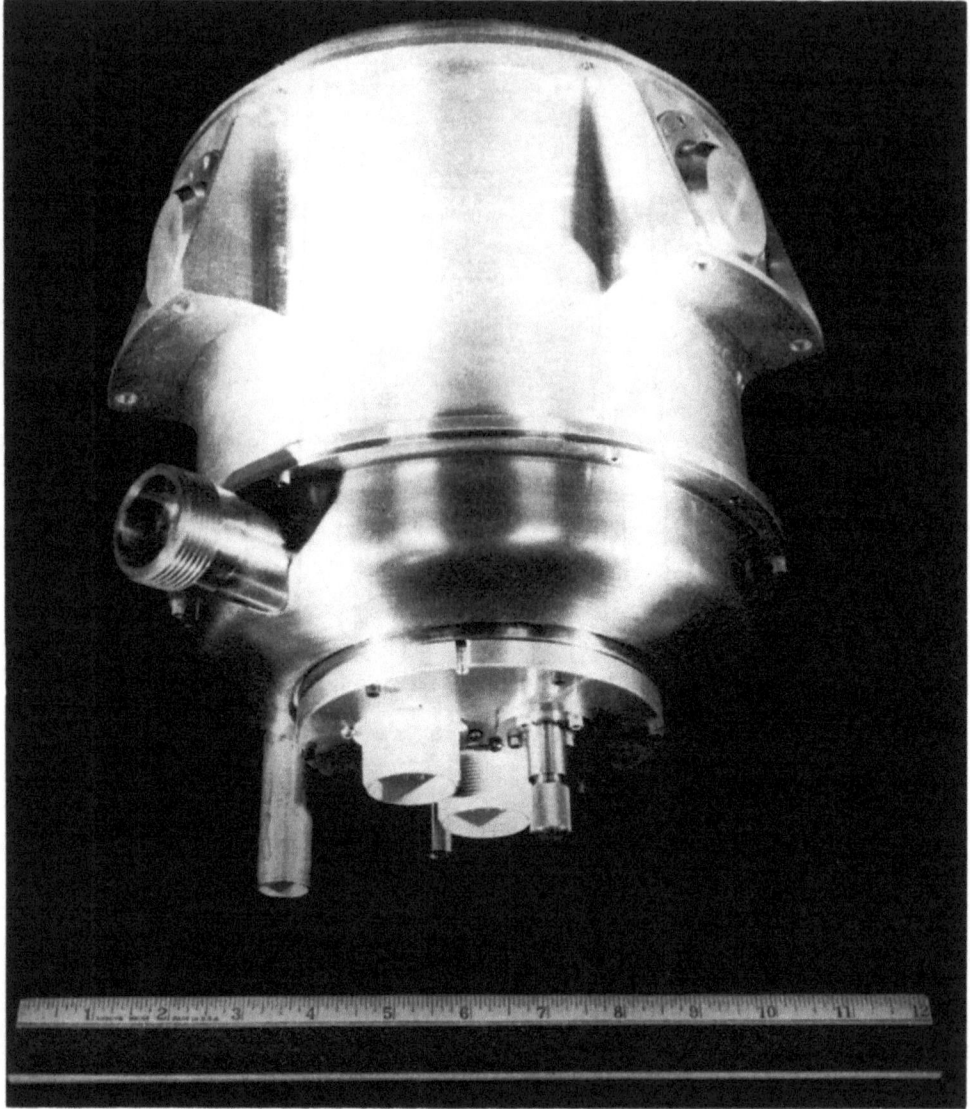

Fig. 7.   Liquid helium dewar for a superconducting magnet.

sublime and the resulting vapor is vented. Figure 8 shows the essential elements of a solid cryogen dewar. It consists of an inner vessel containing the cryogen, an insulation system, and a vacuum shell. A fill and vent tube permits initial filling of the cryogen vessel and venting of a vapor during operation. Auxiliary cooling lines (not shown) are used during the filling operation and during pre-launch storage. A structural support system for the inner vessel (also not shown) is also required.

Depending on the application, a solid cryogen cooler can offer several advantages

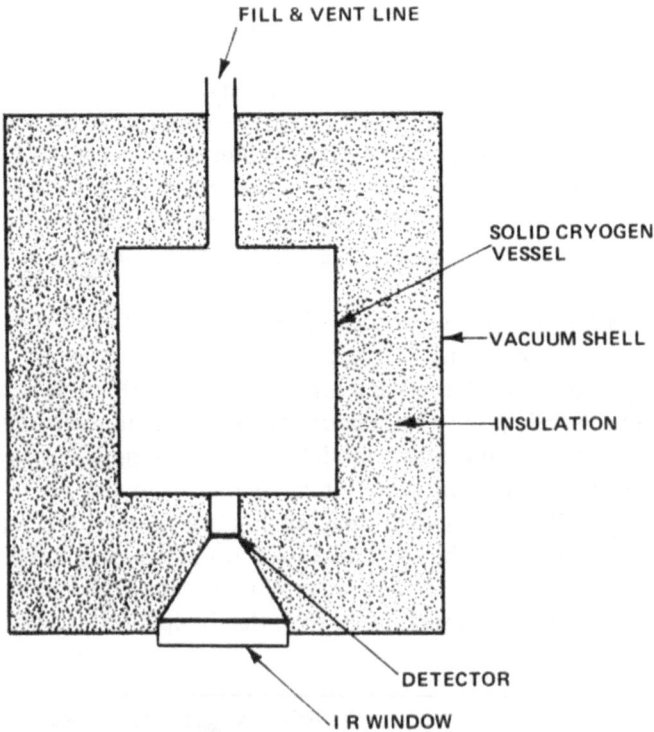

Fig. 8. Schematic representation of a solid cryogen cooler.

over alternative systems [3]. It offers a lower weight system compared with those using stored-liquid cryogens, since the latent heat of sublimation is greater than the latent heat of vaporization. Additionally, solid systems avoid the problems inherent to the uncertain location of liquid under conditions of near-zero-gravity space flight.

Several disadvantages of the subliming refrigerator are that: it imposes restrictions of the detector mounting; and methane and hydrogen, both of which have outstanding properties as solid coolants, are hazardous materials and require special handling.

Essentially, the design of a solid cryogenic cooler is the design of a cryogen storage dewar. The art and technology involved in the development of cryogenic dewars have application to the design of the cooler. Minimum parasitic heat leak is paramount. The service demands that a great deal of ingenuity and technical sophistication be applied to the design of low-heat-leak structural supports and mounting arrangements for the inner vessel and the cooled component. Special operational needs, involving filling procedures, standby requirements, control, and safety are all extremely important to the ultimate design.

To place the weight penalty associated with solid-cryogen coolers in perspective relative to other systems, estimates have been made of the weight of solid-cryogen coolers. Figure 9 shows the estimated weight of solid-cryogen coolers as a function of detector temperature and heat load for a one-year service life. Weights may be scaled from this curve on the basis of the total watt-hours of refrigeration required at the desired temperature.

These curves are intended for first-order determination of cooler weights. They are based on a simple model using solid hydrogen as the cryogen for all temperatures less than 67 K and utilizing both the latent heat of sublimation as well as the specific heat of the gas. For temperatures above 67 K, the curves are based on using solid methane as the refrigerant. Coolers using solid hydrogen are generally lighter (but larger) than coolers employing other cryogens, so the weights cited should be considered as ultimate lower limits.

To reduce the size of the cooler or to avoid the hazards of working with hydrogen, other cryogens must be used. Some of these systems use two cryogens – one with a low sublimation temperature to achieve the temperature required by the detector, and the other with a higher sublimation temperature to surround and shield the lower-temperature cryogen. Several laboratory models of a solid-argon cooler with a solid-carbon dioxide shield have been built for refrigeration in the 50–70 K region [4]. Systems of this type will weigh several times as much as indicated in Figure 9.

The heat load in Figure 9 is the total heat load at the detector and comprises dissipation in the detector (and any cooled preamplifiers) plus heat leak to the detector. The detector heat leak is often much more than the dissipation, so the thermal design

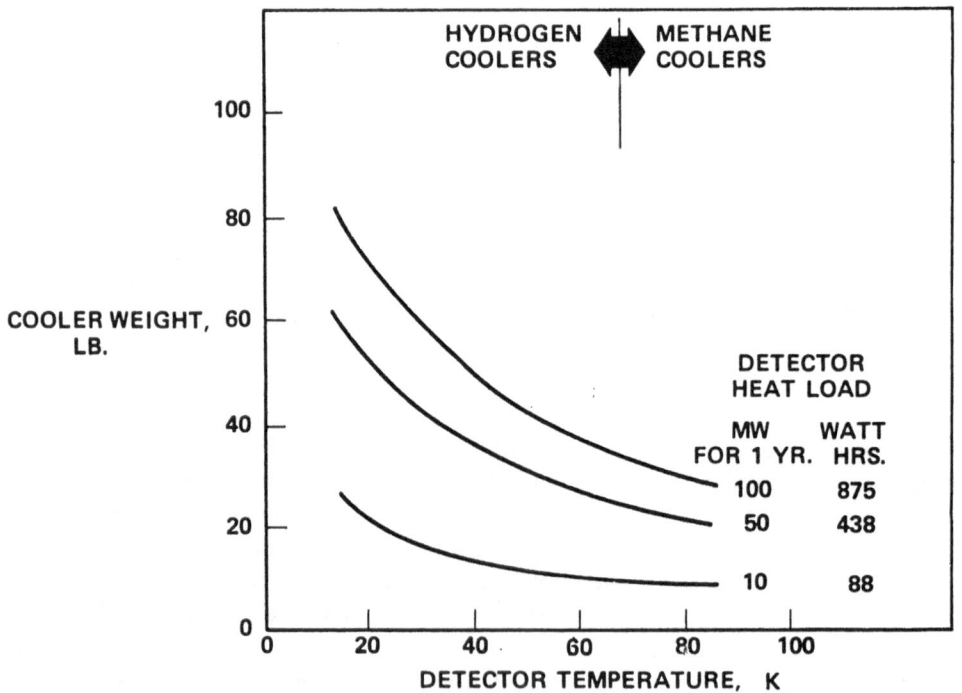

Fig. 9.   Weight of solid cryogen coolers as a function of detector temperature and heat load.

of the detector mounting and integration of the detector with the cooler are critical.

Because they require no power and can be built within the framework of present advanced technology, solid-cryogen coolers will probably see service during the next five to ten years in specialized applications where the heat loads are low; where detector temperature, orbital conditions, or other constraints preclude the use of radiators; and where mission durations are measured in months rather than years. It will be natural to look for efficient, compact, reliable closed-cycle refrigerators to displace them in many applications. Ultimately, their use will probably be confined to short- or intermediate-term missions with very low heat loads.

4. Closed-Cycle Refrigerators

Figure 10 is a block diagram showing the primary elements of a closed-cycle space-borne refrigeration system. The power supply is the basic source of motive power to the refrigerator. For refrigerators designed for electrical power input, the power supply may be solar panels, or, ultimately, closed-cycle electrical power systems deriving their energy from nuclear isotopes or the Sun. Solar panels have been used in

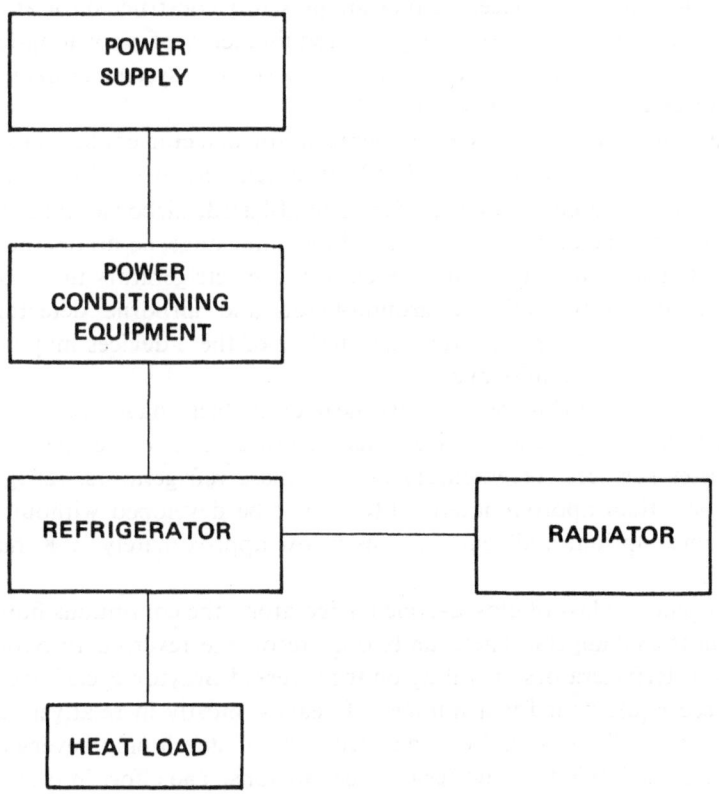

Fig. 10. Block diagram of a closed cycle spaceborne refrigeration system.

most spacecraft to date and will undoubtedly be the primary source of power for some time. Present operational systems have capacities of about 700 W. Work is underway to develop systems to generate up to several kilowatts of electrical power [5].

The power-conditioning equipment converts the raw power to the voltage and frequency required by the refrigerator, and modulates this power according to the needs of the refrigerator. Typically, the equipment is composed of solid-state inverters and control circuitry.

The heart of the system is the refrigerator itself, whose purpose is to pump the heat from the load temperature up to the temperature of the space radiator. The heat load, plus all the energy put into the refrigerator must be rejected by the radiator.

There are two general classes of cryogenic refrigerators. The first of these are the so-called intermittant flow types, which use regenerative heat exchangers. There are three types in this class. They are identified by the thermodynamic cycle on which they operate. They are Stirling cycle refrigerators, Gifford-McMahon cycle refrigerators, and Vuilleumier cycle refrigerators. Stirling cycle and Gifford-McMahon cycle refrigerators have been developed for ground-base and airborne applications, where they are presently being used in operational systems. The maintenance intervals of these systems is from 500 h for the former up to several thousand hours for the latter. It is doubtful whether these devices, as they are presently constructed for airborne use, can be made reliable enough to satisfy the requirements of very long-term space missions. At present no effort is being made to develop a system utilizing either of these cycles for extended space missions.

Development of Vuilleumier cycle refrigerators for detector cooling began several years ago [6]. This refrigeration system is a heat driven derivative of the Stirling cycle refrigerator. Units have been designed for ground-based, airborne, and spaceborne applications. The refrigerators are in the developmental stage, so no operational units are available at this time. When fully developed, the refrigerators may have higher reliability than presently available ground-based and airborne detector coolers. However, the maintenance-free operational lifetime of these devices may not be long enough for extended space missions.

The three systems utilizing regenerative heat exchangers have another limitation involving the minimum practical low end temperature. Due to the characteristics of the heat transfer surfaces (regenerator) used in these refrigerators, refrigeration at temperatures less than approximately 20 K cannot be developed without extremely high power consumption, and temperatures below approximately 10 K may not be possible at all.

The second general class of closed-cycle refrigerators, the continuous flow type, use counterflow heat exchangers. These units operate on the reversed Brayton cycle or derivatives of it. Refrigerators operating on the reversed Brayton cycle have been used in ground-based equipment for a number of years – mostly in relatively large-scale systems. Recently, efforts have been devoted toward developing reversed Brayton cycle refrigerators suitable for long-term space missions. The effort in these programs has been to scale down the size of existing systems, using an approach to the design of the mechanical elements which is inherently very reliable. There are several programs

under way to develop refrigerators of this type. One refrigeration system uses reciprocating machinery [7, 8], and the other uses turbomachinery [9, 10].

Extensive effort has been devoted to developing gas-bearing-supported machinery
for these refrigerators, resulting in equipment which is inherently capable of extremely
long life. The units must be considered as developmental items at this time – but they
appear to be the only ones which are capable of meeting the life requirements of
extremely long-term space missions.

Therefore, in spite of the rather wide range cryogenic refrigerators used today, the
number of systems which can meet the requirements of a long-term space mission is
extremely limited. These systems are closed-cycle refrigerators which operate on
reversed Brayton cycle.

Figure 11 shows the amount of power required by a closed-cycle refrigeration system
as a function of the temperature of the heat load and the refrigeration capacity. These
curves are general curves compiled from data in the open literature. They are intended
to be used only for making preliminary estimates of power. The spread in power
reflects the fact that not much data is available on this type of equipment. The curves
have been drawn for the refrigeration system which results in the minimum power for
refrigeration at a given temperature level (i.e., the curves are composite curves,
reflecting the performance of several types of refrigerators).

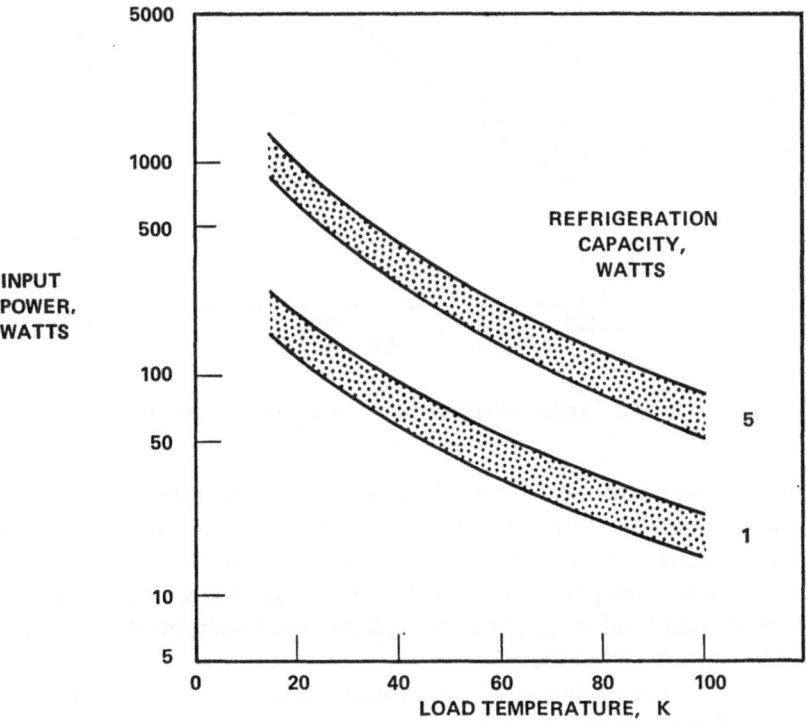

Fig. 11.   Power requirements for spaceborne closed cycle refrigerators.

Figure 12 shows the weight of a closed-cycle refrigerator as a function of the input power to the refrigerator. Again, this is a general, composite curve which indicates the weight of flight-type equipment. It is apparent from Figures 11 and 12 that, to minimize the weight and power consumption of closed-cycle refrigeration systems, the highest possible temperatures and the lowest possible heat loads, consistent with performance requirements, should be used.

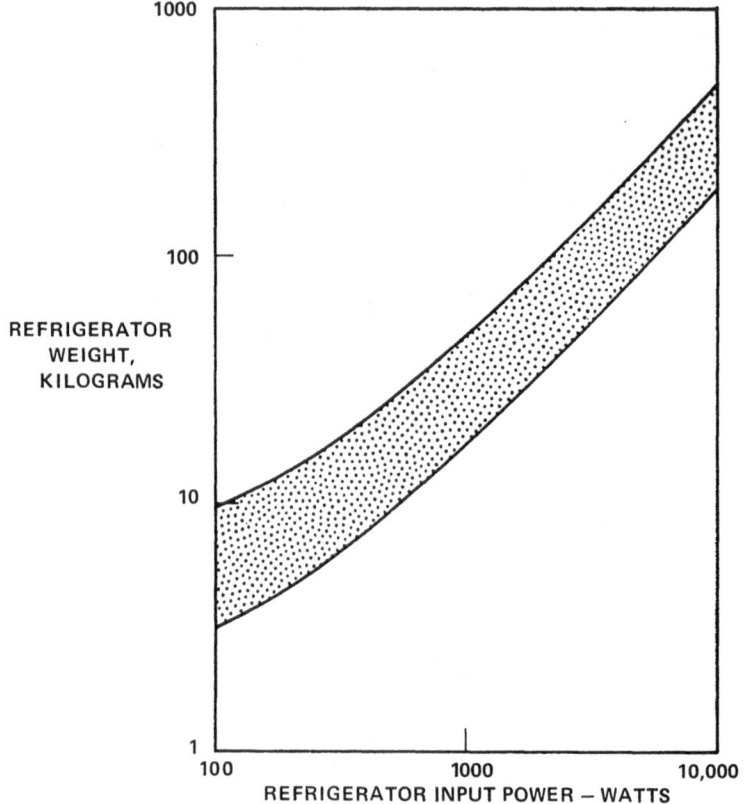

Fig. 12. Weight of spaceborne closed cycle refrigerators.

The previous discussion has covered passive radiators, open-cycle systems, and closed-cycle refrigerators, in ascending order of cost and complexity. The following figures will show the most probable areas of application of these spaceborne cooling systems. The basis of judgment for active systems has been to select the one which yields the minimum total weight. Passive systems have been blocked in where their special merits seem to place them.

Passive radiators are applicable for small heat loads down to their lowest temperature limit (see Figure 13). Solid-cryogen coolers are also applicable for low heat loads, and can be used to reach temperatures below those achievable with radiators, as

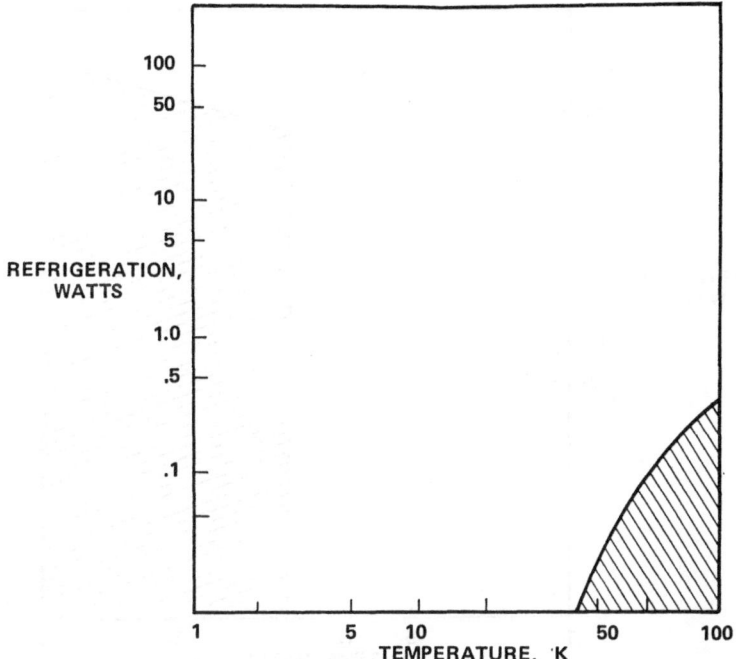

Fig. 13.   Probable area of application for passive radiators.

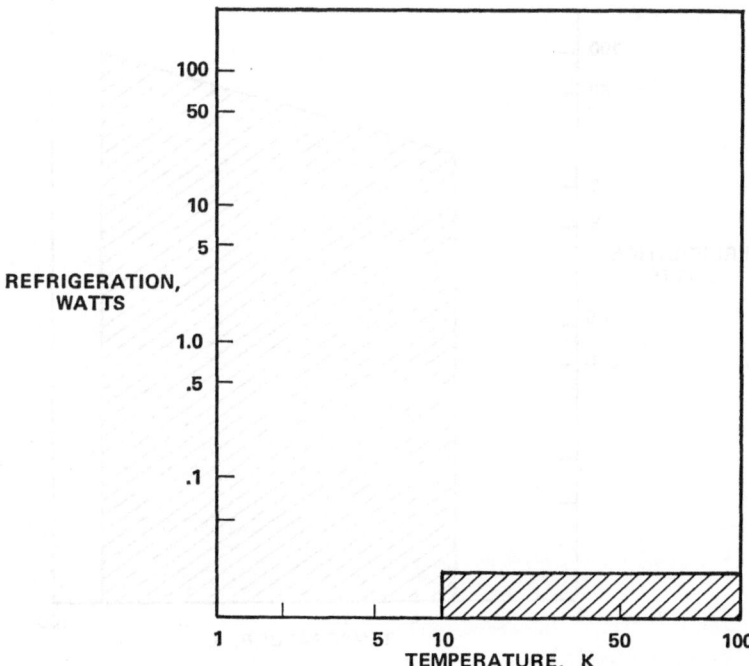

Fig. 14.   Probable area of application for solid cryogen coolers.

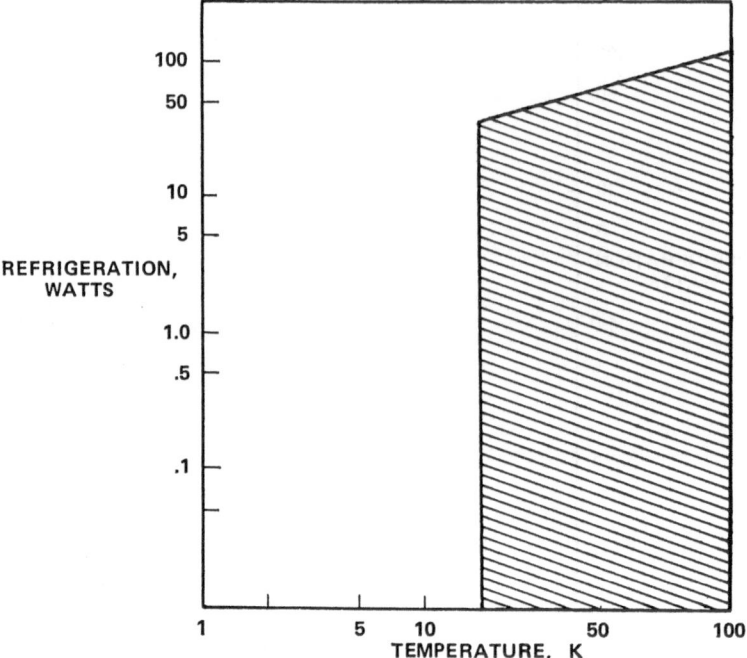

Fig. 15. Probable area of application for Stirling and Vuilleumier cycle refrigerators.

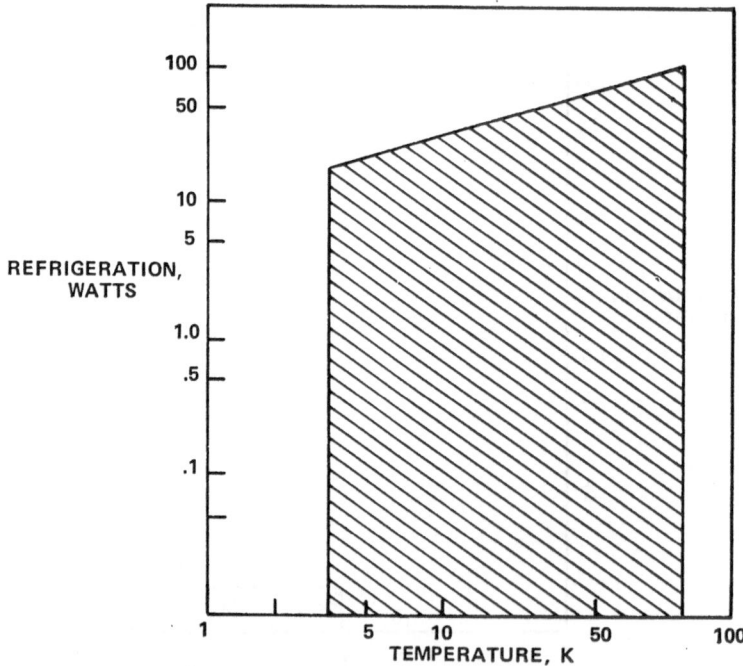

Fig. 16. Probable area of application for Brayton and Claude cycle refrigerators using reciprocating machinery.

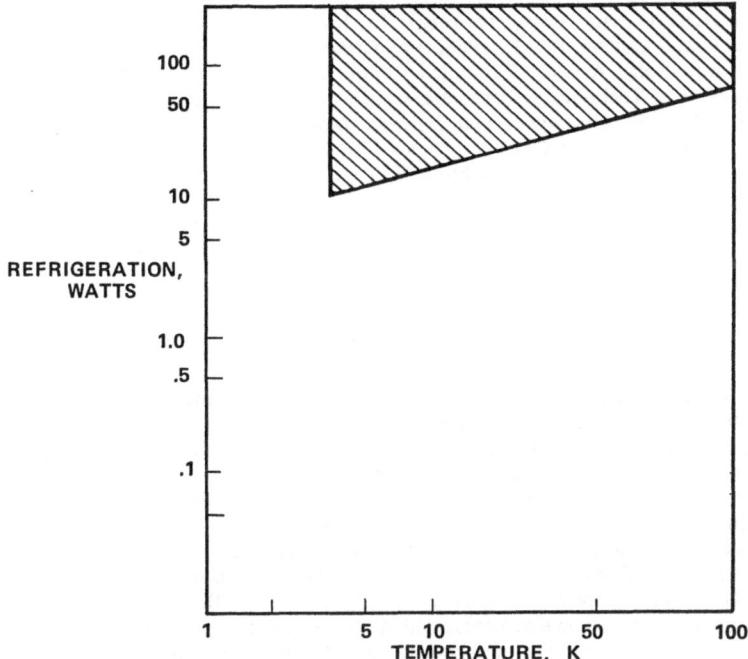

Fig. 17. Probable area of application for Brayton and Claude cycle refrigerators using turbo-machinery.

indicated in Figure 14. Figure 15 shows that Stirling and Vuilleumier cycle refrigerators are suitable in the important region of moderate heat loads at temperatures down to the point where regenerator inefficiencies limit performance. As regenerator technology improves, the low-temperature boundary of this region may move to lower temperatures. The upper boundary of this region is limited by machinery size. Figure 16 shows that reciprocating Brayton cycle or Claude cycle refrigerators are applicable for low and intermediate heat loads at temperatures between liquid helium temperatures and the region of 70 K. The upper capacity boundary of this region is determined by machinery size. Figure 17 indicated that reversed Brayton cycle and Claude cycle refrigerators employing turbomachinery will occupy the region of high heat loads over the entire temperature range.

If one were to superimpose these last five curves, there would be areas of overlap. These areas are due to the developmental nature of the entire technology of spaceborne refrigeration. After systems have been developed further and more have been flown in spacecraft, their capabilities and limitations will be better identified so the areas of application will become more clearly defined.

References

[1] Annable, R. V. *et al.*: 1967, 'A Day-Night High Resolution Infrared Radiometer Employing Two-Stage Radiant Cooling', Final Report on Contract NAS 5-10113, ITT Industrial Laboratories, Ft. Wayne, Indiana.

[2] McNutt, D. P.: 1969, *Astronaut. Aeronaut.* **7**, 44.

[3] Fowle, A. A.: 1966, *Adv. Cryogenic Engin.* **11**, 198.

[4] Caren, R. P. and Coston, R. M.: 1968, *Adv. Cryogenic Engin.* Vol. 13, p. 450.

[5] 'Multikilowatt Solar Arrays', Technical Report 32-1140, Jet Propulsion Laboratory, California Institute of Technology, Pasadena, California, July 15, 1967.

[6] Prast, G.: 1969, in *Cryogenics and Infrared Detection Systems, a Technical Colloquium*, Frankfurt Am Main, West Germany, April 17–18. 1969.

[7] Schulte, C. A., Fowle, A. A., Heuchling, T. P., and Kronauer, R. E.: 1965, *Adv. Cryogenic Engin.* **10**, 477.

[8] Breckenridge, Jr., R. W.: 1968, *Adv. Cryogenic Eng.* **13**, 387.

[9] Gessner, R. L. and Colyer, D. B.: 1968, *Adv. Cryogenic Engin.* **13**, 474.

[10] Maddocks, F. E.: 1968, *Adv. Cryogenic Engin.* **13**, 463.

DISCUSSION

R. D. Joseph: With regard to the solid cryogen system, it seems to me there will be a lot of trouble with keeping the solid cryogen in good thermal contact with the detector.

R. W. Breckenridge: Typically there is a highly conducting mesh inside the solid pack, which provides a conductive path to connect the cryogen to the detector.

P. Léna: From what you said about heat loads, it is clear that any progress made with superconducting junctions, allowing us to work between 15 K and 20 K, will be extremely important.

K. Shivanandan: If we accept the hypothesis that space is at 3 K or 8 K, what sort of endurances can you expect for your closed cycle system once you are already in space?

R. W. Breckenridge: For the closed cycle system, I think that with sufficient development one can talk of 3 to 5 yr endurance. The electronics which control the system may be the ultimate limit.

K. Shivanandan: How does this compare with the situation when man is in space?

R. W. Breckenridge: This is of course an attractive alternative when a man can perform maintenance.

In fact, the only closed cycle refrigerator which I know is being considered for flight is for the manned shuttle programme. It has a total duty cycle of a hundred hours or so.

R. C. Meiner: How does cooling with Peltier effect compare with other systems?

R. W. Breckenridge: The Peltier effect is generally applicable for temperatures on the high side of the temperature range I was discussing. They become inefficient and consume a great deal of power when going down in the 100 K region.

BALLOON-BORNE INFRARED CRYOSTATS

G. CHANIN

Service d'Aeronomie du CNRS, Verrières-le-Buisson, France

Abstract. Far-infrared studies in aeronomy and astronomy are generally not feasible from ground-based observatories because of atmospheric absorption and emission. Stabilized balloon platforms at altitudes superior to 35 km afford a convenient and relatively inexpensive way to avoid most atmospheric interference and permit observation periods in excess of 6 h. Since all sensitive far-infrared instruments need to be cooled to liquid helium temperatures, the development of suitable cryostats was undertaken. Two types of cryostats will be described: a small cryostat where the radiation passes through a warm infrared transmitting window into the cryostat vacuum space to the detection system, and a windowless cryostat in which the incoming radiation traverses the helium gas and liquid before passing through a cold window into the vacuum space. The first type of cryostat has been in use since 1966 and has never been subject to in-flight or laboratory failures. The second is now undergoing laboratory tests for a forthcoming flight.

1. Optical Window Balloon-Borne Cryostat

The cryostat to be described here evolved gradually after use of several different prototypes in infrared experiments using balloon platforms or using optical telescopes in ground-based observatories. Its design is a compromise to satisfy the criteria of performance and convenience under our special operating conditions, low cost, ease of construction and durability. Essentially, the cryostat provides a liquid helium cooled, large (13 cm dia.) optical platform (upon which the optical components of the experiment may be mounted) within a vacuum envelope containing an optical window. Its over-all size is 40 cm height by 17 cm dia. and its weight is 5.1 kg. The rigidity of the optical platform within the cryostat envelope, and its insensitivity to orientation, permit it to be mounted on telescopes in Newtonian or Cassegrain configurations. In an application reported by M. Herse a grating monochromator and bolometer system were installed on the optical platform, producing a gain in detectivity, since cooling the optical equipment decreases the background thermal radiation load on the detector.

The cryostat is shown in its latest form in Figure 1. Three pure aluminum cylindrical radiation shields are used: the outer one anchored to the filling tube, the intermediate shield thermally floating, and the innermost one fixed to the upper part of the helium reservoir. No liquid nitrogen is used: the shields are cooled by the increase in enthalpy of the cold helium gas evaporating from the reservoir. An earlier model of this cryostat used a single radiation shield anchored to a convoluted filling tube but this was abandoned in favour of the present arrangement in order to simplify the construction and decrease the number of welded joints. In both versions, the constructional materials are the same: duraluminum for the outer casing, aluminum for the radiation shields, brass or copper for the optical platform, and stainless steel for the helium reservoir, the filling tube, and the three support tubes between the reservoir and the stainless steel conical top of the cryostat. The difference in thermal expansion between

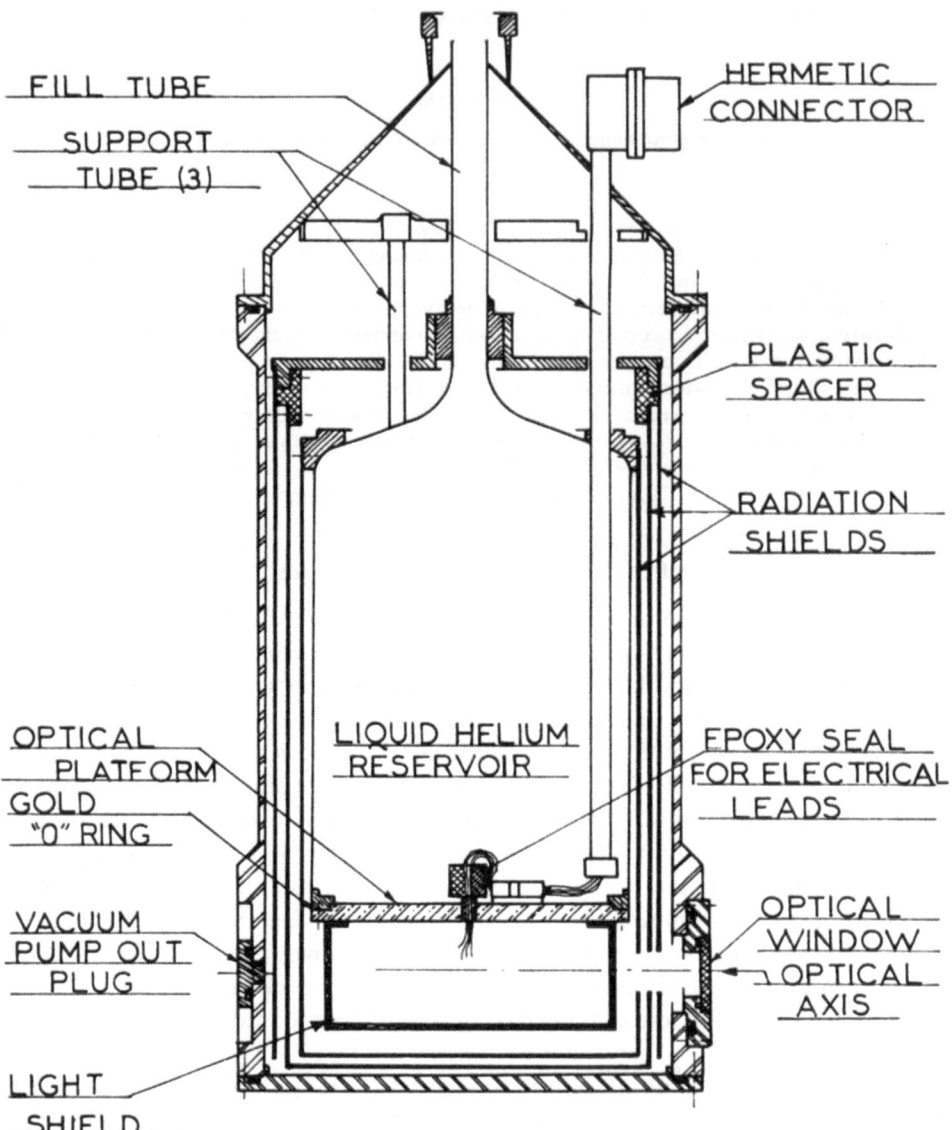

Fig. 1. Lightweight He cryostat.

the filling tube and the three support tubes maintain the reservoir under stress, giving it great mechanical rigidity. One of these support tubes is used to bring the electrical leads from the exterior hermetic connector to within the helium reservoir. Because our experiments generally use a bolometer detector, it is imperative that the measuring wires be thermally shorted to the helium bath if detector performance is not to be degraded. Passing the lead wires through the liquid guarantees that they will not conduct heat to the detector. An epoxy hermetic seal carries the leads through the optical platform to the experimental volume. Because this seal has shown itself rugged and reliable, remaining leakproof after dozens of thermal cyclings, some details of its construction are shown in Figure 2. The principle is that of the familiar Housekeeper glass to copper seal, in which the flexibility of a thin-walled copper tube compensates for the difference in thermal expansion between it and the sealing material, in this case epoxy instead of glass. The copper tube is turned to a taper of about 3° near its extremity, then annealed and chemically etched. A series of fine holes are

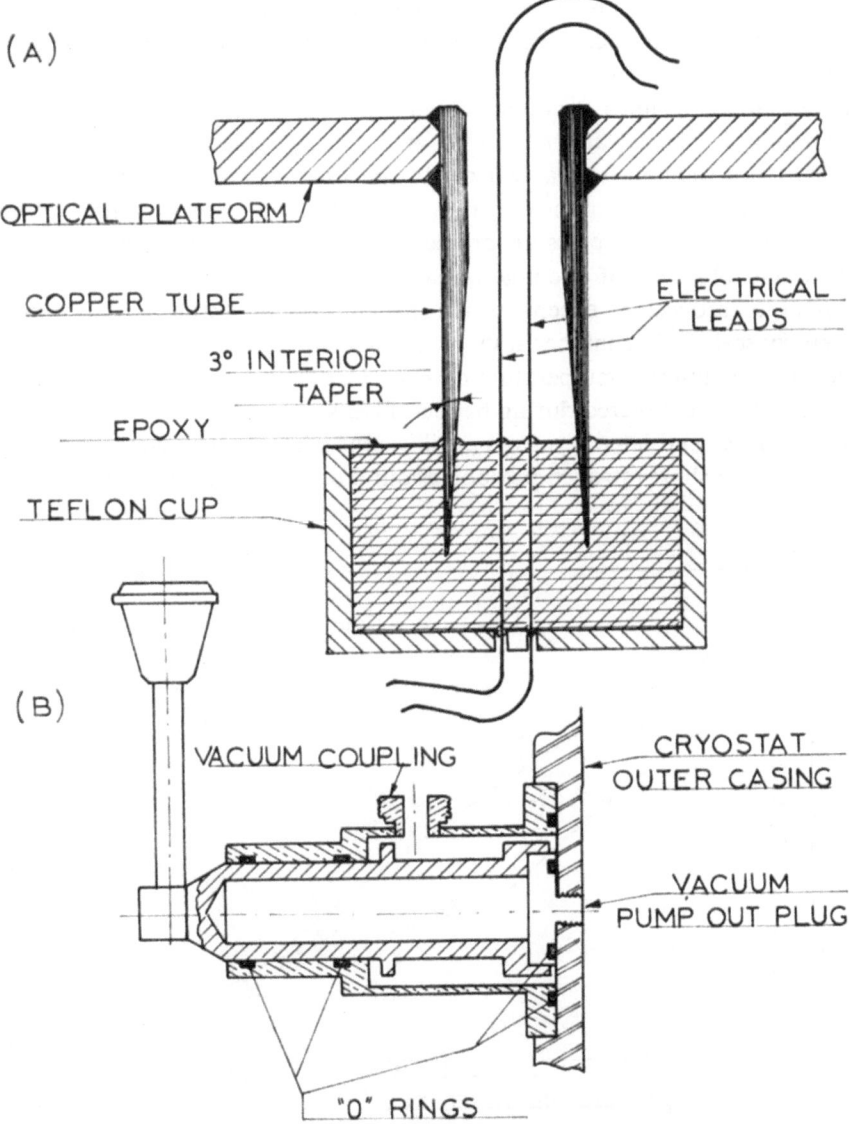

Fig. 2. (A) Hermetic Seal for cryostat Electrical Leads, (B) Vacuum Operator.

drilled in the teflon cup to hold the electrical wires, as shown. The liquified, vacuum outgassed epoxy (both Emerson and Cumings Ecobond 104 and Furane Plastics Co. Epibond 100-A have been used) is poured into the prewarmed cup and the wires gently agitated for a few moments to avoid the possibility that a chain of air bubbles may form along the length of wire and produce a leak. An oven cure which follows the epoxy manufacturer's recommendations completes the seal.

The optical platform is sealed into its flange at the bottom of the helium reservoir with a gold 'O' ring: 0.7 mm diameter gold wire was fused into a ring of proper size and annealed to a dull red colour just before mounting on the flange, and was then compressed to about 0.3 mm thickness. The seal was leakproof even to superfluid helium, and leaks due to thermal cyclings were not encountered. Indium gaskets have also been successfully used.

The cryostat is put under vacuum using an external pumping system. The vacuum pump-out plug is opened and closed by an 'O' ring sealed operator as shown in Figure 2b. Once vacuum is established and the vacuum pump-out plug sealed into place, the operator may be removed, thus eliminating the excess weight and bulk of a vacuum valve. No over-pressure protective devices are needed since the optical window would blow out if a dangerous over-pressure occurred. For reliability all joints are argon arc-welded except for the silver-brazed mounting of the copper tube for the epoxy seal. Cold chamber tests of the cryostat in all orientations revealed no leaks down to an ambient temperature of −45 °C, thus ensuring vacuum reliability at the temperatures encountered during balloon flights.

Both the earlier and later cryostats exhibit approximately the same operating performance. After a short nitrogen precool, about 4 litre of helium are needed to fill the

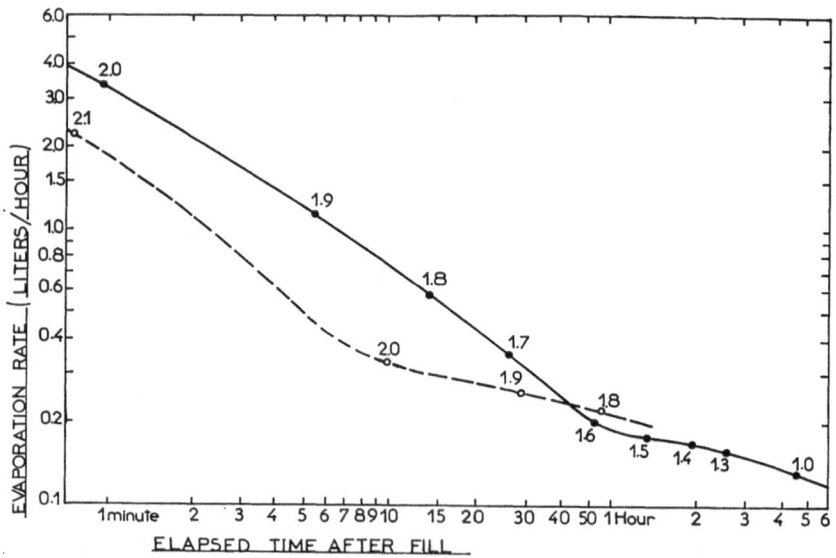

Fig. 3. Evaporation rate of lightweight Cryostat.

2.15 litre capacity cryostat. The liquid helium evaporation rate as a function of time after filling is shown in Figure 3. The initially high evaporation rate decreases during the first hour after transfer, as the radiation shields approach their equilibrium temperature. Subsequent topping up of the level thus benefits from a lower radiation load with a consequent improvement in the evaporation rate for a given helium level. After topping up, helium hold-times of about 20 h are obtained.

A pump of relatively modest capacity (35 m³/h) allows temperatures down to 1.5 K to be obtained. Operation of the cryostat with its axis tilted has little effect upon the evaporation rate; the cryostat has even been used within several degrees of the horizontal! The centre filling tube and the three supporting tubes provide exceptional rigidity without complicating the assembly or disassembly by transverse support struts. Less than five minutes are needed to obtain full access to the optical platform. From vertical to horizontal orientation the platform moves only 0.07 mm with respect to the outer casing, so that the optical alignment with a telescope is unimpaired.

2. Windowless Balloon-Borne Cyrostat

For measurements of the absolute radiated power level of weak extended sources in the far infrared the optical window of the cryostat must be eliminated: no two beam modulation procedure is possible so that the chopper itself must be at liquid helium temperature if its thermal emission is to be reduced to acceptable levels. The thermal emission of any warm optical components before the chopper would then be modulated and drown the signal from the weak source. To avoid these problems, a wide mouth cryostat was designed in which the incoming radiation passes through the helium gas and liquid before encountering the chopper and optical system, then traverses a cold window for its detection by a bolometer in the vacuum space. The cryostat is shown schematically in Figure 4. Its dimensions are: height: 50 cm; diameter: 36 cm; weight: 31 kg. The helium compartment has a volume of 9 litre, the nitrogen/oxygen reservoir, a volume of $9\frac{1}{2}$ litre. Constructional materials are stainless steel throughout except for pure aluminum for the radiation shields and brass or copper for the base-plate of the liquid helium compartment. All joints were argon arc-welded. The base-plate seal used an indium gasket. The entire optical system is within the helium compartment and cooled by film flow from the superfluid liquid at 1.5 K. Electrical leads to the detector are thermally shorted to the helium bath by the device of Figure 2a. The envelope may be pumped using the operator shown in Figure 2. A field of view up to 12° total included angle can be accepted by a light collector of 5 cm diameter without collecting the thermal emission of the cryostat inner walls.

The filling of the cryostat requires approximately 19 litre of liquid helium after a two hour nitrogen precool. Its subsequent evaporation rate is shown in Figure 5. Total hold time is about 25 h. The useful operating time (including pump-down) before the liquid level drops too far to ensure proper cooling of the optical components is approximately 10 h. During the balloon flight, the use of liquid oxygen is to be preferred to nitrogen in order to avoid solidification with consequent loss of thermal

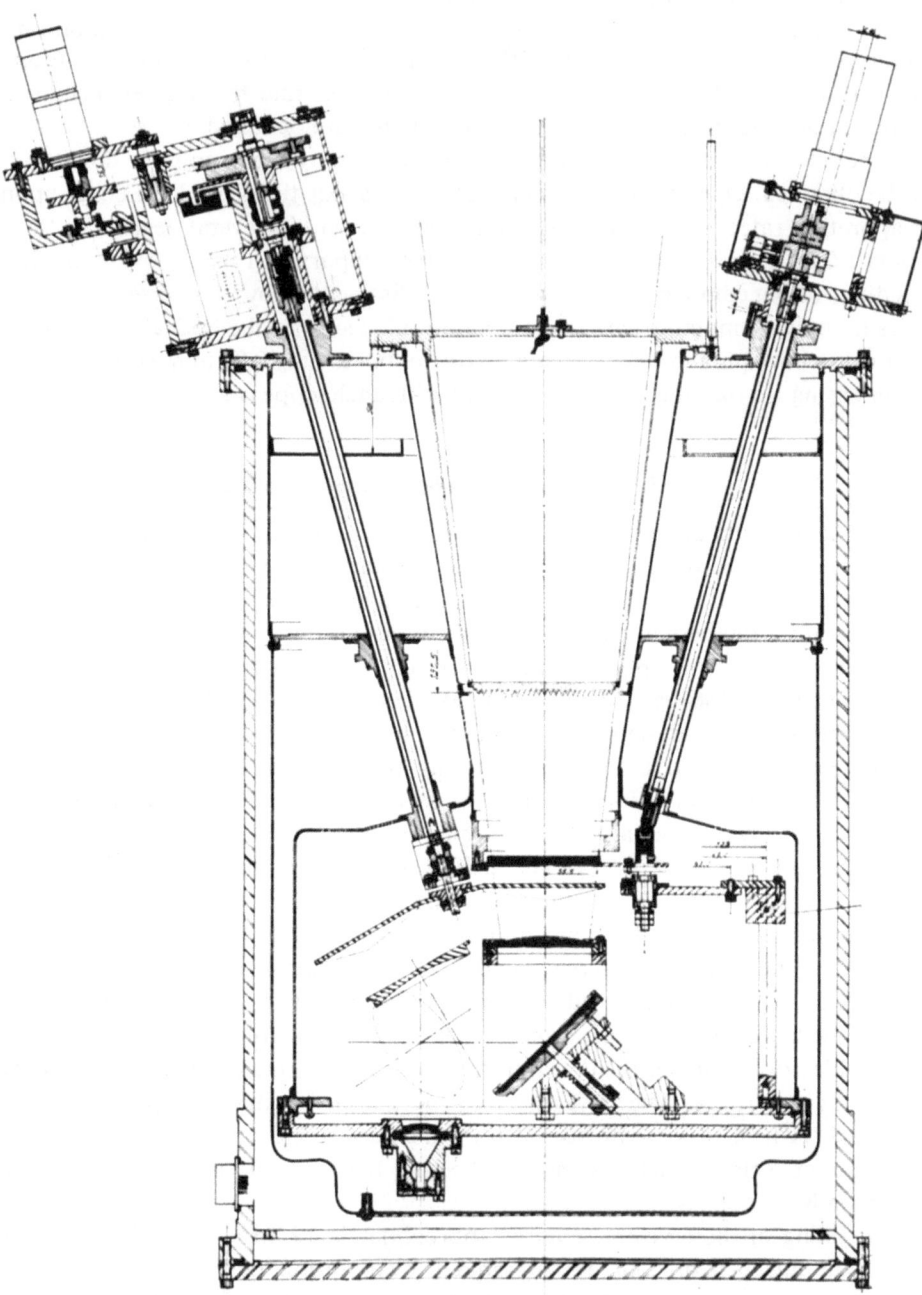

Fig. 4.   Windowless Cryostat.

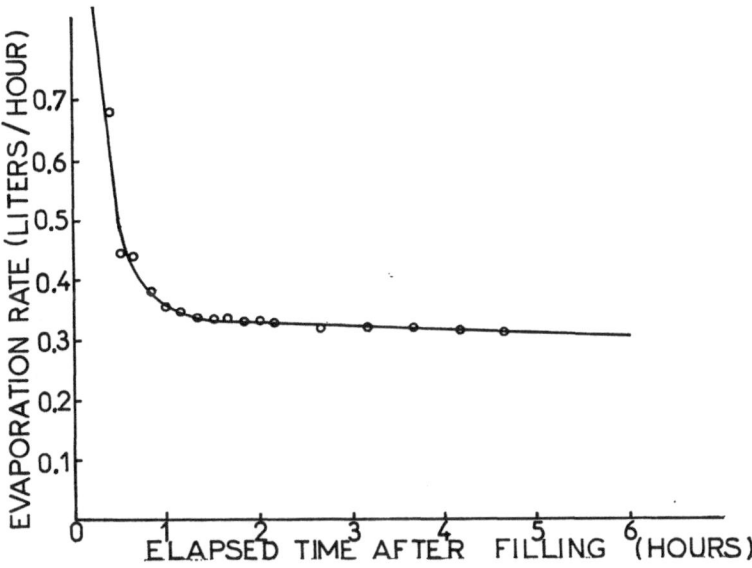

Fig. 5. Evaporation Rate for Windowless Cryostat.

contact to the cryostat walls. The use of oxygen also gives a lower temperature at the balloon height (35 km): 55 K as against 63 K for nitrogen. To further decrease the evaporation of helium during the standby period and during the ascension to operating altitude, a conical liquid nitrogen/oxygen cryostat can be introduced into the mouth of the cryostat. It remains in place long enough to permit its use as a black-body calibrator for the experiment, then is withdrawn by a motor driven winch to permit an unobstructed view of the sky.

DISCUSSION

M. Douma: Can you tell what was the electrical resistance of the epoxy which you used and also what are the heat losses of your first cryostat?

G. Chanin: The epoxy has a high electrical resistance, between 10^8 and 10^{10} Ω.

As far as the heat losses, the major one is due to the conductivity of the supporting structure.

5. FILTERS

SURVEY OF THE PRESENT STATE OF ART OF INFRARED FILTERS

S. D. SMITH and G. D. HOLAH

Heriot-Watt University, Edinburgh, U.K.

and

J. S. SEELEY, C. EVANS, and R. HUNNEMAN

Reading University, U.K.

1. Introduction

Since 1960 there has been a steadily increasing demand for high performance filters for the wavelength region $\lambda > 2$ μm and extending to $\lambda \sim 1$ mm. This demand has been due to a number of causes, the most important ones being:

(a) The steady development of infrared techniques including Fourier transform spectrometers and the emergence of the far infrared lasers, e.g. CO_2, H_2O, HCN.

(b) A special stimulus has come from space requirements, e.g. the Nimbus projects of Oxford/Heriot-Watt Universities, NOAA, NASA (GSFC) and JPL, and generally from ground and balloon-based infrared astronomy as well as the requirements of deep space probes.

This paper describes how various techniques have been used to produce good filters for this spectral region and discusses how the filter development is likely to continue.

Radiation filters, like electrical filters, may be classified according to the spectral profile. Accordingly the type of filters discussed will include narrowbandpass, wide-bandpass, low-pass and high-pass filters, the latter two cases refer to the frequencies which are passed. The methods other than interference filters which have been used to produce some of these types of filters include ground aluminium scatter plates, natural absorption, e.g. semiconductor edges, lattice vibration absorption (Reststrahlen bands) and molecular absorption in various polymers, e.g. PTFE, Yoshinaga filters, i.e. fine powders of alkali halides in polyethylene film, and selective chopping. For lowpass filters many of these methods have had to be used at the same time. In general these techniques are not capable of producing such high performance filters as can be made using interference methods, except for the semiconductor edge filters. Natural absorption is very useful in helping to reject short wave radiation. The interference filters discussed here use either multilayer dielectric stacks of differing refractive indices to produce reflecting interfaces or alternatively metal mesh for these reflecting elements. The design techniques are similar in the two cases and these will be discussed first.

The design techniques of interference systems have been discussed for example, by Smith [1], Pidgeon and Smith [2], Seeley and Smith [3], and Seeley [4]. Smith [1] has shown that a non-absorbing system of reflecting elements can be considered analytically in terms of two effective interfaces. For this system the power transmission after summing the multiply reflected and transmitted beams is given by

$$T = \frac{T_1(\omega)T_2(\omega)}{[1 - R(\omega)]^2} \times \frac{1}{1 + F \sin^2 \theta} \tag{1}$$

Manno and Ring (eds.), Infrared Detection Techniques for Space Research, 199–218. All Rights Reserved
Copyright © 1972 by D. Reidel Publishing Company, Dordrecht-Holland

T_1 and T_2 are the frequency dependent transmission of the two interfaces.

$$T_1 + R_1 = 1 \qquad F = \frac{4R(\omega)}{[1 - R(\omega)]^2}$$

$$T_2 + R_2 = 1 \qquad R(\omega) = \sqrt{R_1(\omega)R_2(\omega)}$$

$\theta = \phi - \beta$, ϕ is a frequency dependent phase term between transmitted and reflected beams, β constitutes a phase term dependent on the separation d of the interfaces, $\beta = 2\pi d/\lambda$.

In Equation (1), if $\theta = m\pi$, $\sin^2 \theta = 0$ and the transmittance is therefore a maximum. For small values of $\theta \sin^2 \theta$ varies rapidly and consequently the transmission is a rapidly varying function of θ. If $\phi \approx 0$, then the transmission is highly peaked near $\beta = 2\pi d/\lambda = m\pi$ (normally $m = 1$), i.e. the transmission is peaked at $\lambda = 2nd$ (n is the refractive index), the half-wave point. A high value of $R(\omega)$, and consequently F, leads to a sharpening of the peak, i.e. the condition is a critical one. The half-wave system is the basic building block of bandpass filters. A better bandpass shape can be obtained by using more than one half-wave system, e.g. a Double-Half-Wave. For this system $R_1(\omega)$ and $R_2(\omega)$ may for example, vary as shown in Figure 1, the overall transmission

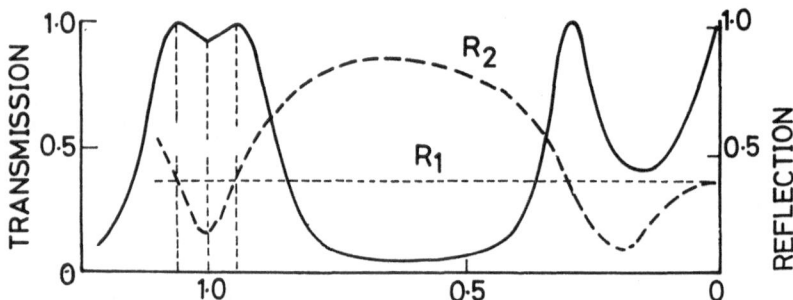

Fig. 1.   Reflection and transmission characteristics for effective interfaces in a DHW filter.

can be seen to have twin peaks which occur at $R_1 = R_2$. The maximum transmission is given by:

$$T_{max} = \frac{T_1(\omega)T_2(\omega)}{[1 - R(\omega)]^2} \qquad (2)$$

In Figure 1 R_1 is assumed to be constant for simplicity of discussion, if R_1 is adjusted such that it meets R_2 only at one point, there is only one peak with $T_{max} = 1$, if R_1 is always less than R_2 again only one peak occurs but this time $T_{max} < 1$.

In Equation (1) F is sometimes called the effective number of interfering beams. From a consideration of the turning points of Equation (1) it may be shown that F is a minimum at the centre of the passband for a DHW whereas for a single half-wave, a Fabry-Pérot, it is always relatively large. This determines that the effect of losses is much less in the passband for a DHW than for a FP. Further, F is a maximum for the

DHW in the stopband therefore aiding rejection. Half-wave systems of varying complexity have been used to construct the narrow and wide bandpass filters described in this paper. The lowpass filters are based upon the quarter-wave which will be discussed now.

Returning to Equation (1), if $\sin^2 \theta = 1$, $\sin^2 \theta$ will vary only slowly with θ. This leads to the condition where the transmission and reflectance are relatively flat functions of frequency. If $\phi = 0$ then the wavelength at which the transmission reaches a maximum is given by $\lambda = 4nd$. This constitutes a quarter-wave layer and has the useful property that at wavelengths greater than $\lambda = 4nd$ the transmission rises and never falls again.

These design concepts can be applied to filters utilizing either multi layers or meshes to produce a wide variety of spectral profiles. The two techniques divide themselves conveniently in wavelength. For wavelengths less than 40 μm the multilayer filter is used in either bandpass or lowpass form, above this wavelength material problems have led to the development of mesh filters.

The short wavelength region will be considered first.

2. Development in Multilayer Interference Filters

A. SUBSTRATES AND SEMICONDUCTOR EDGE FILTERS

In the region 1–9 μm extensive use is made of the natural electronic absorption edges of various semiconductors, e.g. Si, Ge, Fan and Becker [5]. These materials constitute lowpass filters in themselves and are often used as substrates for multilayer lowpass and bandpass filters. Other materials which have been used include PbS, InAs, PbTe and InSb. When these materials are used by themselves, the thickness is a very important parameter in determining a rapid transition from stop to passband, too thin a layer results in a slower cut-on.

From Table I it can be seen that there are certain gaps in the ability of natural electronic absorption to produce continuous blocking up to 20 μm at octave intervals. These gaps have to be filled using multilayer techniques and occur approximately at 2.6 μm, 5 μm, and 12 μm.

TABLE I
Characteristics of semiconductor edge filters

Material	Si[a]	Ge[a]	PbS[b]	InAS[a]	InSb (n-type)[c]	InSb[a]
Location of absorption edge (μ)	1.1	1.8	3.0	3.8	5.5	7.0
Best thickness (mm)	1	1	< 0.006	< 0.2	< 0.05	< 0.1
Refractive index (beyond edge)	3.5	4.0	4.2	3.4	3.6	4.0
Antireflected with	ZnS	ZnS	ZnS	ZnS	ZnS	ZnS

[a] Excellent blocking filter.
[b] Prepared by evaporation. Inadequate, too thin to give sharp edge.
[c] Inadequate transmission. Edge degraded by impurities.

B. MULTILAYER LOW-PASS FILTERS

The electronic absorption of the aforementioned semiconductors is such that the rejection is usually better than 10^6. To produce equivalent rejection of shortwaves in multilayer filters is a difficult problem, the solution of which requires filters with materials of high index ratio and relatively many layers. The strength and breadth of rejection is determined by the low refractive index n_L, the high refractive index n_H, and the number of layers x. The breadth of the primary rejection period is given by

$$\frac{400}{\pi} \text{ arc sin} \left[\frac{n_H - n_L}{n_H + n_L}\right]\%$$

and the rejection strength by

$$\left[\frac{n_H}{n_L}\right]^x.$$

Due to the difficulty of depositing many layers of relatively thick films, it is highly desirable to keep the number of layers to a minimum which for good rejection requires a large index contrast ratio, about 2.4:1, with 15 layers. Detailed filter profile calculations require a design system which optimises the transmission in the passband whilst maintaining good rejection. Various systems have been used, e.g. Baumeister [6] and a useful method has been found to be based on Tchebyshev equal ripple functions. The details of multilayer manipulation using these exact functions have been discussed elsewhere, Seeley [4] and Seeley and Williams [7]. It is sufficient here to quote the conclusions which may be drawn as to the relative values of the refractive indices and phase thicknesses.

If the refractive index of the substrate is denoted by n_S, the first layer on the substrate should have the high refractive index, n_H, if $n_H/n_S > n_S/n_L$ and a low index if $n_S/n_L > n_H/n_S$. The fractional ripple in the passband is given by

$$[(n_H/n_S - n_S/n_L)/(n_H/n_S + n_S/n_L)]^2$$

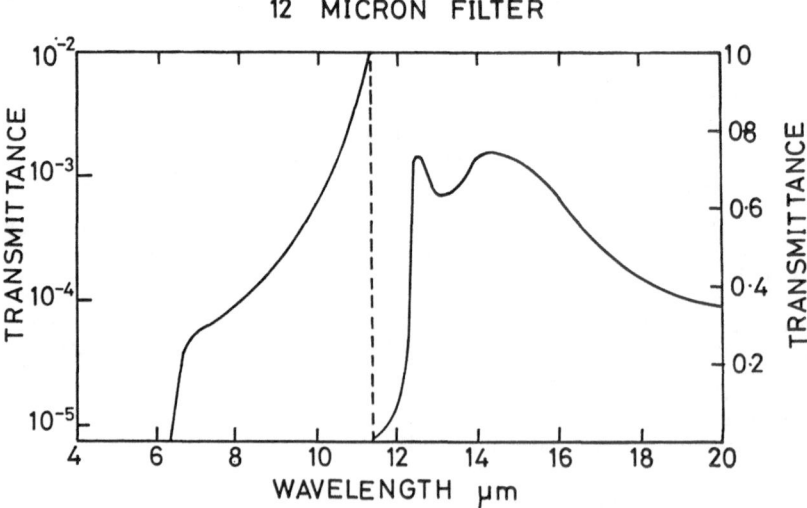

Fig. 2.   Spectral performance of 14 layer 12 μm lowpass filter.

The phase thicknesses at cut-off should be less than $\pi/2$ for low-pass and conversely should be between $\pi/2$ and π for high-pass filters. In order to antireflect the substrate by means of a suitable thickness of n_L, the front layer should be n_H.

The materials which have found greatest favour in the fabrication of low-pass filters with edges at 2.6 μm, 5 μm, and 12 μm have been ZnS as the low index material ($n_L = 2.2$) and either PbTe or Ge as the high index ($n_H = 5.1$ or 4.0). An experimental profile is shown in Figure 2. The construction is given in Table II. This design is backed up by a thin layer (100 μm) of InSb since the rejection bandwidth is just sufficient to link with the InSb absorption edge at 7 μm. Using 14 layers of PbTe/ZnS a rejection of greater than 10^4 with Ge as the substrate, assuming ideal antireflection the optimum ripple is 0.04.

TABLE II

Construction of 12 μm edge filter. Optimum thicknesses of fourteen-layer lead telluride/zinc sulfide filters

Layer	Relative thickness	
	Low pass	High pass
1 and 14	0.55	1.25
2 and 13	0.82	1.11
3 and 12	0.92	1.05
4 and 11	0.96	1.025
5 and 10	0.98	1.015
6 and 9	0.99	1.01
7 and 8	1	1
15 (antireflection)	2.0	0.5

Substrate Ge, Odd layers ZnS, Even layers PbTe

These materials, i.e. PbTe and ZnS, and the semiconductors enable a complete range of blocking filters up to 20 μm to be produced. The extension to longer wavelengths with these materials becomes increasingly difficult due to free-carrier and lattice absorption which increases the absorption beyond about 20 μm to an unacceptable level. Materials which may however be useful for this range include the heavy alkali halides, i.e. those with lattice absorption peaks above 100 μm. After a series of evaporation and transmission tests Caesium Iodide (CsI) became a possibility as a low index material ($n \sim 1.7$) for filters beyond 20 μm. Accordingly a low-pass filter using a total of 30 layers of PbTe/CsI was fabricated with an edge near 20 μm, the spectral performance is shown in Figure 3. Unfortunately CsI is hygroscopic and to overcome possible peeling of the layers the final filter is sealed using a thin film of polystryrene. Using this combination of materials filters with edges out to 30 μm have been constructed. Auxiliary blocking filters to remove the shortwave passbands of the main filter are used.

Although multilayer techniques have produced good blocking filters with edges out to 30 μm this probably represents a reasonable limit in the light of present-day tech-

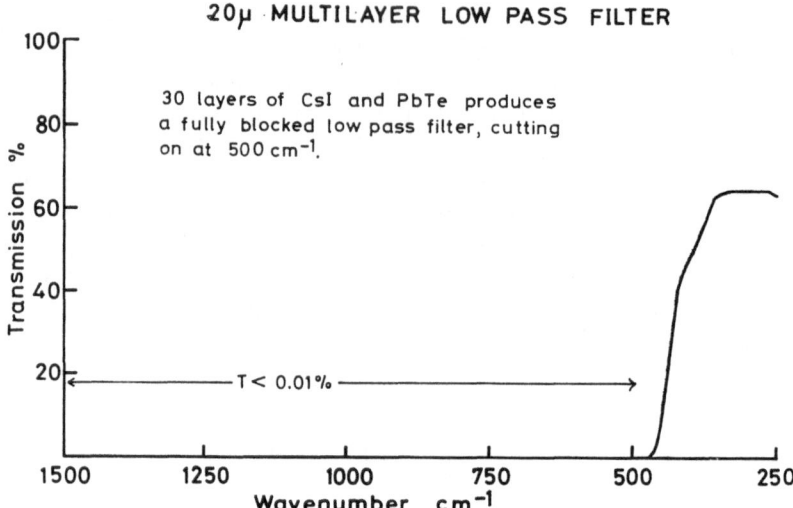

Fig. 3.   Spectral performance of 30 layer 20 μm lowpass filter.

niques. The extension to longer wavelengths is complicated not only by absorption but also by the difficulty in depositing the thicker films with good stability.

The longwave range can be more efficiently covered using new developments in metal mesh filters.

C. MULTILAYER BANDPASS FILTERS

Before the manufacture of longwave lowpass filters is discussed the production of bandpass filters for 1 μm $< \lambda <$ 20 μm will be discussed. The specific filters which will be considered as examples form part of the optical components for the Nimbus 4 and 'E' satellite vertical temperature sounders.

The heart of the Nimbus 4 experiment lies in the narrowband interference filter at 668.0 cm^{-1} which selects the 'Q' branch of the ν_2 vibration-rotation band of atmospheric carbon dioxide. The choice of this band and further details of the Selective Chopper Radiometer successfully flown on Nimbus 4 have been discussed elsewhere and will not be considered here [8, 9]. It is sufficient to say that for temperature sounding of high accuracy a resolution of 200 is required for the narrowband filter, this is equivalent to a bandwidth of 3 cm^{-1} at 668 cm^{-1}, i.e. $\frac{1}{2}$%. Such a bandwidth implies that the evaporated layers have to be controlled to better than $\frac{1}{6}$% and also be uniform to within $\frac{1}{10}$% over the filter aperture. To manufacture filters with layers deposited to this degree of control needs considerable development in thickness monitoring, substrate temperature control and source-substrate conditions.

The spectral profile of a 10 layer Fabry-Pérot is shown in Figure 4, it was constructed to the following design.

$$L/\text{Ge}/LHLHL\ \underline{HH}\ LHLH$$

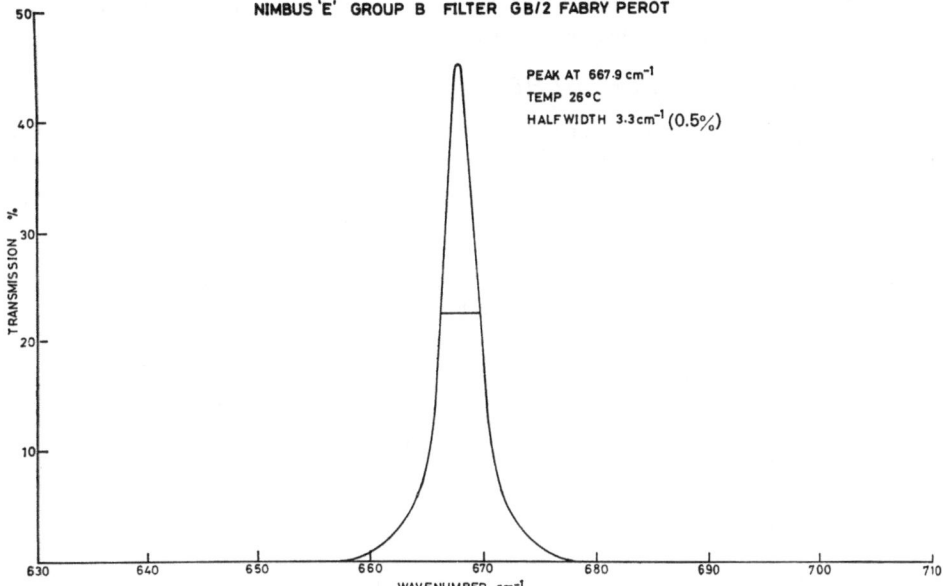

Fig. 4.   Narrowband 'Q' branch Fabry-Pérot at 668.0 cm^{-1}.

L is a quarter-wave thickness layer of ZnS.

H is a quarter-wave thickness layer of PbTe.

The index ratio for these materials is 2.5.

In order to produce the required reflectivity and hence achieve a similar half-width using Ge and ZnS, which have an index ratio of 1.6, would require twice the number of layers.

Such filters may be centred to about ± 1 cm^{-1} in practice, but may be tuned by tilting, i.e. changing the angle of incidence of the incoming radiation. From Pidgeon and Smith (1964) the shift in the frequency of the transmission peak from that of normal incidence may be written, for this design.

$$\frac{\varDelta\nu}{\nu} = \frac{1}{2}\left[\frac{\sin\phi}{n^*}\right]^2$$

where ϕ is the angle of incidence and n^* is an effective refractive index, i.e. an equivalent monolayer, ν is the peak frequency for normal incidence. For this design a tilt of 10° is allowable with a tolerable drop in transmission and degradation of half-width. Since n^* is approximately 3.3 and the energy gathering power is proportional to n^{*2} this leads to an energy gathering advantage of 10 with respect to an air-spaced Fabry-Pérot. Tuning can also be achieved by a variation in the filter operating temperature, for this design an increase in temperature of 1 °C gave an increase in peak frequency of 0.07 cm^{-1}, i.e. 10 °C will tune by 0.7 cm^{-1}.

Both Nimbus 4 and 'E' radiometers required a rather wider narrow-band filter pitched near 15 μm with a half-width of 10 cm^{-1}. This is an ideal application for DHW design, the following design was used in the profile shown in Figure 5.

$$L/Ge/LHL\ \underline{HH}\ LHLHLHL\ \underline{HH}\ LH$$

The blocking of shortwave radiation, <12 μm, was achieved by the 12 μm lowpass filter already discussed, integrated rejection measurements have shown that over 99% of the transmitted energy of the combined filter lies within the passband of the filter using a 300 K source.

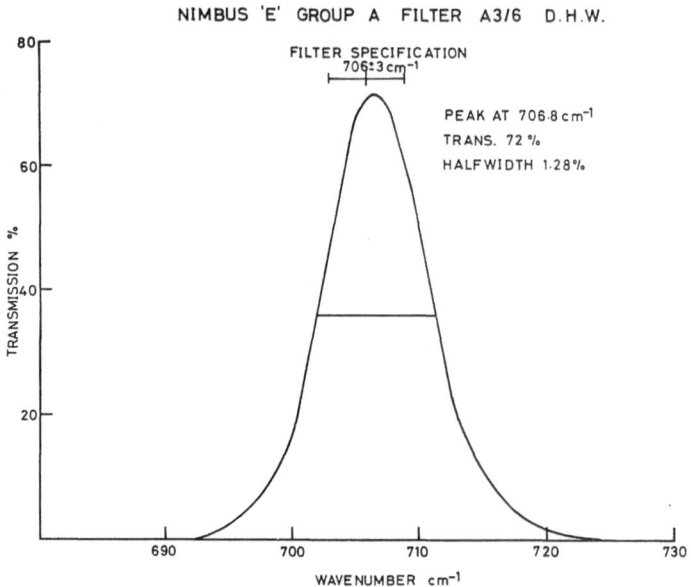

Fig. 5.   DHW filter near 15 μm.

These two types of filters have already been flown successfully on Nimbus 4. The Nimbus 'E' experiment incorporates further channels in order to measure, in addition to the vertical temperature profile, the distribution of cirrus cloud, water-vapour distribution and reflected sunlight. To achieve all these aims a 16 channel radiometer has been constructed in which eight channels are equipped with the aforementioned narrowband filters. Some of the additional channels represented an advance in the state of filter technology.

The first to be considered are very narrow, 1.5% halfwidth, multilayer filters for the region 2–3.5 μm. At 2.59 μm water vapour absorbs so strongly that the lower atmosphere is opaque and hence no radiation from the lower atmosphere can reach the satellite, specifically this wavelength supplies information regarding the particle size of the cloud using reflected sunlight off the high-level cloud. The results of this filter are shown in Figure 6, made according to the degenerate DHW design [1].

$$Ge\ HLH\ \underline{LL}\ HLH\ L^1HLH\ \underline{LL}\ HLH$$

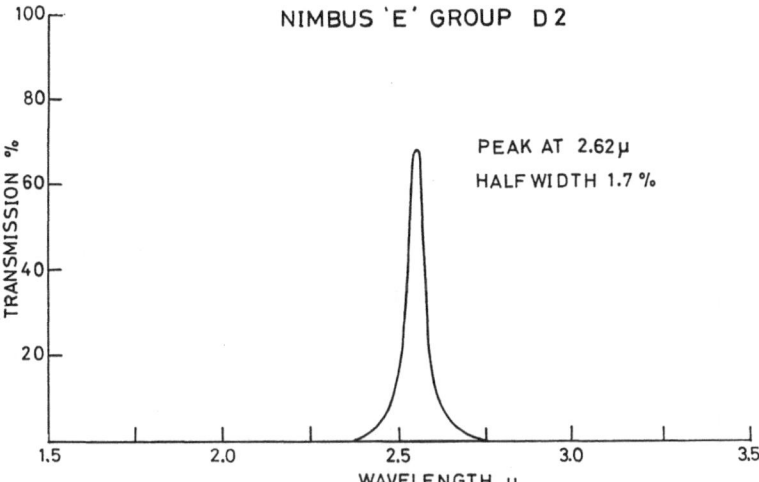

Fig. 6. DHW multilayer filter at 2.59 μm.

H is a quarter wave of Germanium

L is a quarter wave of Silicon Monoxide

L^1 is a reflectance matching layer of ZnS

The specification for this filter is derived from a consideration of the water-vapour absorption lines and in order that accurate measurements from cloud may be made, the cut-on and cut-off of the filter was required to cut out certain H_2O and CO_2 absorption lines and the need for edge steepness resulted in the DHW design. The passband is restricted to 2.5% which is in itself a tight specification. The position of the edges, in this case defined as the 50% of maximum transmission, has to within 3760 cm^{-1} and 3880 cm^{-1} with a minimum width of 45 cm^{-1}. In order to achieve such high performance a large index ratio and a large number of layers is required. The rejection is approximately given by $(n_H/n_L)^x$ is, for this design using Ge (4.0) and Si O (1.7), $(2.35)^{17}$, i.e. 2×10^6. Similar filters with peaks at 2.65 μm and 3.5 μm were also produced.

A bandpass filter peaking at 18.6 μm was required in order to obtain further information on water vapour distribution. The specification of this filter was a halfwidth of 10 cm$^{-1} \pm 5$ cm^{-1}. Such a specification can be ideally met by using a DHW of suitable materials, however problems were anticipated since the transmission of ZnS, which has a restrahl wavelength of 33 μm, becomes intolerably low towards 20 μm. However some filters were made to the following design.

Ge/L $\underline{HH}$ LHLHL $\underline{HH}$ LH

with $L=$ZnS and $H=$PbTe. The result is shown in Figure 7, it can be seen that the transmission reaches almost 40% and therefore becomes acceptable without needing a different low-index material. This probably represents the longwavelength limit of the use of ZnS. It is generally accepted that a material may be used up to wavelengths half that of its restrahl wavelength without the absorption losses becoming intolerable.

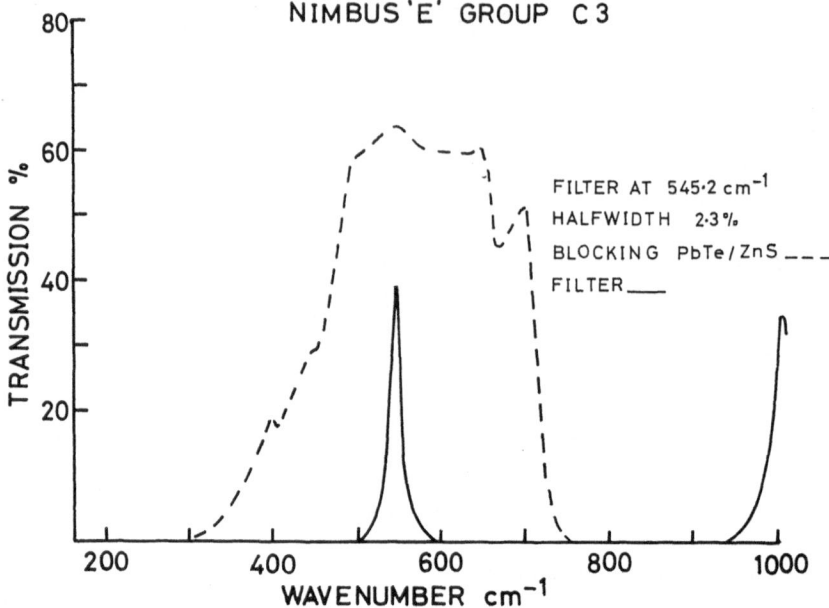

Fig. 7.   DHW multilayer filter at 18.6 μm.

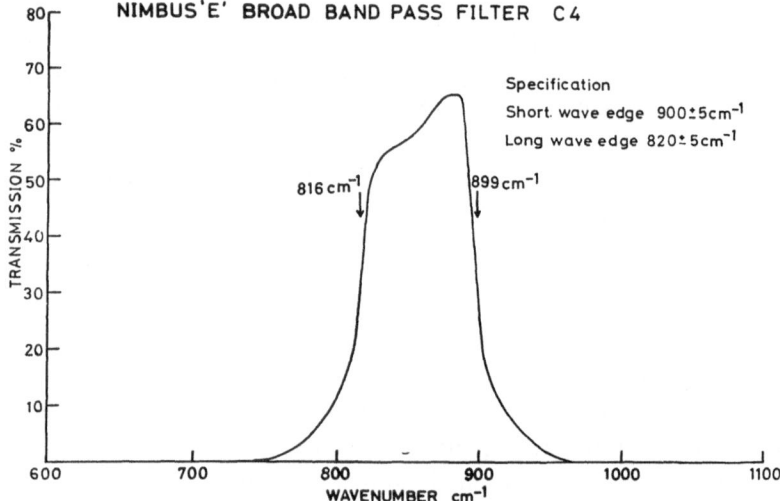

Fig. 8.   Widebandpass triple-half-wave multilayer filter at 11.1 μm.

D. WIDEBANDPASS FILTERS

To produce rather wider bandpass filters, say 10% halfwidth a variety of techniques may be used. These include multiple half-wave filters, repeated period filters and a combination low-pass/high-pass system. For Nimbus 'E' a bandpass filter with a peak at 11.1 μm and a half-width of 80 cm^{-1} ± 10 cm^{-1}. To satisfy this specification a triple-half-wave filter of the following design was used

$$Ge/HL \; \underline{HH} \; LHL \; \underline{HH} \; LHL \; \underline{HH} \qquad L = ZnS$$
$$H = PbTe$$

The result is shown in Figure 8.

These multilayer lowpass and bandpass filters probably represent the limit of present-day multilay technology. The extension to longer wavelengths is restricted by the lack of suitable high and low index materials which can be deposited in thick stable films.

3. Beamsplitters

In order to achieve simultaneous monitoring at a wide range of spectral intervals, the Nimbus 'E' radiometer incorporates time multiplexing and wavelength separation. The incoming radiation is first separated by a reflecting/transmitting chopper. Each of these beams is then incident on a beamsplitter which separates by reflection and transmission according to wavelength.

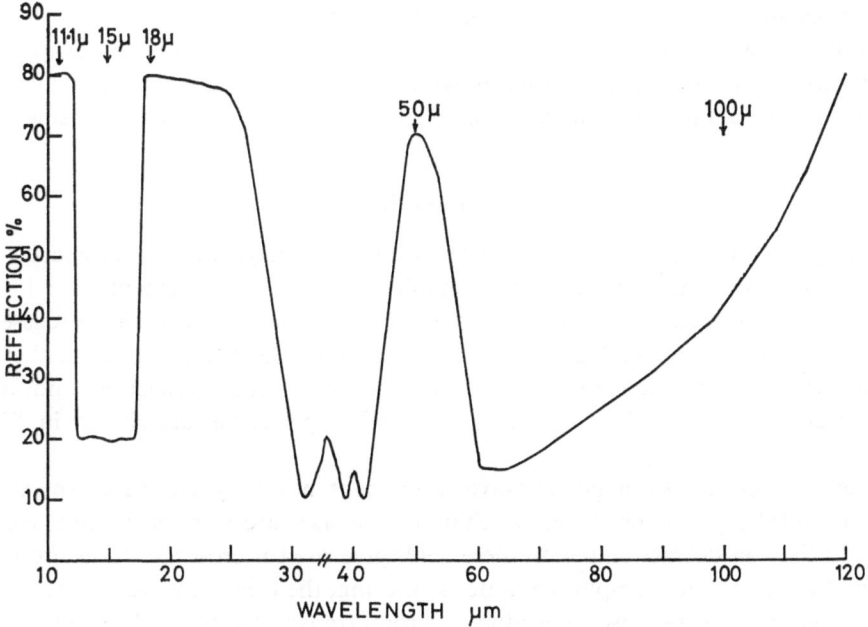

Fig. 9.   Reflection spectra of B/C beamsplitter.

The shortwavelength, 2 μm → 3.5 μm, channel is taken together with the 15 μm DHW channel. A slice of sapphire is sufficient to achieve separation by reflecting the 15 μm channel and transmitting the shortwavelength channel. To achieve separation of the other channels is a more serious problem. This required a beamsplitter which reflects strongly at 11.1 μm, 18.6 μm, 48 μm, and 100 μm and transmits at 15 μm. The 11.1 μm, 18.6 μm and 15 μm bandpass filters have already been discussed, the long-wave filters will be discussed later.

The beamsplitter, used at an angle of incidence of 45°, which satisfied all these requirements is shown in Figure 9.

The shortwavelength performance, $\lambda < 25\ \mu$m, is achieved using germanium ($n = 4.0$) as the substrate with alternate layers of half and quarter wave optical thicknesses of ZnSe and PbTe. The beamsplitter has a diameter of 2″. The design is

Sub.	0.52	0.252	0.52	0.252	0.52	0.252	0.52	0.252	0.26	AIR
Ge.	ZnSe	PbTe	ZnSe	PbTe	ZnSe	PbTe	ZnSe	PbTe	ZnSe	

$$n = 4.0 \quad 2.6 \quad 5.6$$

These thicknesses are computed for $\lambda_0 = 15\ \mu$m. The selective reflection at 47 μm arises from the strong lattice vibrational absorption of ZnSe whose restrahlen wavelength is 46.5 m. The strong reflection near 100 μm has been achieved from the free-carrier plasma absorption of PbTe with the appropriate number of carriers being obtained by reduction of the PbTe in a nitrogen/hydrogen mixture.

Further reflection at 100 μm was achieved by using CsBr (restrahl wavelength 135 μm) as an antireflection film, $nd = 3\lambda/2$ at $\lambda = 15\ \mu$m.

The combination of these materials coupled with control over the evaporation conditions, i.e. substrate temperature, pressure of oxygen in the evaporation plant and source temperature together enabled the result shown in Figure 9 to be obtained.

4. Mesh Filters

The design methods which have been discussed for multilayers can be extended to the design of longwave filters using metallic mesh as the reflecting elements.

Two types of metal mesh have been used in the longwave filters discussed in this section. From an analogy with transmission line equivalent circuits they are called inductive and capacitative and have complementary reflection and transmission characteristics. Their structure and optical properties are shown in Figure 10 a, b.

When a plane electromagnetic wave is incident normally on the mesh, surface currents and charges are produced which in turn act as sources of electric and magnetic fields. In the non-diffraction region these fields give rise to zero-order transmitted and reflected waves of equal amplitude. After scattering, the original wave and zero-order transmitted waves superpose to produce the total transmitted wave. The development of the application of metallic mesh in interference filters is due mainly to Ulrich [10, 11, 12].

For a loss-free system the transmission and reflection of the two types of filters are related by

$$T_c + T_i = 1, \quad R_c + R_i = 1$$
$$T_c = R_i, \quad R_c = T_i$$

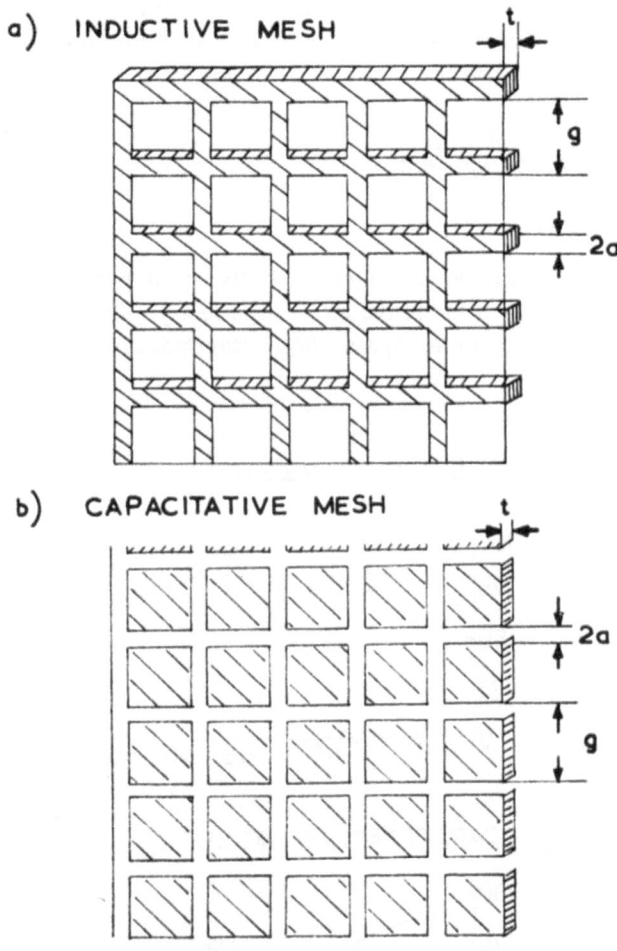

Fig. 10a.   Geometrical parameters of inductive and capacitive meshes.

The transmission of a thin capacitive mesh in the non-diffraction region, $a \gg t$ and $\lambda > g$ may be written, including loss,

$$T = \frac{r^2 + Z_0^2(\omega/\omega_0 - \omega_0/\omega)^2}{(1 + r)^2 + Z_0^2(\omega/\omega_0 - \omega_0/\omega)^2}$$

where r is an ohmic loss term due to surface currents flowing in the mesh, and is related to the absorption by

$$A = |\Gamma|^2 2r \quad \text{where} \quad r = (c/\lambda\sigma)^{1/2}\zeta$$

where $|\Gamma|^2$ is the reflectivity and ζ, for a capacitive mesh is given by

$$\zeta = \frac{1}{1 - 2a/g}$$

σ is the bulk dc conductivity, ω is a normalized frequency g/λ and ω_0 is the frequency where the transmission of the capacitative mesh is a minimum. Z_0 is a characteristic impedance which for the capacitative mesh is

$$Z_0^c = \frac{1}{2 \log_e(\text{cosec } \pi a/g)}$$

$$Z_0^c \cdot Z_0^i = 1.$$

From Figure 10a,b it can be seen the capacitative mesh has maximum transmission at low frequencies whilst an inductive mesh is perfectly reflecting, consequently only capacitative mesh is used for low-pass filters. Bandpass filters may be made using either type of mesh.

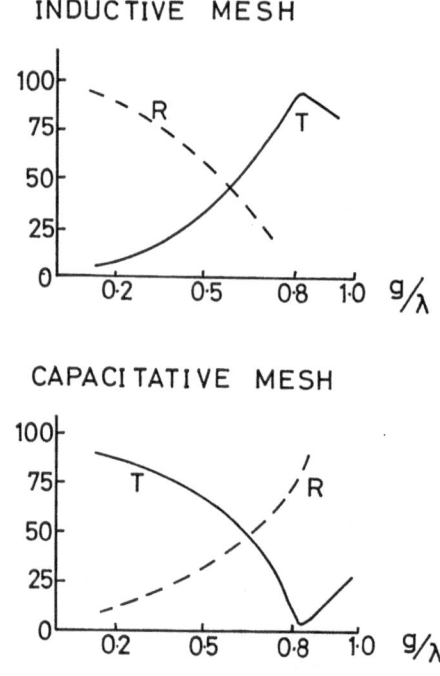

Fig. 10b. Optical performance of inductive and capacitative meshes.

For capacitative meshes the dielectric loss in the substrate must be considered in addition to the ohmic loss. This loss is not simply the dielectric loss of the substrate alone but an increased loss, by a factor $g/2a$, due to the concentration of the electric field energy into the gaps between the metal squares.

In using Eqation (1) the frequency dependence of the reflection and transmission terms is due to the geometry of the meshes, this replaces the dispersion of multilayer filters.

5. Bandpass Mesh Filters

Three specific bandpass filters were required at wavelengths varying from 50 μm to 337 μm. A filter with a half-width of less than 10% with a peak wavelength near 50 μm was required for helping to measure the cirrus cloud distribution in the atmosphere, in the Nimbus 'E' project. A similar filter with a peak wavelength between 85 μm and 100 μm was required for interstellar hydrogen emission experiments.

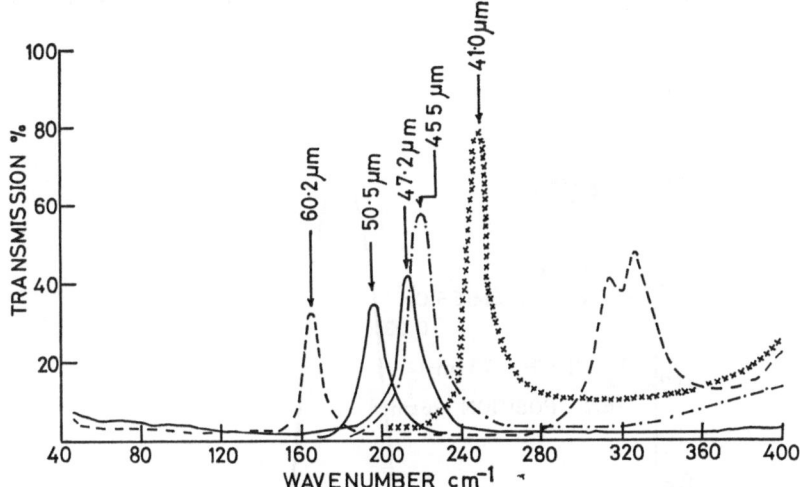

Fig. 11.   Mesh Fabry-Pérots near 50 μm.

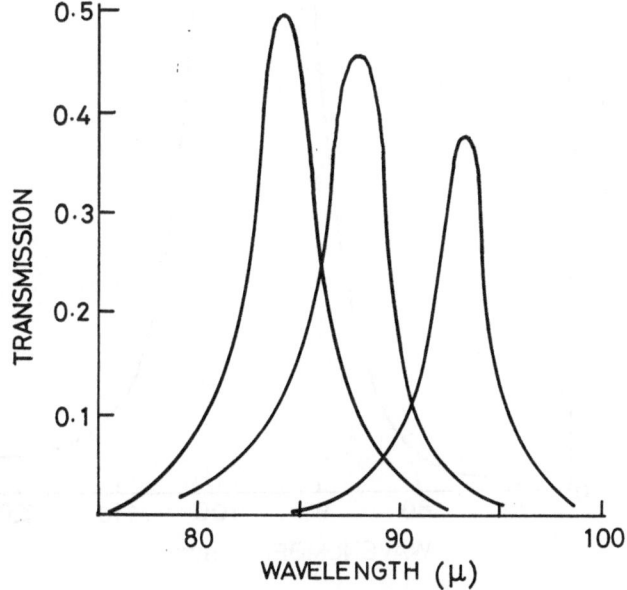

Fig. 12.   Mesh Fabry-Pérots between 85 μm and 95 μm.

The HCN laser lines require separation as mentioned previously and a bandpass filter is needed with a peak wavelength at 337 μm.

The results for the Nimbus filter are shown in Figure 11. They are Fabry-Pérot filters with two inductive meshes of $g = 37.5$ μm and a spacing of between 20 μm and 30 μm. The fall-off in peak transmission towards longer wavelengths is due to the fall in transmission of the constituting meshes.

For the bandpass filters between 85 μm and 100 μm two types of systems have been used. The Fabry-Pérot results are shown in Figure 12 using inductive meshes with a

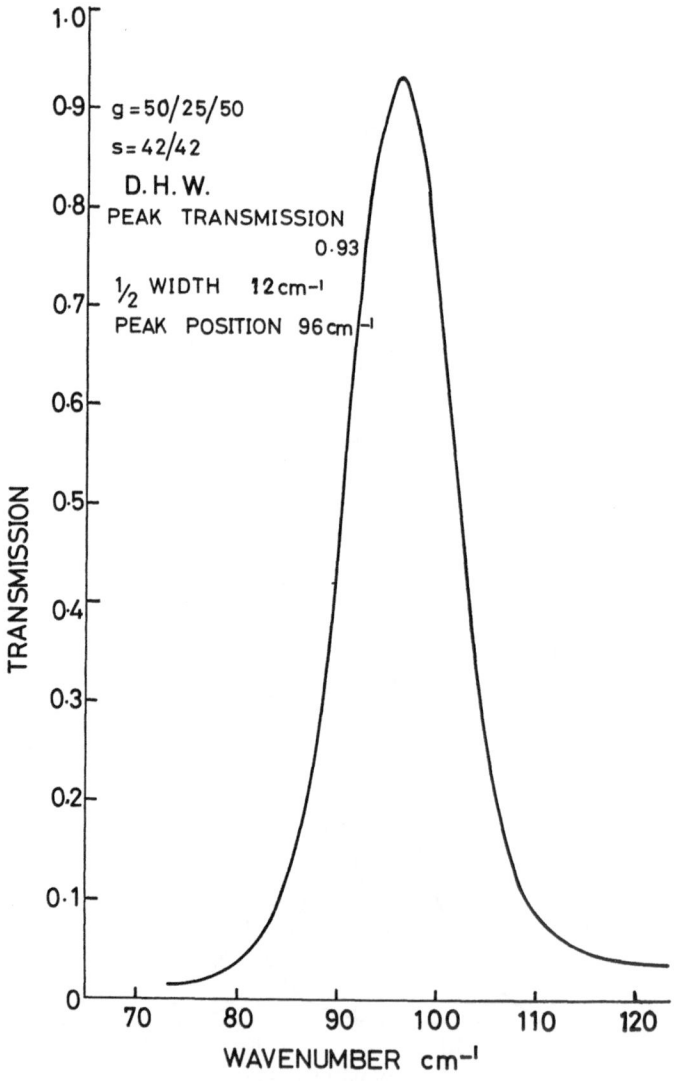

Fig. 13.   Mesh DHW filter at 100 μm.

g factor of 25 μm and spacings between 42 μm and 50 μm. A Double-Half-Wave system is shown in Figure 13. The design is 50 μm/25 μm/50 μm for the g values and equal spacers of 42 μm.

The HCN laser line filter is shown in Figure 14 and from the presence of the twin-peaks it is clear that the design is a DHW. This design is $g = 100$ μm/50 μm/100 μm and spacing $s = 150$ μm/150 μm.

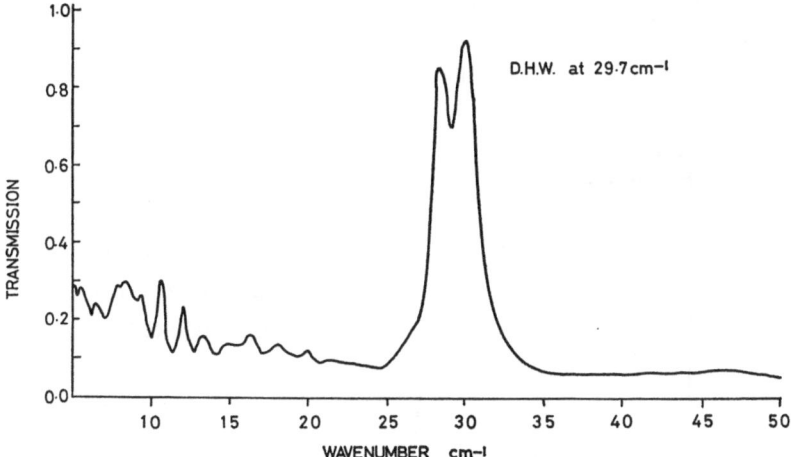

Fig. 14. Mesh DHW filter at 337 μm for HCN laser.

These three filters show that good performance bandpass filters of both Fabry-Pérot and DHW types are now possible for the wavelength region $\lambda > 40$ μm.

6. Lowpass Mesh Filters

One of the requirements for the Nimbus 'E' project was a lowpass filter with an edge near 100 μm, this coincides with the filtering needs of vacuum grating and Fourier transform spectroscopy.

Capacitative meshes are not commercially available and this led to development work on substrate material (metal squares need supporting) and type of metal film.

Since aluminium is easy to evaporate and has a high conductivity it was chosen, but copper, gold, and silver are also usable. The pattern was produced by evaporating a thin film, ~0.1 μm, on to a substrate and then using photo-etching techniques to cut the pattern.

The substrate materials which were possible were polyethylene, polypropylene and Mylar. Polyethylene has good transmission but is too flexible so it was not suitable although indications are that it might be useful for short-wave filters. Polypropylene has also good transmission and has high mechanical rigidity, unfortunately it was only readily available in thick films (~12.5 μm physical thickness) which, since this would constitute a major part of the spacing (optical thickness 20 μm), would require metal annular spacers of 5 μm which is intolerably thin for easy handling and availability. Consequently 2.5 μm thick Mylar was chosen since it had high mechanical rigidity, however its relatively large absorption at $\lambda < 100$ μm, whilst helping low-pass filters with edges near 100 μm, means that shorter wavelength filters may suffer from dielectric loss.

In order to achieve a sharp transition from pass to stop band needs at least four meshes. The rejection is also aided by the number of meshes, this rejection region can be extended and improved if one or two of the meshes have their g factor coincident with twice the spacing. The lowpass mesh filters have superior performance if the spacings are all equal. At $\lambda = 2s$ there will be a peak in transmission, this would

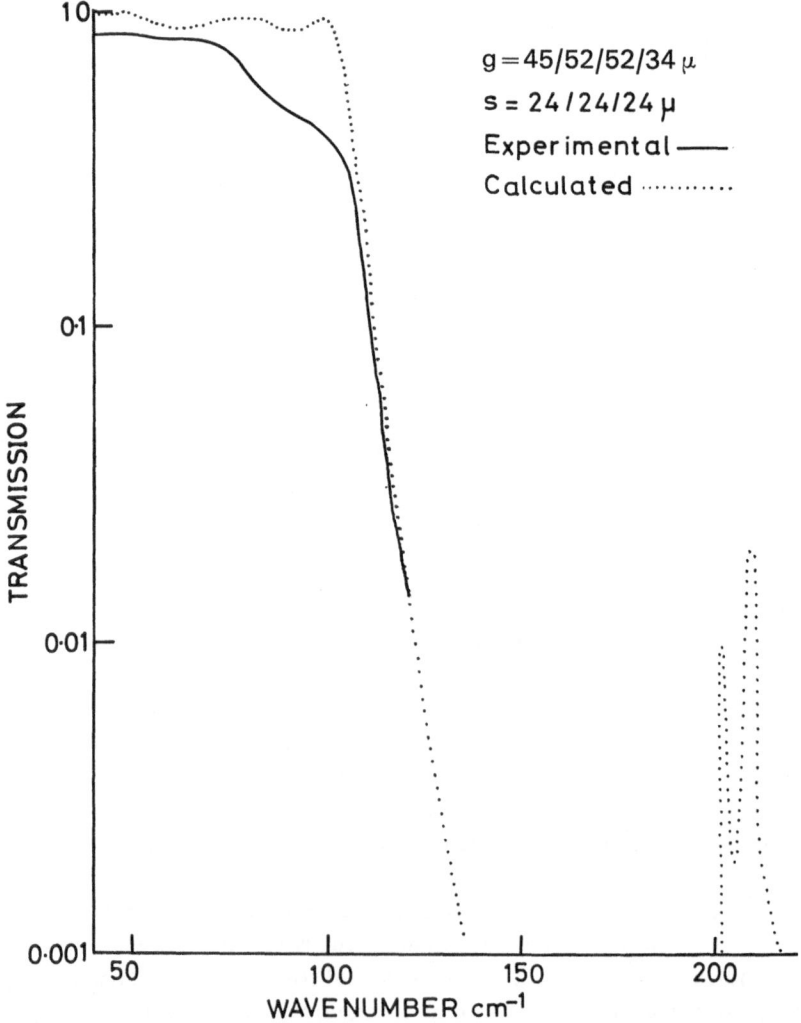

g = 45/52/52/34 μ

s = 24/24/24 μ

Experimental ———

Calculated ············

Fig. 15. Lowpass 4 element mesh filter with an edge at 100 μm.

normally limit the rejection width, as in multilayer filters, however if $g=2s$, the transmission at $\lambda=2s$ is reduced to a low value since the transmission of a capacitative mesh is a minimum at $\lambda=g$. It may not always be possible to produce exactly the correct g factor and some way of extending the design latitude must be arranged. This may be done by removing the critical nature of the previous $g=2s$ condition. The Nimbus 'E' filter required the 50% transmission point to be at 100 μm ± 3 μm. This was done by using two meshes in the filter which have their transmission minima at each side of the wavelength $\lambda=2s$. This overlapping of minima gives an added latitude in moving the edge simply by changing the spacer. The theoretical and experimental curves for the Nimbus 'E' filter is shown in Figure 15. The rejection is improved in the shortwavelengths below the diffraction wavelength $\lambda=g$ where the mesh theory as given here does not hold, by the addition of a piece of PTFE and in fact the rejection was $>10^3$ with transmission of $>70\%$ in the passband with PTFE.

7. Conclusions

The production and design of bandpass and lowpass filters for the infrared region 1 μm $\rightarrow$ 1000 μm has been discussed and it has been shown that such filters are now commercially available. The availability of high performance interference filters has enabled progress to be made in infrared spectroscopy and particularly the application of infrared techniques to astronomy.

The limitations of both multilayer and mesh filter techniques have been discussed and it has been shown how the two methods are complementary. Multilayer methods are usually more efficient at $\lambda < 40$ μm whilst at $\lambda > 40$ μm meshes are to be preferred. It may be found that for the region 30 μm $< \lambda < 40$ μm hybrid filters of multilayer and mesh methods will prove to be the most efficient technique, this is the subject of present development.

Acknowledgements

We would like to acknowledge the financial support of the Science Research Council and the assistance of Grubb Parsons in the manufacture of some of the multilayer filters. The capacitative meshes were constructed by Cambridge Consultants whom it is a pleasure to thank. The computational assistance of Miss C. Gregory of Heriot Watt University is gratefully acknowledged.

References

[1] Smith, S. D.: 1958, *J. Opt. Soc. Am.* **48**, 43.
[2] Pidgeon, C. R. and Smith, S. D.: 1964, *J. Opt. Soc. Am.* **54**, 1459.
[3] Seeley, J. S. and Smith, S. D.: 1966, *Appl. Opt.* **5**, 81.
[4] Seeley, J. S.: 1961, *Proc. Phys. Soc.* (*London*) **78**, 998.
[5] Fan, H. Y. and Becker, M.: 1951, *Proceedings of the Reading Conference on Semiconducting Materials* (Butterworths, London) pp. 32, 120, 138.
[7] Seeley, J. S. and Williams, J. C.: 1962, *Proc. I.E.E.* **109B** Suppl. 23, 827.
[6] Baumeister, P.: 1958, *J. Opt. Soc. Am.* **48**, 955.
[8] Houghton, J. T. and Smith, S. D.: 1970, *Proc. Roy. Soc. London* **A320**, 23.

[9] Abel, P. G., Ellis, P. J., Houghton, J. T., Peckham, G., Rogers, C. D., Smith, S. D., and Williamson, E. J.: 1970, *Proc. Roy. Soc. London* **A320**, 35.
[10] Ulrich, R.: 1967a, *Infrared Phys.* **7**, 37.
[11] Ulrich, R.: 1967b, *Infrared Phys.* **7**, 65.
[12] Ulrich, R.: 1968, *Appl. Opt.* **7**, 1987.

DISCUSSION

F. Kneubühl: If you use metallic mesh filters in the far infrared, around 100 μ or slightly more, why did you not use it in higher orders.

If you take a Fabry-Pérot and a submillimetre wave laser, you can adjust them to an order of two hundred or so and then you would have a better percentage. What is the disadvantage of this?

S. D. Smith: Just the free range. It depends on what you try to separate out. The simplest filter is in the first order. Afterwards one can always put something else on. You lose nothing by going to higher orders. The half width to spacing of the filter remains a function only of the absorptive and degrading properties of the system not of the order, which is always useful. But of course the first steps of this business are to make them in first order with the biggest range which is free and then find ways of rejecting the rest, particularly if you are looking for a continuous filtering, from a continuous background. But putting the order up in a multicavity filter is a little more difficult than in the Fabry-Pérot.

J. Ring: I would like here to add a comment. It would be a mistake to assume that 1% is the best one can do, because you can easily envisage adding a small scanning Fabry-Pérot stage to an interference filter. In fact my colleague, Dr Selby, is building such a thing which looks as though it would get to 10^{-3} ($\lambda/\Delta\lambda \sim 10^3$) at about 10 μ. Of course the snag is the limited wavelength range you can scan. If you are working at 10 μ and start off with a narrow-band filter you can then scan through the pass band of that filter very easily, and I do not think it would be profitable to go to more stages than that, because then you may start having problems.

H.-J. Bolle: Do you have experience of the temperature stability of these various filters at 4 K also?

S. D. Smith: For most of the multi layers the temperature coefficient is of the order of 0.07 cm^{-1}/°C. You can cool them all to liquid He temperature without problems. For the mesh filters there are tests going on and the results are encouraging.

METALLIC GRID FILTERS

G. CHANIN

Service d'Aeronomie du CNRS, Verrières-le-Buisson, France

Abstract. The problem of defining and isolating a wavelength band in the far infrared and sub-millimetre region (45 to 1000 μ) is particularly difficult because of the large thermal energy at shorter wavelengths that must be rejected. Metallic grid filters have been extensively studied during the past several years and are particularly suitable for high-sensitivity instruments because they function well in liquid helium. For use as filters in our balloon-borne experiments we have been studying the transmission and reflection characteristics of three types of metallic grids in the sub-millimetre wavelength region. While our results are not yet complete, they have enabled us to construct long-wave pass filters suitable for our experiments.

The grids studied include: woven wire cloth, electroformed nickel, and gold grids and photo-etched copper grids on quartz and mular substrates. The general features of their transmission spectra are similar: a very low short wavelength transmission, a sharp increase to a peak and a gradual roll-off at longer wavelengths to essentially zero transmission. The geometric parameters that characterize grid performance are summarized in Figure 1, representing an electroformed metallic grid. Woven wire

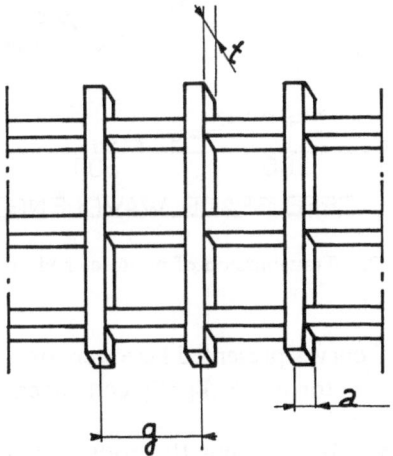

Fig. 1. Grid parameters.

cloth, of course, departs significantly from planinity. The photo-etched grids are supported on a quartz or mylar substrate and are of course, rigorously flat. Other parameters that affect grid performance are the index of refraction of the substrate (if any) and surrounding medium and the electrical conductivity of the metal. Since the latter is strongly temperature dependent the performance of the grids at liquid helium temperatures might be expected to be different than at room temperature. However, the room temperature conductivity is already sufficiently high for ohmic losses to be negligible compared to diffraction and scattering: no significant differences have been noted for those tests made in liquid helium.

Typical transmission characteristics for the three general types of grids are shown in Figures 2, 3, and 4. The results obtained for a given grid are subject to considerable variation if the incident radiation is strongly convergent or if the grid is placed at the

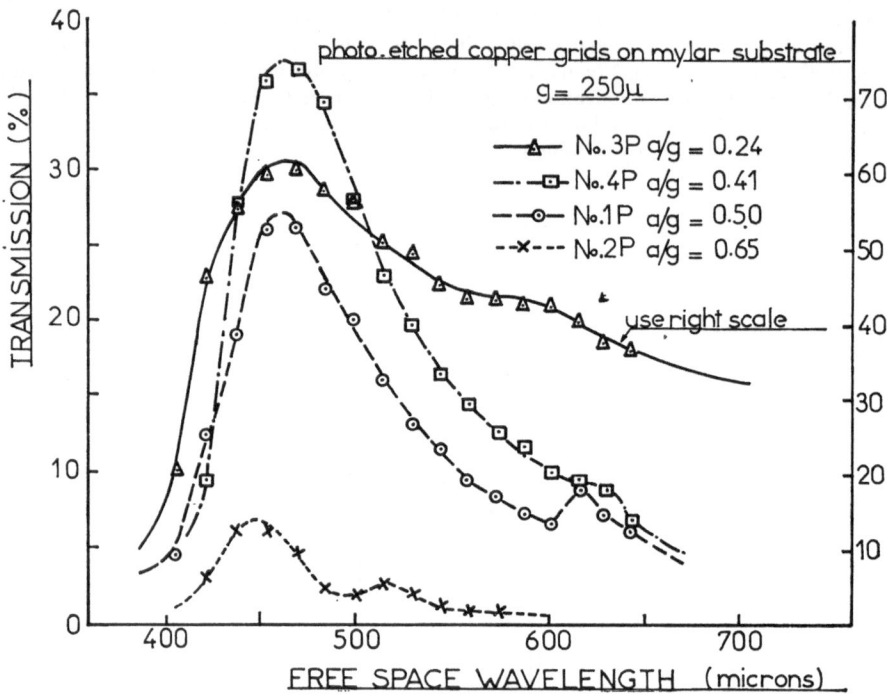

Fig. 2.   Transmission of photo-etched grids.

entrance of a light cone. All curves presented here were obtained at room temperature with a 'Cameca' monochromator using slightly convergent light and without a light cone.

The photo-etched grids may be conveniently produced with different values of a (the width of the metallic strips) for the same periodicity, g. Results for a series of such grids, identical except for a, are shown in Figure 2. Transmission maxima for all grids is dependent only on g and is constant at $1.9\,g$. We note that for extreme values of a/g (grids No 3 P and No 2 P) either the bandwidth of the transmitted radiation is too large or its amplitude is too small for most applications. The optimum appears to be in the neighbourhood of $a/g = 0.4$. The rather large shift (90%) between grid periodicity and transmission peak wavelength may be attributed to the change in wavelength in the substrate material from the free space value used as the abscissa in Figure 2.

Transmission curves of woven wire cloth and electro-formed gold and gold-plated nickel grids are presented in Figure 3 and Figure 4. The maxima are higher than for the photo-etched grids and are shifted only 20% with respect to the grid period, g, because of the absence of substrate. The bandwidth appears to depend upon a/g in the same way as for the grids of Figure 2. To further accentuate the band-pass characteristics of these grids and to improve their short-wavelength rejection, two or more

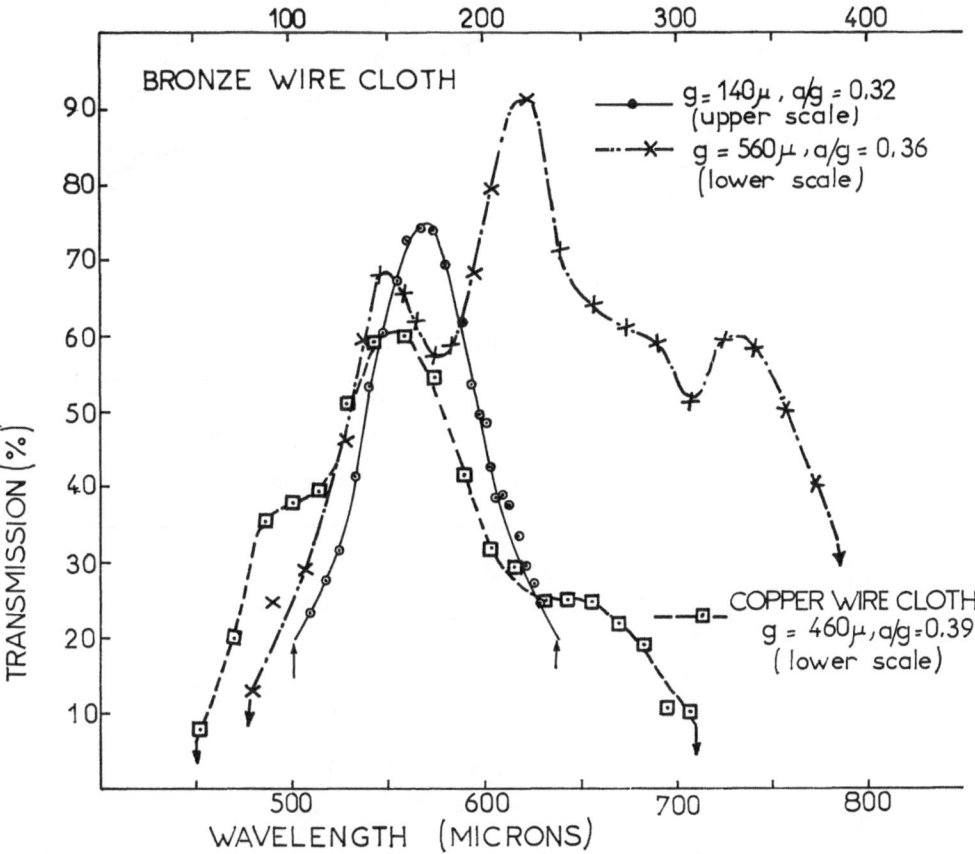

Fig. 3.   Transmission of commercial bronze wire cloth.

may be stacked in superposition (Figure 5). Similar results are obtained for the woven wire cloth grids. The stacking of grids with a non-parallel spacer between them leads to very similar results. The superposition of substrate-supported grids is not useful because of the large reflection loss resulting from the multiple substrate surfaces.

Long-wavelength pass filters have been obtained by using woven wire cloth in reflection. Reflection spectra for two such grids are shown in Figure 6. Reflection is essentially perfect at wavelengths greater than $2\,g$. If care is taken to assure the flatness of the grids (by stretching them over metal rings) they may be used to replace all the

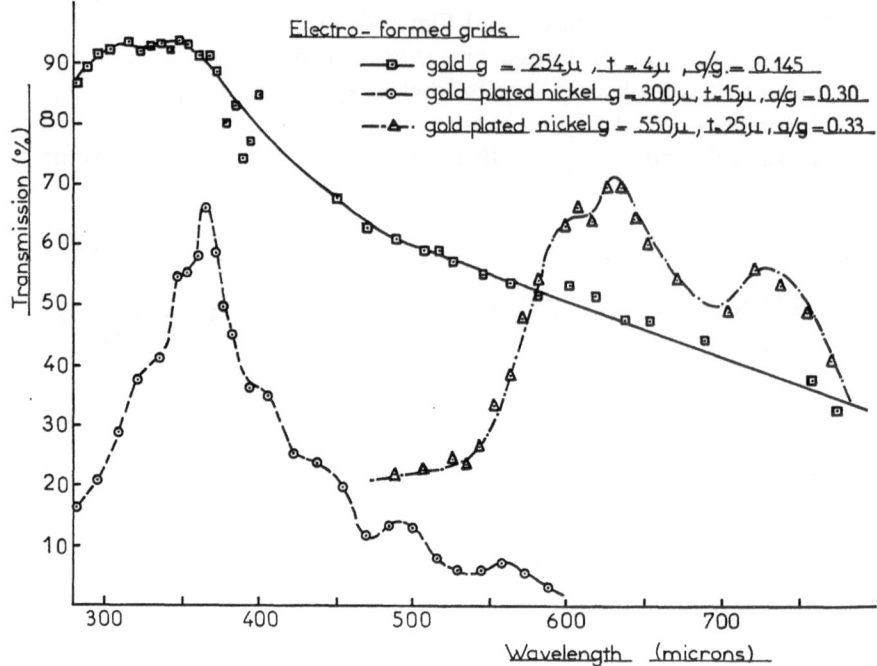

Fig. 4.   Transmission of electro-formed grids.

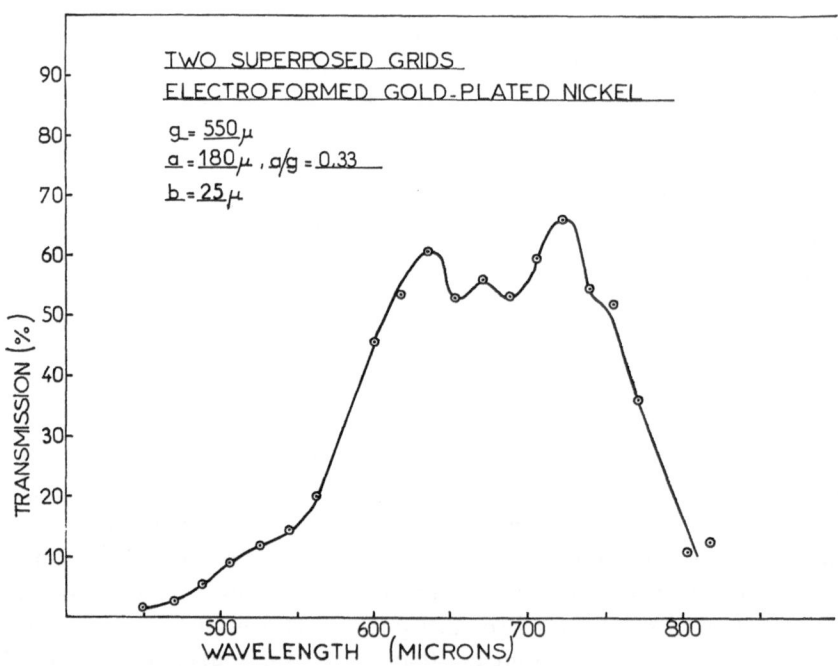

Fig. 5.   Transmission of two superposed electro-formed grids.

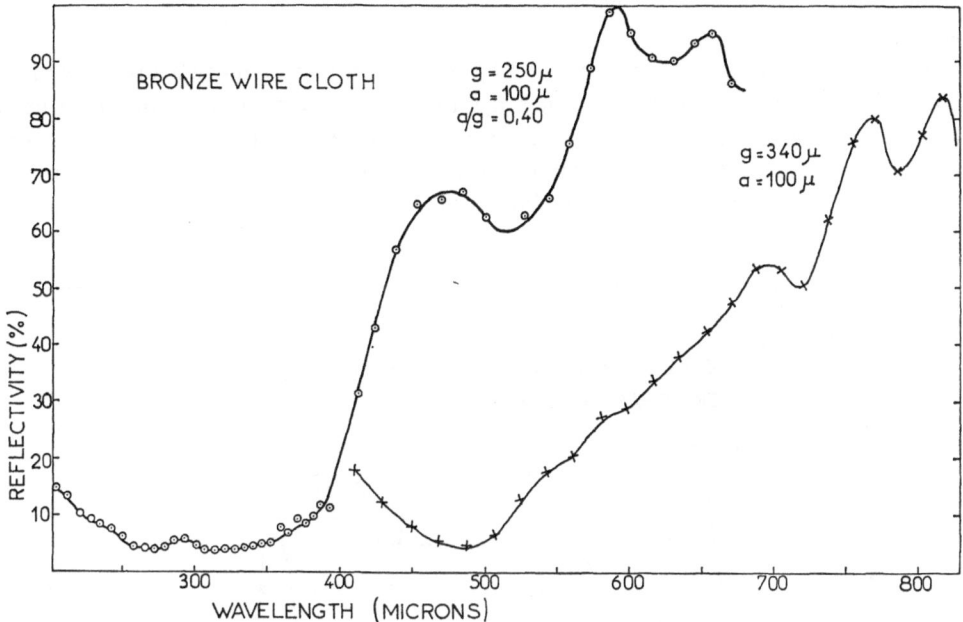

Fig. 6.   Reflection of commercial bronze woven wire cloth.

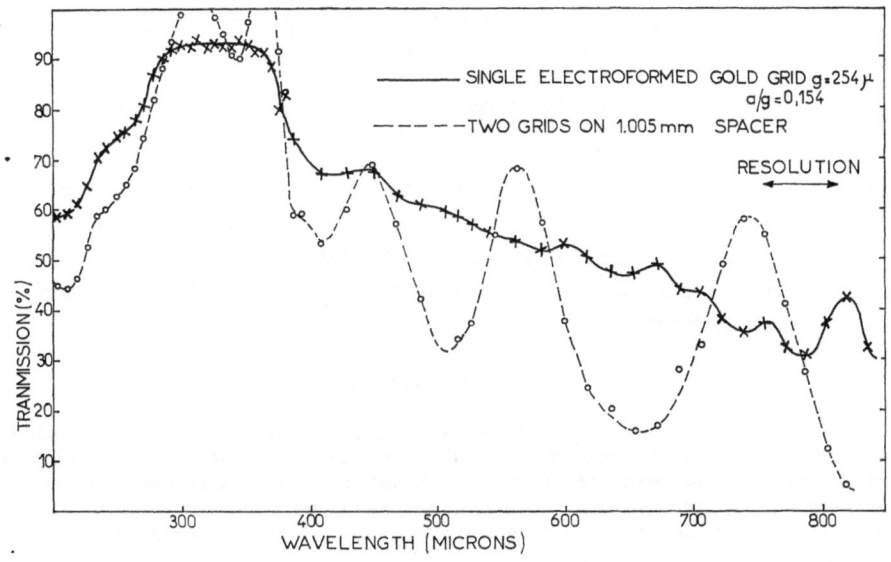

Fig. 7.   Transmission of a Fabry-Pérot consisting of two electro-formed grids
of low reflectivity.

plane mirrors in an infrared instrument and thus greatly improve rejection of short-wavelength radiation.

Narrow band filters may be obtained by separating two or more grids by parallel-faced spacer rings. The fringes of such a Fabry-Pérot interferometer obtained using two electro-formed grids are shown in Figure 7. Note that as the single grid transmission approaches 30% (70% reflectivity) the fringes become narrower with minima approaching zero transmission. The fringes of Figure 8 show how increasing the grid

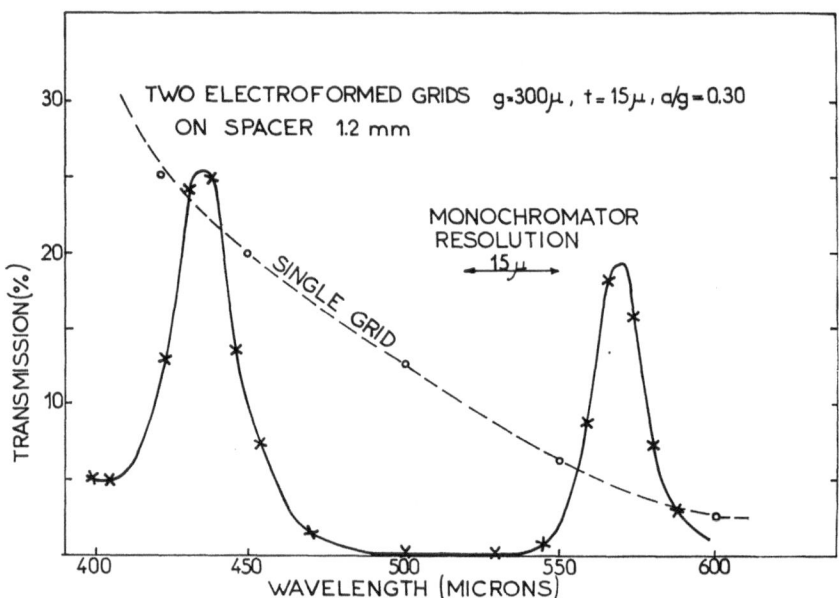

Fig. 8.   Transmission of a Fabry-Pérot consisting of two electro-formed grids
of high reflectivity.

reflectivity (by increasing a/g) narrows the fringes still further. The true transmission and bandwidths are far superior to the curve given here because the monochromator bandwidth limited the spectral resolution. Combining this filter with that of Figure 7 or with the grid of Figure 4, would allow isolation of a single order and thus produce a very narrow-band filter.

DISCUSSION

S. D. Smith: I would like to point out that we also did some studies of the effective thickness and different metals, and thus your conclusion that the conductivity will not make much difference is correct. All these metallic mesh filters must be sufficiently thick that the boundary conditions are properly obeyed, that's to say that the wave does not really penetrate the thickness of it. We found the most convenient material to be aluminium at about 0.1 μ, which was contrary to expectations. But I would like to point out that the design variables you've got with thickness and spacing certainly exist, but don't neglect the much greater freedom of action you've got by combining three or four grids, because this really gives the possibility of very sharp transitions between on and off.

NEW NARROW-BAND METALLIC MESH FILTERS
FOR THE 50 μ REGION

MARTIN HEIDEMANN

Meteorologisches Institut der Universität München, München, G.F.R.

Abstract. Metallic mesh filters were made for the 50 μ region with a halfwidth of less than 5 cm^{-1} and a peak transmission of 40%. Peak wavenumber and transmission were reproduced within ± 2 cm^{-1} respectively $\pm 2\%$, this limit being given by the accuracy of the used spectrometer.

There have been attempts in the past [1] to build dielectric interference filters for the 50 μ region, but the success was not very convincing with respect to bandwidth and stability. Another alternative are Fabry-Pérot type filters made of metallic meshes.

For a Fabry-Pérot filters [2, 3] the peak transmission is given by

$$\tau_{\max} = k\left(1 - \frac{A}{1 - R}\right)^2$$

the halfwidth by

$$H = \frac{1}{\pi h \sqrt{F} \cos \theta} \sqrt{\frac{2}{k} - 1}$$

and the usable spectral range by

$$D = \frac{1}{2h \cos \theta}$$

where A is the absorption of the meshes, R the reflection, h the distance of the two meshes, θ the angle of incidence, $F = 4R(1 - R)^{-2}$, and k is a factor which accounts for the inequality of the Fabry-Pérot mirrors.

$$k = \frac{\tau_{\max}}{\tau_{\max \text{ ideal}}}$$

k as a function of R is shown in the Figure 1. The computation of k is based upon the following assumption: We can say that one interferometer consists of many part interferometers. The transmission of each part for the wavenumber ν is given by Born and Wolf [2].

$$\tau' = \frac{1}{1 + F \sin^2 2\pi\nu(h + \Delta h)}$$

Δh is the error of the part interferometer. The transmission of the whole interferometer is

$$\tau = \sum a_i \tau' = \sum a_i \frac{1}{1 + F \sin^2 2\pi\nu(h + \Delta h)}$$

Manno and Ring (eds.), Infrared Detection Techniques for Space Research, 171–188. All Rights Reserved

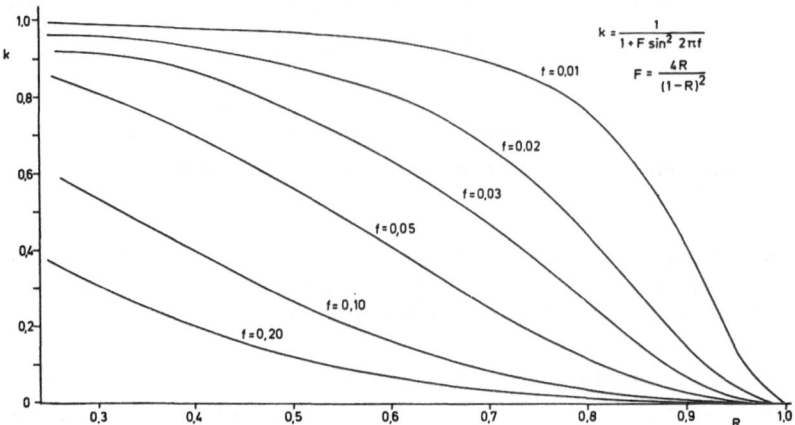

Fig. 1. k-factor as a function of the reflectivity R and the inequality of distance.

for the peak, $\sin 2\pi h = 0$ and

$$k = \sum a_i \frac{1}{1 + F \sin^2 2\pi v_0 \Delta h}$$

Now there must be a mean value Δh for the error of distance so that

$$k = \sum a_i \frac{1}{1 + F \sin^2 2\pi v_0 \Delta h} = \frac{1}{1 + F \sin^2 2\pi \overline{\Delta h}}$$

with $v = 1/\lambda$ and $\overline{\Delta h} = f \cdot \lambda$

$$k = \frac{1}{1 + F \sin^2 2\pi f}.$$

One can see that for a desired reflection of more than 60% the average inequality should be much smaller than 0.03λ and for $R > 90\%$ it should be better than 0.01λ.

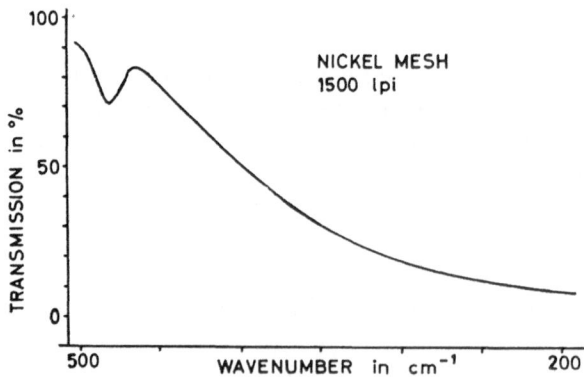

Fig. 2.   Transmission of a nickel mesh (1500 lines per inch).

Figure 2 shows the transmission of a nickel mesh with 1500 lines per inch ($g=$ 16.8 μm). Such a mesh can be used as a Fabry-Pérot mirror for wavelengths longer than 30 μ. For shorter wavelengths the reflection is too poor.

In the region where the reflectivity is 50–90% the absorption can be neglected. For copper meshes it is ∼1% and for nickel meshes ∼2% [4, 5]. So one can say $R = 1 - T$ and can later on calculate k from a transmission measurement of the filter:

$$k = \frac{2}{H^2\pi^2h^2F + 1}.$$

The construction of a filter is illustrated in Figure 3. The mesh is fixed on the ring with

Fig. 3.   Construction of the filter parts.

an adhesive and will be stretched by putting the ring with the mesh over the cylindrical part of the flange. It is then fixed with three screws on the base of the flange.

It is important that the front end of the cylindrical part is exactly plain, because it defines the inequality of the Fabry-Pérot mirror. Two such parts as shown in Figure 3 give one filter. With three springs and screws the wanted distance between the meshes can be tuned. This is shown in Figure 4.

The chosen construction maintains a good stability of the distance, also in case of temperature variations, if the frame and the screws have the same thermal expansion coefficient.

The variation of the screw lengths will then compensate the variation of the frame. For nickel meshes and aluminum frame there is no marked change in distance, when the temperature of the filter is changed between +30° and −30 °C. However, at lower temperatures, there was a considerable decrease of flatness.

The filters are always adjusted in a spectrometer to maximum transmission. From the distance of the centre frequency of the transmission peaks the distance of the mirrors can be determined, and from peak transmission and halfwidth the value of k.

When all values are in the desired range, the filter can be tuned on the wanted wave-number. A result can be seen in Figure 5. Here the second order of the transmission is at 254 cm^{-1}, which is the wavenumber of a water vapour line. The distance of the metallic meshes was computed to be $h = 1/2D = 39.4$ μm, the halfwidth is 8 cm^{-1}, the peak transmission 40%. With a reflection of 87% (Figure 2) k is 0.66 and the inequality

Fig. 4.   The ready filter.

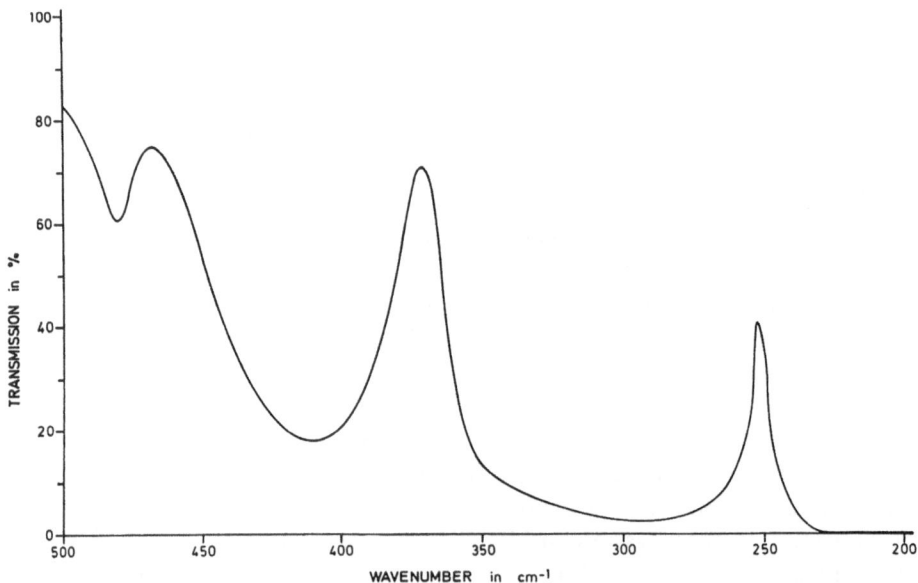

Fig. 5.   Transmission of the filter for 254 cm^{-1}. Halfwidth 8 cm^{-1}, peak transmission 40%.

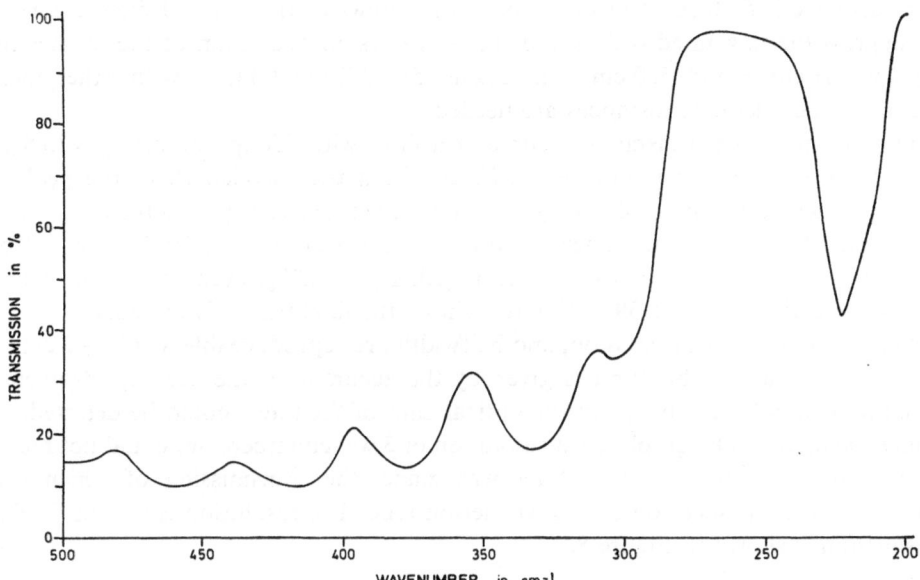

Fig. 6.   Transmission of one prefilter.

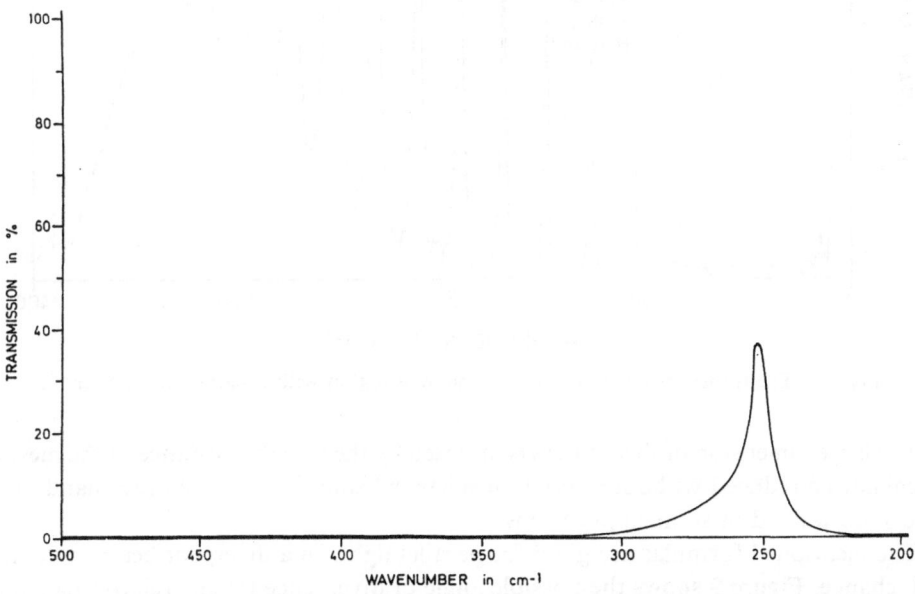

Fig. 7.   Transmission of the four filters.

of the distance is $f=0.007\,\lambda$ or $0.3\,\mu$. Now the screws of the finished filter are fixed. To suppress the unwanted orders and the transmission maximum of the meshes for $\lambda=g$ (at 381 cm^{-1} and 500 cm^{-1} in Figure 5) additional filters with other mesh constant g and adequate distances are needed.

Figure 6 shows the transmission curve of a filter with 750 lpi meshes ($g=33.8\,\mu$). Since the mesh constant is not very different from the wavelength of the peak at 254 cm^{-1}, there is a transmission of 97%. With three filters to suppress the not wanted transmission for $\mu>300$ cm^{-1} there is still a transmission of 90% for 254 cm^{-1}. The resulting transmission of all four filters together is <0.7%, even at 381 cm^{-1} and 500 cm^{-1}; at 254 cm^{-1} it is 39%. Figure 7 shows the final transmission curve.

Peak wavenumber, transmission, and halfwidth are reproduceable within ±2 cm^{-1}, ±2% resp. ±1 cm^{-1}. This limit is given by the accuracy of the used spectrometer. Within four months no change in the optical data of the filters could be detected.

Since the used spectrograph has a resolution of 3 wavenumbers, we could not measure a lower halfwidth. Another filter was made, the transmission of which was measured in the Iris spectrometer, a Michelson type. The resolution is 0.1 cm^{-1}. The transmission is shown in Figure 8.

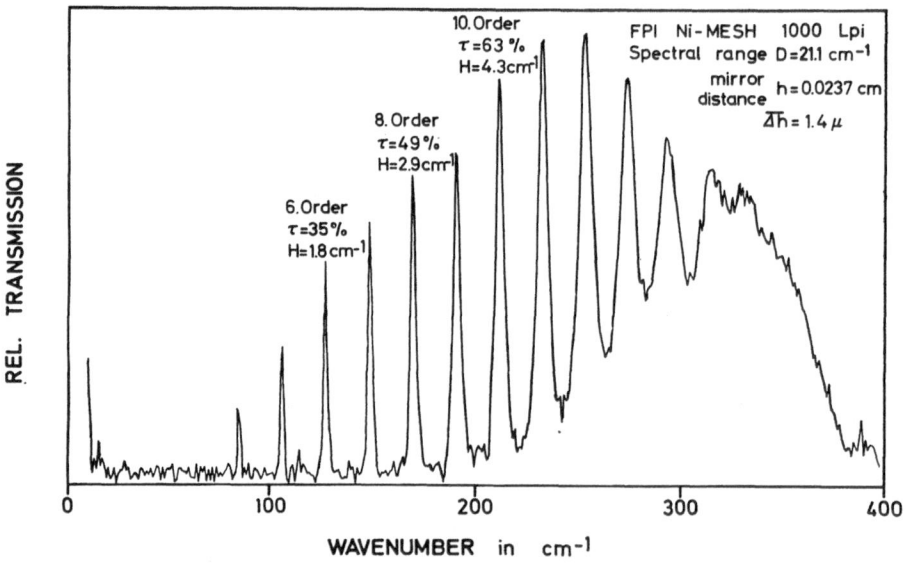

Fig. 8.   The transmission of a filter for the 50 μ region with a halfwidth of 5 cm^{-1}.

The large dimension of these filters is imposed by the variable distance of the meshes. It cannot be reduced without a loss in mechanical stability. On the other hand, large filters are wanted in some applications.

The mentioned formulas are good for parallel light. In a divergent beam, halfwidth will change. Figure 9 shows the possible angle of divergence for the relative halfwidth $n=H/D$.

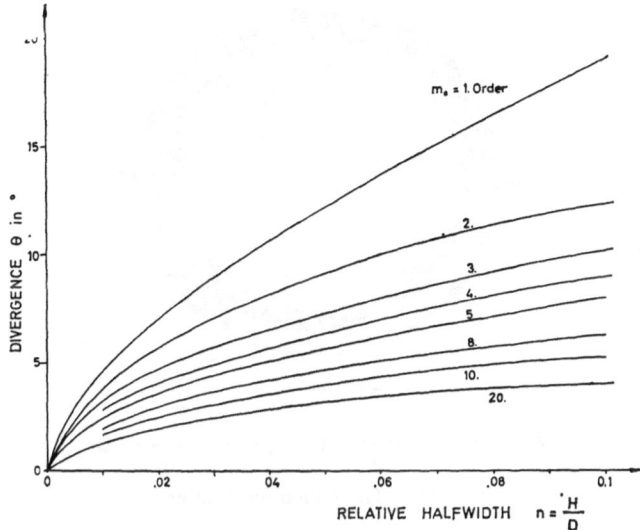

Fig. 9. Possible divergence in beam for Fabry-Pérot filters as a function of order and relative halfwidth.

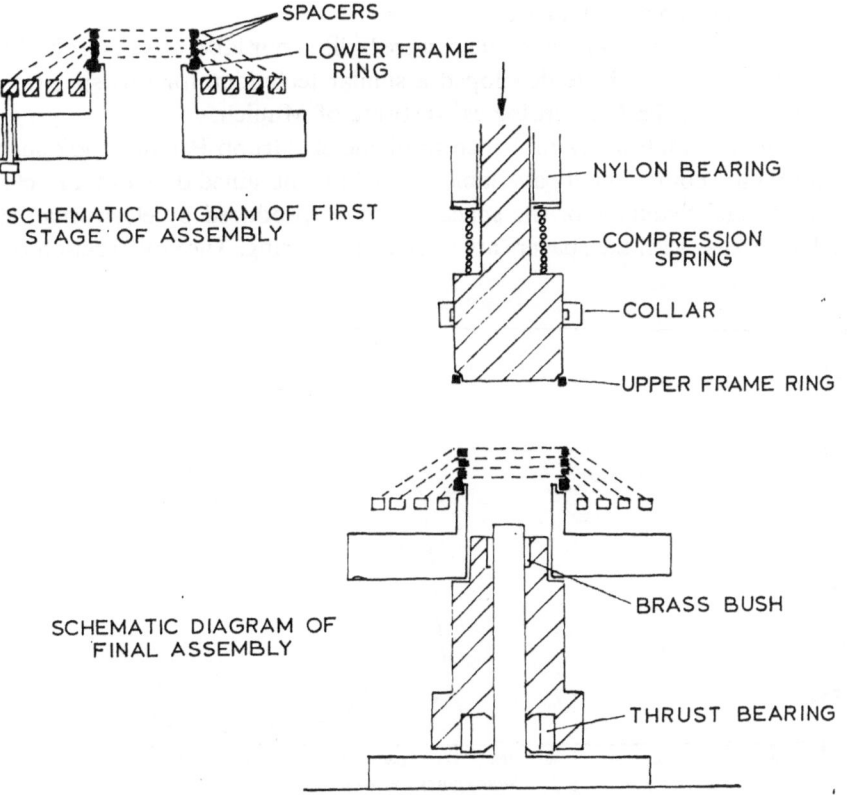

Fig. 10. The construction of the Grubb Parsons filters.

Fig. 11.   The 1 inch mesh filter.

It is calculated by assuming that the radiation within this angle has its transmission maximum within the theoretical halfwidth of the filter for parallel light. So the change in halfwidth is not greater than 1.5.

During the same time when these filters were built in our laboratory, Grubb, Parsons & Co., Ltd., Newcastle, have developed a similar technique for this spectral region under contract with the Meteorological Institute of Munich.

In order to get small filters of 1 inch in diameter, Grubb Parsons took spacers of gold leaf that are commercially available, and can be obtained down to a thickness of 10 μ. The wanted thickness of the spacer could be produced by electroplating these leaves. The meshes with the spacers are stretched on a ring. The construction is shown

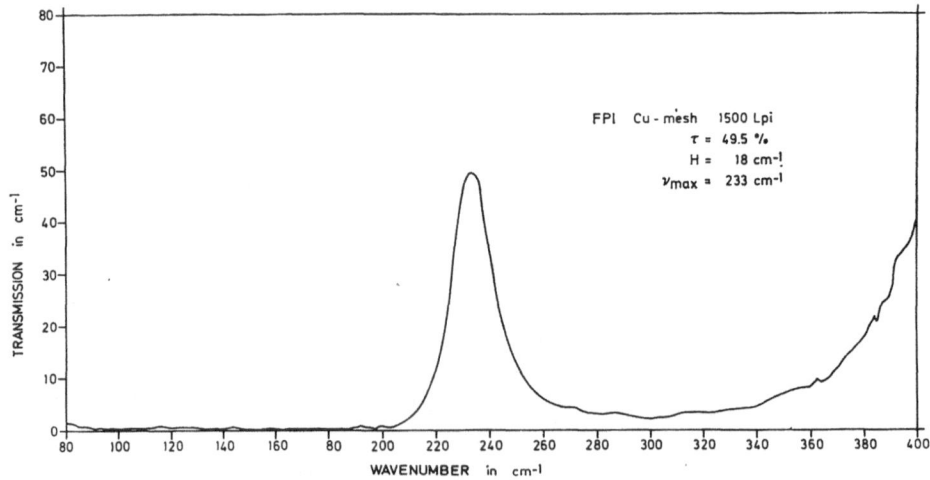

Fig. 12.   The transmission of the filter (Figure 10).

in Figure 10; technique that has to our knowledge first been applied by Ulrich (1967) [4].

In the first stage of assembly, the optical data of the filter can be measured. If they are satisfactory, the upper frame ring is pressed on the filter, with the nylon bearing, the collar is mounted, and the mounting rings are severed from the meshes. The finished filter is shown in Figure 11. The transmission of this filter (Figure 12) is 49% at the peak wavenumber of 233 cm^{-1}, the halfwidth is 18 cm^{-1} in the first order. So the inequality of distance is about 1/100. With a relative halfwidth of $n=0.07$ the possible angle of divergence is $\pm15°$. In order to suppress the unwanted transmission, a blocking filter has to be used together with the metallic mesh filter, the transmission curve of which is shown in Figure 13.

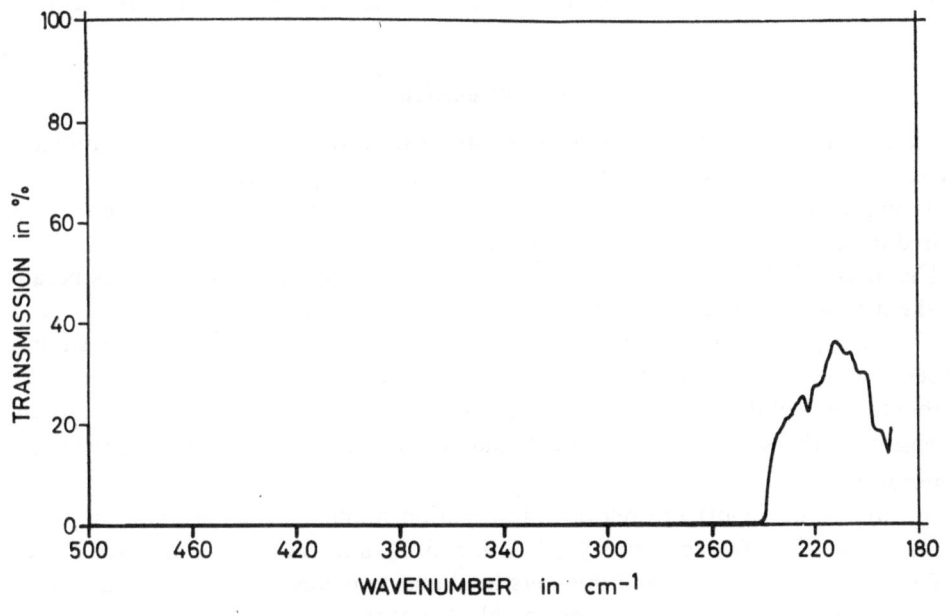

Fig. 13.   Blocking filter.

We are now working on an extension of the temperature range for mesh filters down to lower temperatures.

References

[1] Smith, S. D. and Seeley, J. S.: 1968, Multilayer Filters for the Region 0.8 to 100 Microns. Final Sci. Report. AF 61(052)-833.
[2] Born, M. and Wolf, E.: 1959, *Principles of Optics*, Pergamon Press, pp. 322 f.
[3] Heidemann, M.: 1966, Studie über die Verwendung von Fabry-Pérot-Interferometern zur Auflösung des infraroten Wasserdampfspektrums in der hohen Atmosphäre. Forschungs- bericht BMwF-FB W 66-02.
[4] Ulrich, R.: 1967, *Infrared Phys.* **7**, 37; *Infrared Phys.* **7** (1967), 65.
[5] Vogel, P. and Genzel, L.: 1964, *Infrared Phys.* **4**, 257.

HIGH-PASS FILTERS FOR FAR INFRARED
ASTROPHYSICAL INVESTIGATION

MARIA GRAZIA BALDECCHI, BIANCA MELCHIORRI, and
GIOVANNA M. RIGHINI

CNR GIFCO-Te.S.R.E. Laboratory, Florence Section Space Physics,
University of Florence, Florence, Italy

Abstract. The preparation of high-pass inclusion filters for the far-infrared spectral range is described. Rejection factors ($1/T$) better than 10^3–10^4 in the visible and near infrared together with a high transmittance ($T = 70$–80%) in the far infrared up to $\sim 300\,\mu$, have been obtained.

These filters are used in high altitude experiments on far-infrared detection of extraterrestrial sources.

1. Introduction

In recent years the search and study of extraterrestrial sources of far infrared radiation (50–1000 μ) greatly developed through high altitude flying instrumentation.

As many customarily used detectors have a 'flat' response with λ, filters are required in order to eliminate unwanted radiation.

The most suitable filters for high altitude research are transmission filters because of the mechanical advantages that make them easy to handle.

The main problem of making a far infrared filter is to find a basic composition in order to eliminate the visible and near infrared radiation with a high rejection factor, without seriously decreasing the transmittance in the range of interest (the expected intensity of the sources is very weak indeed and near to the sensitivity limit of detectors).

In Table I we report the relevant information about recent high altitude experiments in the far infrared region [1, 2, 3, 4, 5, 6, 7], and the filters employed therein.

One can see that many experimenters used quartz plates (which cut infrared down to about 40 μ) coupled with sheets of black polyethylene (to remove the visible); white polyethylene (high and nearly constant T in the infrared range beyond 15 μ) or teflon (similar behaviour beyond 50 μ) are used for windows. However, it is difficult to find in the references an explicit evaluation of the overall transmittance and of the rejection factor in the visible and near infrared.

The same kind of filtering is currently used in laboratory experiments, where sources such as Hg lamps and globars often require rejection factors as large as 10^3–10^4 in the visible and near infrared region: blackened quartz plates of suitable thickness are usually employed. In this case, however, the far infrared emission of the sources is rather strong: so we can use filters without a good T in this range.

One can see (Figure 1) that a fused quartz plate 1 mm thick shows only 40% transmittance in the interesting range; even a thickness of 0.5 mm does not give much more than 60%. If we consider that the quartz thickness used in high altitude experiments

TABLE I

Experiment	Altitude	Detector	Filtering system	Transmission band (filters and detector)	Remarks	References
Far infrared survey of the night-sky	27 km (balloon)	Ge bolometer (1.8 K)	f/1.2 crystal quartz lens (1.8 K) black polyethylene (1.8 K) No. 80 wire mesh (1.8 K) thallium bromide (1.8 K)	300–450 μm		[1]
Far infrared galactic center observation at 100 μm	27 km (balloon)	Ge bolometer (1.8 K)	f/1.2 crystal quartz lens (1.8 K) (2.3 mm) black polyethylene (1.8 K) (0.1 mm) No. 300 electroformed metal mesh (1.8 K) BaF in polyethylene (cold radiation shield) Black polyethylene (cold radiation shield) No. 300 electroformed metal mesh (cold radiation shield) Crystal quartz dewar window (210 K) (1 mm)	80–120 μm	peak transmittance 12%	[2]

Table I continued

Experiment	Altitude	Detector	Filtering system	Transmission band (filters and detector)	Remarks	References
Far infrared observations of the night-sky and background radiation	higher than 130 km (rocket)	Ge:Ga photoconductive detector (2.4 K)	Wire mesh Ge interference filter	5.2–130 μm		[3]
		Ge:Ga photoconductive detector (2.4 K)	Wire mesh Ge interference filter NaCl crystal (2 mm)	5.2–130 μm	opaque at 63 μm (atmospheric atomic oxygen emission)	
		InSb detector (2.4 K)	Wire mesh Ge interference filter Black polyethylene	400–1300 μm		[4]
Detection of celestial sources at far infrared wavelengths	28 km (balloon)	Ge bolometer (1.8 K)	Black polyethylene (1.8 K) (0.19 mm) Crystal quartz (1.8 K) (1 mm) Crystal quartz (210 K) (2 mm)	45–250 μm		[5]
Far infrared observations of the galactic center	15 km (aircraft)	Ge bolometer (1.8 K)	Crystal quartz (2 K) (0.46 mm) Fabry lens intrinsic silicon (2 K) (0.46 mm)	40–350 μm	upper limit due to diffraction limit of the telescope	[6]

Table I *continued*

Experiment	Altitude	Detector	Filtering system	Transmission band (filters and detector)	Remarks	References
			Black polyethylene (100 K) (0.15 mm) High density polyethylene (300 K) (3 mm)	40–350 μm	half-peak transmission at 65 μm	
		Ge bolometer (1.8 K)	Crystal quartz (2 K) (0.46 mm) Fabry lens intrinsic silicon (2 K) (2 mm) Black polyethylene (100 K) (0.15 mm) High impact polysytrene (300 K) (1.5 mm)	50–350 μm	upper limit due to diffraction limit of the telescope half-peak transmission at 75 μm	
Observations of galactic and extragalactic sources in the far infrared	15 km (aircraft)	Ge bolometer (1.8 K)	Teflon (2 K) (1.6 mm) Black polyethylene (2 K) (0.3 mm) Crystal quartz	50–300 μm	upper limit due to diffraction limit of the telescope rejection λ shorter than 50 μm better than 10^3	[7]

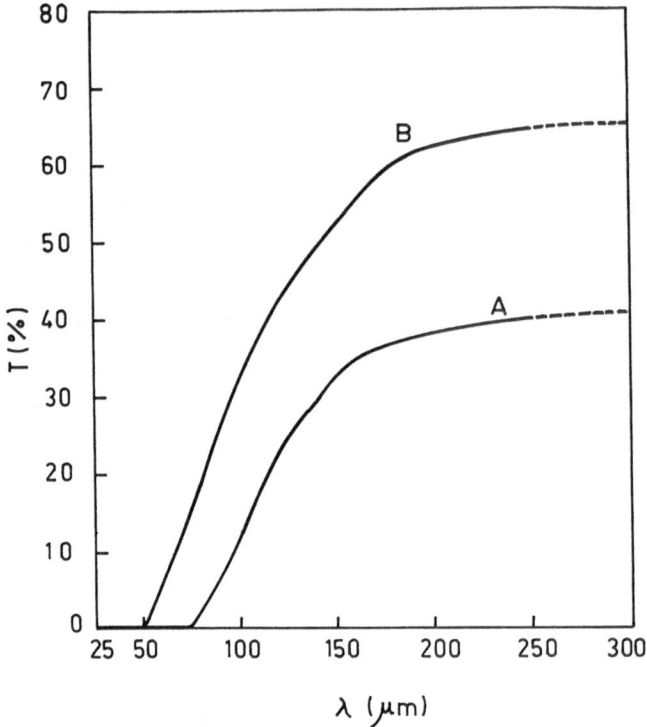

Fig. 1. Transmittance of fused quartz plates (A) 1 mm thick; (B) 0.5 mm thick.

is often greater than 1 mm, we conclude that the results are not satisfactory. Moreover the filters thus obtained are easy to break and difficult to cool, when required, because they are made of several different parts.

2. Inclusion High-Pass Filters

The method used by us is that of including opaque materials in a transparent support, following Möller *et al.* [8], who modified the original method proposed by Yamada *et al.* [9].

The first step has been to establish the minimum quartz quantity capable of eliminating the near infrared, without affecting seriously the transmittance in the long wavelength range.

We found that this is easily obtained using powdered fused quartz. Quartz is pounded into very thin powder and mixed carefully with soot and polyethylene powder (Rigidex,* Nathene**). The mixture is then pressed between the heated plates of a hydraulic press, the plate temperature being kept at about the polyethylene melting point. With a moderate pressure (5×10^{-3} tonn/cm^2) we obtain a sample whose

---

* Rigidex: Grubb, Parsons, Walkergate, Newcastle upon Tyne 6, U.K.
** Nathene: Pechiney Saint Gobain, 47 Rue de Villiers, (32) Neuilly sur Seine (Paris), France.

thickness can be chosen in a wide range making use of variable thickness gauges: the homogeneity can then be improved by cutting this first product into several parts, heating and pressing them together again many times. After several trials we found a satisfactory quartz powder amount, which gives to the sample a high transmittance together with a good rejection in the undesired wavelength range (while a direct comparison with a quartz plate is not fully justified, the amount of the powder is equivalent to a few tens of microns).

Soot provides complete opacity in the visible; a small residual transmission band centered at about 17–18 μ is cut off by adding BeO to the mixture [9].

Figure 2 shows the transmission spectrum of a typical filter. The spectrum was

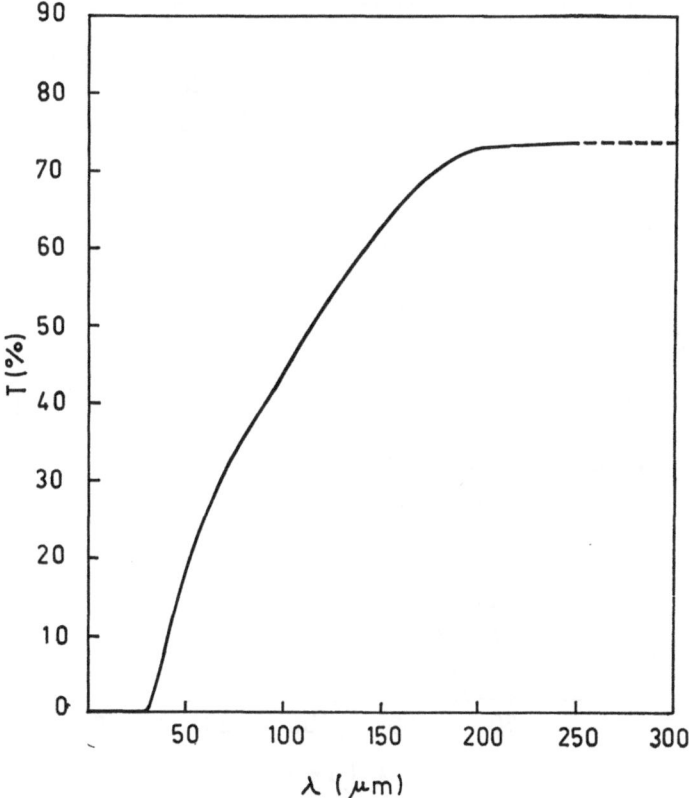

Fig. 2.   Typical transmission spectrum of a basic high-pass filter. Percent powder (by weight) with respect to polyethylene is as follows: soot 3%, BeO 25%, fused quartz 20%: thickness 340–360 μm.

carried out with two spectrophotometers: a Perkin Elmer 225 (2.5–50 μ) and a Hitachi Perkin Elmer FIS-3 (25–333 μ).

As reported by many experimenters (see for instance [8, 9, 10, 11]), powders of ionic crystals (oxides, halides, carbonates) show transmission spectra with a distinctive steep absorption band that comes from the selective reflectivity that these crystals show in the 10–10^3 μ range [12]. This feature has been widely used to obtain band stop filters for laboratory purposes (a set of this kind of filters is used in some far infrared monocromators, such as the Hitachi FIS-3).

The addition of these compounds to the basic filter mixture allowed us to obtain high pass filters with cut-on point in a wide wavelength range and with very steep slope. Actually, the basic composition can be slightly modified when other constituents are added, because their presence is often effective in lowering the transmittance also in the wavelength range shorter than 40 μ. In this way we obtained rejection factors in the short wavelength region better than 10^3–10^4, with a transmittance after the cut-on point increasing steeply up to 80%, such as the ones shown in Figure 3.

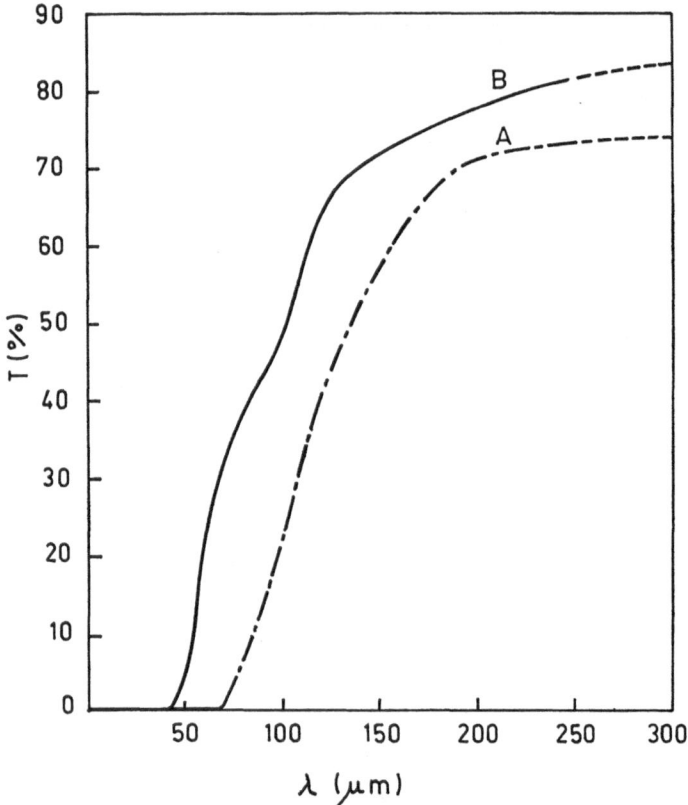

Fig. 3. Spectra of filters obtained by adding ionic crystal powders to the basic composition (rejection factor in the range of the visible and near infrared better than 10^3–10^4). (A) soot 3%, BeO 25%, fused quartz 20%, ZrO_2 30%: thickness 200–280 μm; (B) soot 3%, BeO 25%, fused quartz 20%, NaF 10%: thickness 280–330 μm.

3. Final Remarks

Our filters appear to be particularly suitable for airborne and balloon experiments, because they are mechanically strong; only one element constituted, our filters are easy to cool and to handle.

In Figure 3, A and B indicate the filters used by us in balloon experiments intended for intensity and polarization measurements of far infrared extraterrestrial sources. It is an intrinsic feature of this kind of filters that they do not change the polarization state of the incident radiation: this makes them preferable in our case to meshes [13], which do not give full assurance on this point.

When using liquid helium refrigerated detectors, it is necessary to cool filters in order to minimize their thermal emission. As ionic crystals slightly change their optical properties at low temperature [14], the next step will be to investigate the behaviour of these filters at liquid helium temperature [15].

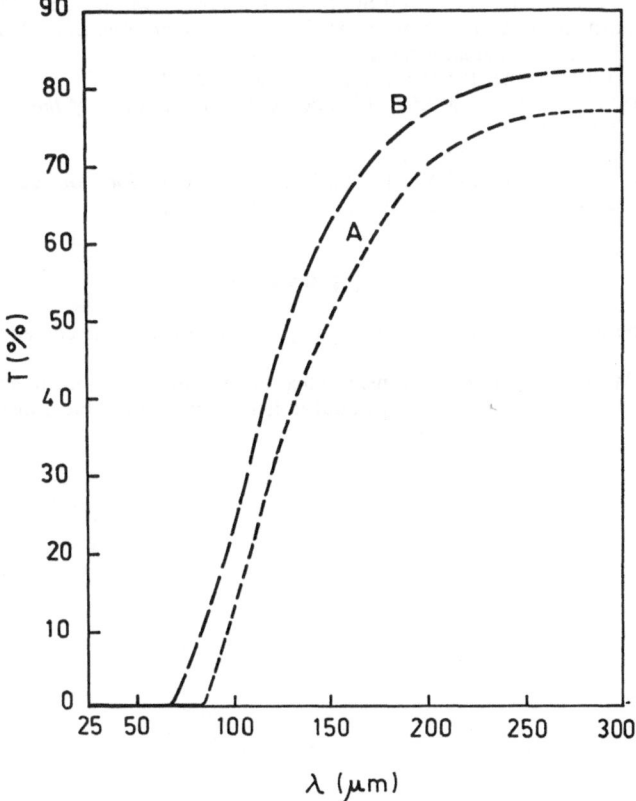

Fig. 4. Spectra of filters for detectors sensitive beyond 50 microns (residual transmittance in the visible and near infrared $T \leqslant 5\%$). (A) ZrO_2 30%, BeO 5%: thickness 500 μm; (B) BeO 25%, fused quartz 20%, ZrO_2 30%: thickness 180–250 μm. *Note:* Beyond 250 μm curves are extrapolated because the instrumental error in T becomes larger than 2%.

Finally we note that when the detector is sensitive only in definite wavelength regions (this is the case of some of the experiments quoted in Table I, [3, 4]) one can remove the unnecessary amounts of opaque constituents and still ensure a sharp cut-on and a steep slope with a good transmittance in the region of interest (Figure 4).

Acknowledgement

We would like to thank Dr F. Melchiorri for helpful discussions.

References

[1] Hoffman, W. F., Woolf, N. J., Frederick, C. L., and Low, F. J.: 1967, *Science* **157**, 187.
[2] Hoffmann, W. F. and Frederick, C. L.: 1969, *Astrophys. J.* **155**, L9.
[3] Houck, J. R. and Harwit, M.: 1969, *Science* **A64**, 1271.
[4] Shivanandan, K., Houck, J. R., and Harwit, M.: 1968, *Phys. Rev. Letters* **21**, 1460.
[5] Friedlander, A. W. and Joseph, R. D.: 1970, **162**, L87.
[6] Aumann, H. H. and Low, F. J.: 1970, **159**, L159.
[7] Low, F. J. and Aumann, H. H.: 1970, **162**, L79.
[8] Möller, K. D., McMahon, D. J., and Smith, D. R.: 1966, *Appl. Opt.* **5**, 403.
[9] Yamada, Y., Mitsuishi, A., and Yoshinaga, H.: 1962, *J. Opt. Soc. Am.* **52**, 17.
[10] Ackermann, F. W.: 1940, *Ann. Phys. Leipzig* **37**, 442.
[11] Quintard, P. and Delorme, P.: 1967, *Rev. Phys. Appl.* **2**, 215.
[12] Hadni, A.: 1967, *Essential of Modern Physics Applied to the Study of the Infrared*, Pergamon Press, 1967, Chap. VI.
[13] Ulrich, R.: 1967, *Infrared Phys.* **17**, 65.
[14] Mitsuishi, A., Yamada, J., and Yoshinaga, H.: 1962, *J. Opt. Soc. Am.* **52**, 14.
[15] Zwerdling, S. and Theriault, J. P.: 1968, *Appl. Opt.* **7**, 209.

DISCUSSION

G. Chanin: Did the transmission curves which you presented refer to the cooled filters or at room temperature?

F. Melchiorri: We've measured the transmittance of the filters at room temperature at liquid nitrogen and liquid helium temperatures, without noticing any appreciable difference.

FILTERS FOR FAR INFRARED ASTRONOMY

J. J. WIJNBERGEN, W. H. MOOLENAAR, and G. DE GROOT

Dept. of Space Research, University of Groningen, The Netherlands

1. Introduction

In developing far infrared photometers for astronomical purposes one of the main problems is the efficient spectral filtering of the radiation. As a consequence of the unknown spectral distribution of the usually weak radiation of the objects, the filter-bands must have a high transmission with steep edges; in addition a specified spectral band and bandwidth should be attainable. To lower the background radiation, which limits the performance of our broadband detector low temperature operation is required. From literature (Hadni) it appears that such problems can be solved by applying the reststrahlen reflection of crystal materials: the remarkable high reflection of crystal materials: the remarkable high reflection at low temperatures permits one to use multiple reflections, thus obtaining a high rejection specially for the short wavelengths, which are the most troublesome.

For our balloon photometer, which is launched 7 June 1971 in Tallard in France, this method is applied. Two telescopes of 20 cm diameter collect the light for two filtersystems in a dewar which also contains the condensing optics and detectors. Each filter system has four reflecting surfaces of the same material. KJ is chosen for the 105–140 cm^{-1} and CsBr for the 75–125 cm^{-1} region. In a second flight the KJ filter is combined with a TlJ filter for the 57–85 cm^{-1} region. The results of a calculation of the reflection, based on a measurement of the reflection at one surface at liquid helium temperatures is shown in Figure 1.

A photograph of the dewar optics with the filter systems and detectors is shown in Figure 2.

For laboratory purposes and in practice for interferometry transmission filters of the type developed by Yamada et al. [1] seem very attractive in spite of the poor bandpass characteristics: weak edges and a transmission of not more than 80%. Thus using more filters in series or a higher concentration of the crystal powder results in a less efficient filter than the reststrahlen filter.

In Figure 3 the characteristic features of a crystal reflection filter (CsJ) and a powder filter (LiF) at different temperatures can be compared. The influence of concentration and thickness is demonstrated by the behaviour of a CuCl powder filter at room temperature. In first instance the transmission curves seem to be governed by the law of Beer.

For the third type of filters – metallic mesh – we have obtained some interesting data with commercially available sieves for the chemical industry. These filters appear very promising for narrow-band photometry.

Manno and Ring (eds.), Infrared Detection Techniques for Space Research, 243–256. All Rights Reserved
Copyright © 1972 by D. Reidel Publishing Company, Dordrecht-Holland

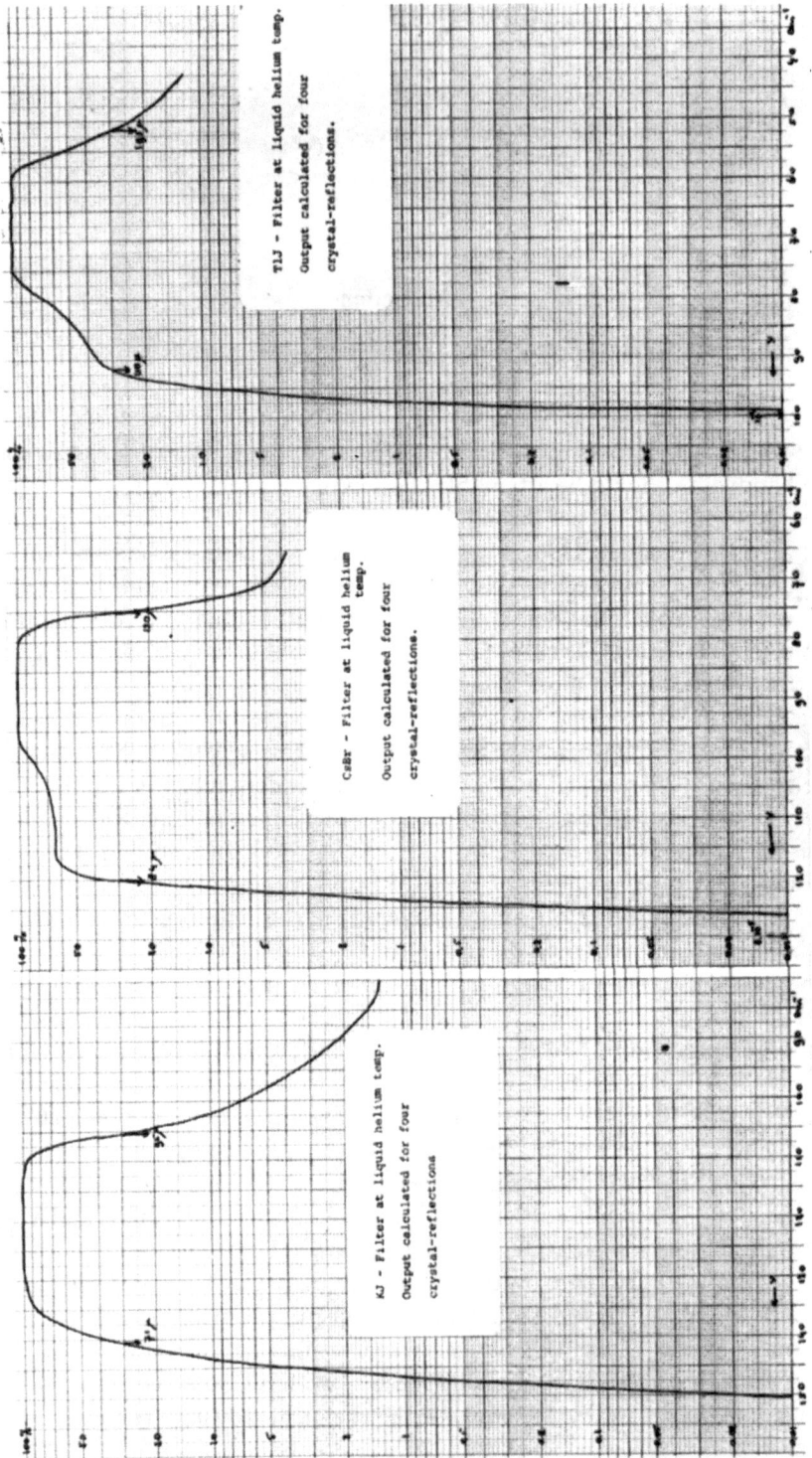

Fig. 1.   Reflection filters for the 1971 balloon photometer.

Fig. 2. Dewar optics with two filter systems and detectors.

2. The Apparatus

In our laboratory we use two Hitachi double beam spectrometers to cover the wavelength region from 2.5–50 μ. Above ca 30 μ we use a Grubb-Parsons cube interferometer. Both instruments can be adapted for reflection and transmission measurements at low temperatures.

In the Hitachi instrument the complete dewar (a Texas Instruments CLF-1) is placed in the sample beam after removing the cover of the source compartment. In the reference beam a tube is placed filled with dry air and provided with two polythene windows as used in the dewar. The sample holder designed for the interferometric work is mounted on a slide and contains two diaphragms one of which is used for the sample. In practice the possibilities of the reference are not utilized. We attenuate the reference to some extent and make a run with and without the sample obtaining a more accurate result.

For reflection measurements the dewar is placed on the adapted reflection attachment of the instrument and the beam passes through a window in the bottom. Now the sample slide is provided with the sample and a gold plate. The reference beam is attenuated to reach nearly full scale output for the gold plate and in fact a reflection of the sample is measured with respect to gold. The angle of reflection has to be ca. 12°.

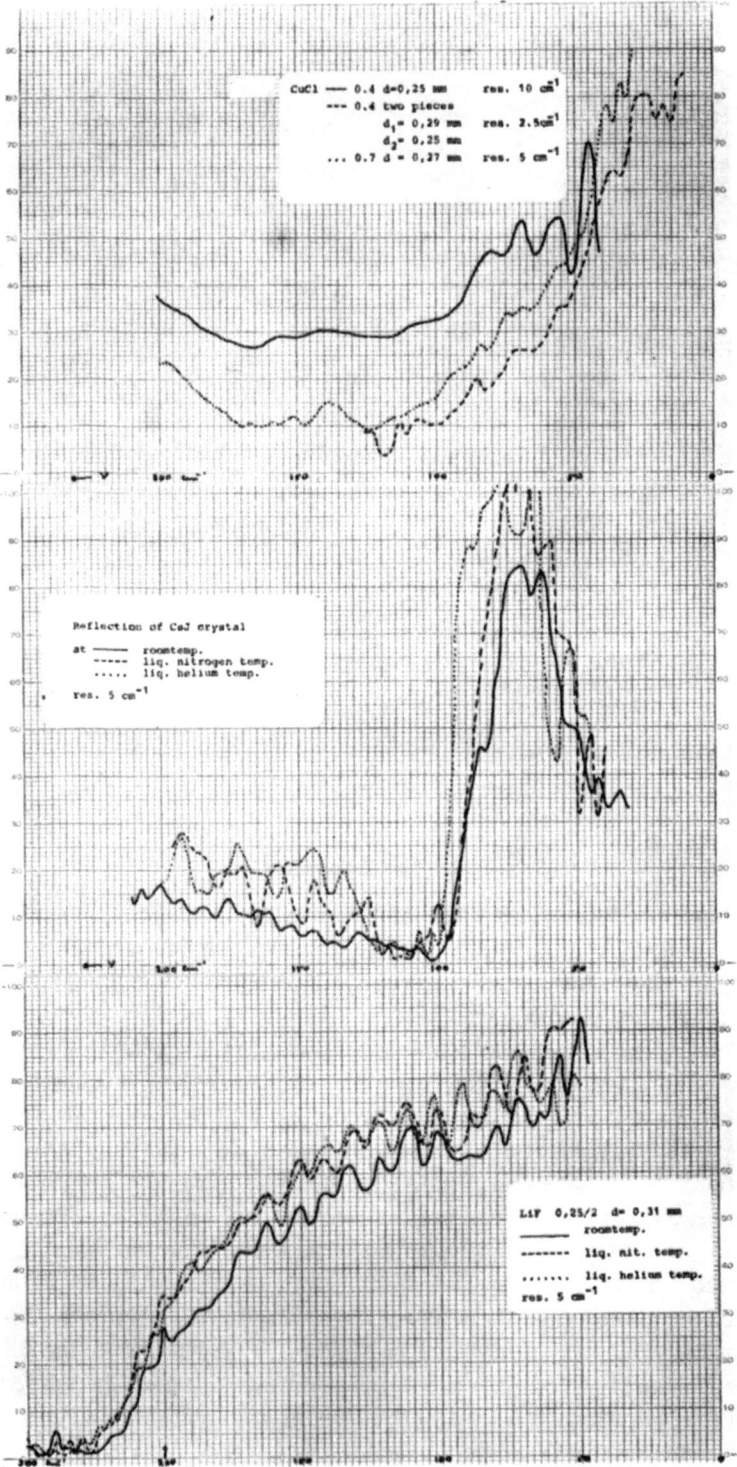

Fig. 3.   Typical curves of crystal reflection and powder-filters.

The interferometer is adapted for cold work by mounting the dewar between the cube and the detector and by changing the polythene lens to obtain a higher transmission of the instrument.

To perform the transmission measurement a slide with two diaphragms, one of which contains the sample, can be moved perpendicular to the beam. For the reflection measurements, which are performed at 45°, the detector is moved over 90° whereas another slide is mounted inside the dewar containing sample and goldplate. The set-up as arranged for reflection measurements can be seen in Figure 4.

Fig. 4. Laboratory set-up arranged for 45° reflection measurement.

In the standard interferometer the mirror is driven by a stepping motor over distances of 2.5 μ per step. So with one step in two seconds and a mirror travel of ± 0.5 mm around the zero path centre, 800 s are needed to perform a measurement with a resolution of 10 cm^{-1} (for our two-side analysis). This resolution is far too low for the longer wavelength region around 200 μ in which we are interested. Without spending more time at a larger mirror travel a higher resolution is easily obtained by driving the mirror over much larger distances per step. Therefore we developed an electronic multiplier, which can be switched into the circuit and which multiplies the trigger-pulses actuating the steppermotor circuit by a factor between one and ten. These pulses are applied within a very short time with respect to the original triggerpulse interval of the system.

The larger steps require an additional filtering to eliminate interfering short wavelengths. In principle with steps of $n \times 2.5 \mu$ wavelengths shorter than $\lambda_s = n \times 10 \mu$ should be completely blocked. This is accomplished by applying a suitable powder filter in addition to a 0.1 mm sheet of black polythene. Because of the relatively weak front edge of these filters oversampling is required. The spectrum calculation is designed to start at a λ_s where the radiation contribution can be neglected, but a reasonable energy output is only to be expected for much longer wavelengths.

After some tests a simple operation manual was arranged, given in Table I. The resolution, when spending the same measuring time, is proportional to the mirror step, in casu, using step 8 eight times higher than when using step 1.

TABLE I

Mirror step	λ_s	Beamspl.	Filter	Wavelength region
1–2.5 μ	10 μ	3.75 μ	Bl.Pol. (0.1 mm)	450–60 cm^{-1}
2–5 μ	20 μ	6 μ	Bl.Pol. + BeO 0.25/2	330–40
4–10 μ	40 μ	25 μ	Bl.Pol. + LiF 0.25/2	125–25
8–20 μ	80 μ	38 μ	Bl.Pol. + LiF 0.25/2 + PbF2 0.5/2	60–20 cm^{-1}

Except the black polythene of 0.1 mm (Bl.Pol.) the thickness of all filters has been ca. 0.25 mm. The concentration of the filter material has been given in terms determined by Yamada; for LiF: 0.25 g powder mixed with 2.0 g polythene indicated as LiF 0.25/2.

3. Reststrahlen Reflections

The crystals of which we measured the Reststrahlen reflection are partly reported in the literature (see e.g. [2] and [3]). As we were looking for attractive filters we tried to measure many materials and to select them for their spectral band, steepness of the short wavelength edge and the value of the maximum reflection at low temperatures. Until now we measured AgBr, AgCl, BaF2, CaF2, CsBr, CsJ, KCl, KJ, LiF, TlBr, and TlJ at cryogenic temperatures. Partly the spectra were measured on commercially available crystals, partly on pellets pressed under high temperatures from the powder. In Figure 5 we see a number of spectra of CsBr at different temperatures. As usual, the reflection becomes higher at lower temperatures whereas the short wavelength edge becomes very steep. The same features are more distinct in the curves of AgCl and AgBr reproduced in Figure 6. In Figure 7 spectra of some pellets (CsCl, PbJ_2, and TlCl) at room temperature are given.

Until now the temperature of the crystal surface could not be determined very precisely. Carbon resistors stuck behind the crystal surface during measurement indicate low temperatures but as the heat conductivity of the crystals and the heat supply of the filtered light of the mercury lamp is unknown, we cannot give a precise value.

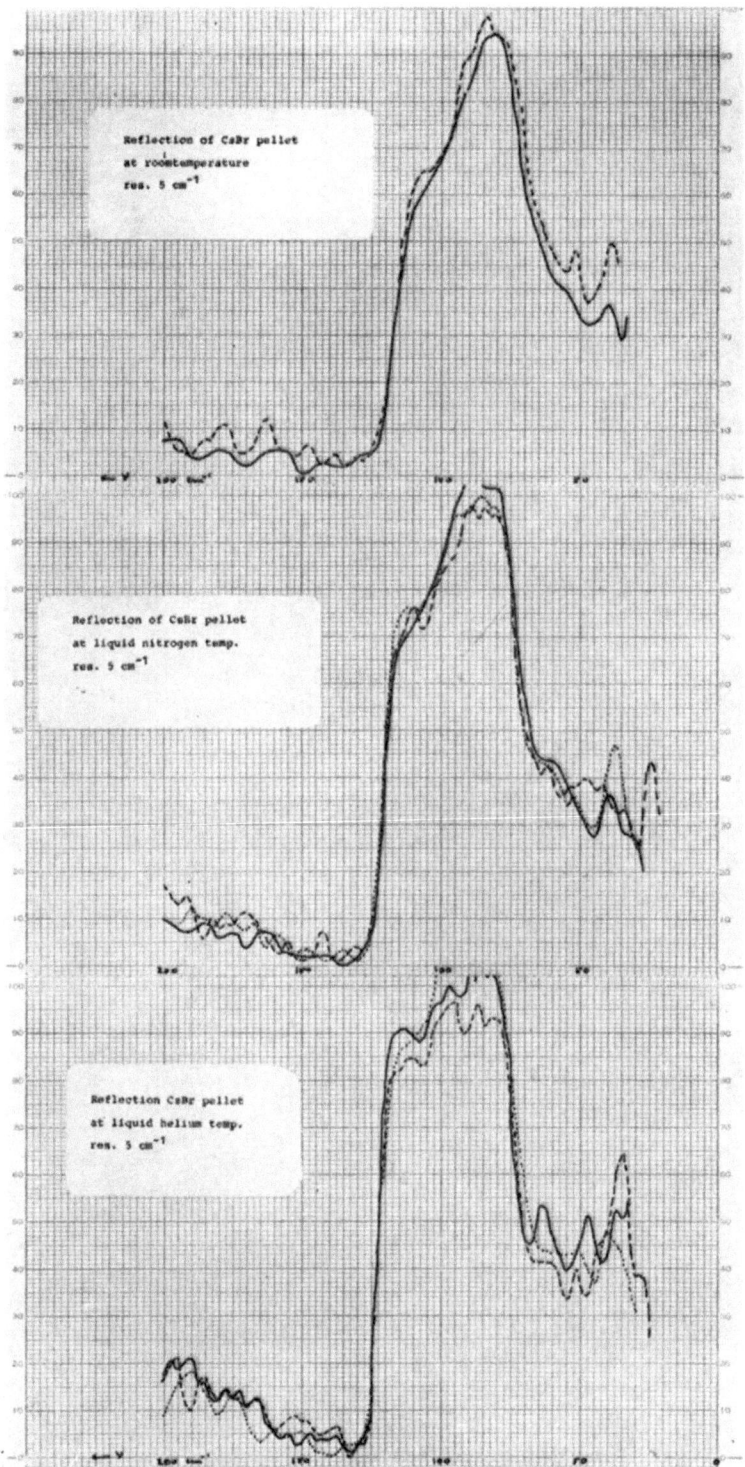

Fig. 5.   Reflection of CsBr pellet as a function of temperature.

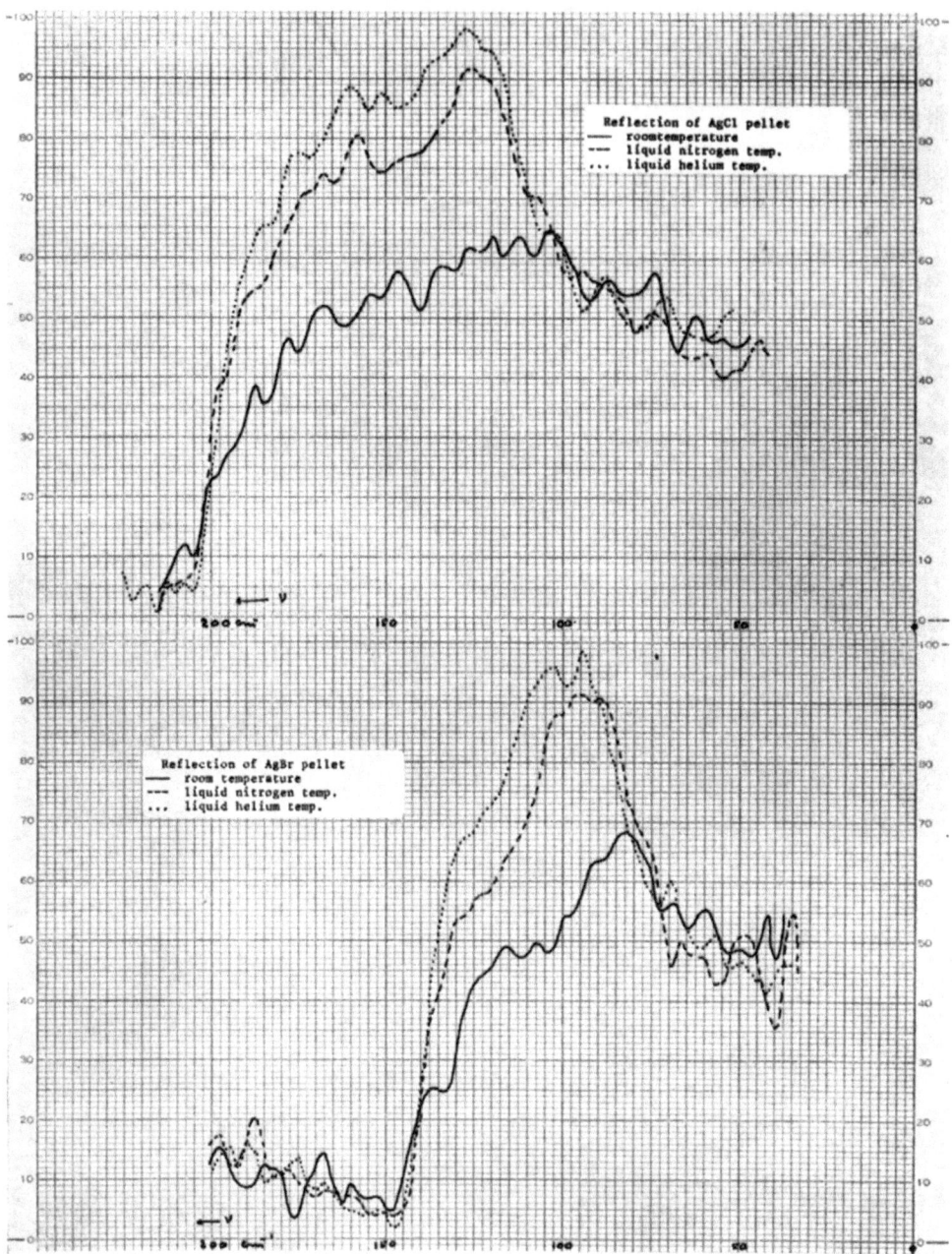

Fig. 6. Reflection of AgCl and AgBr as a function of temperature.

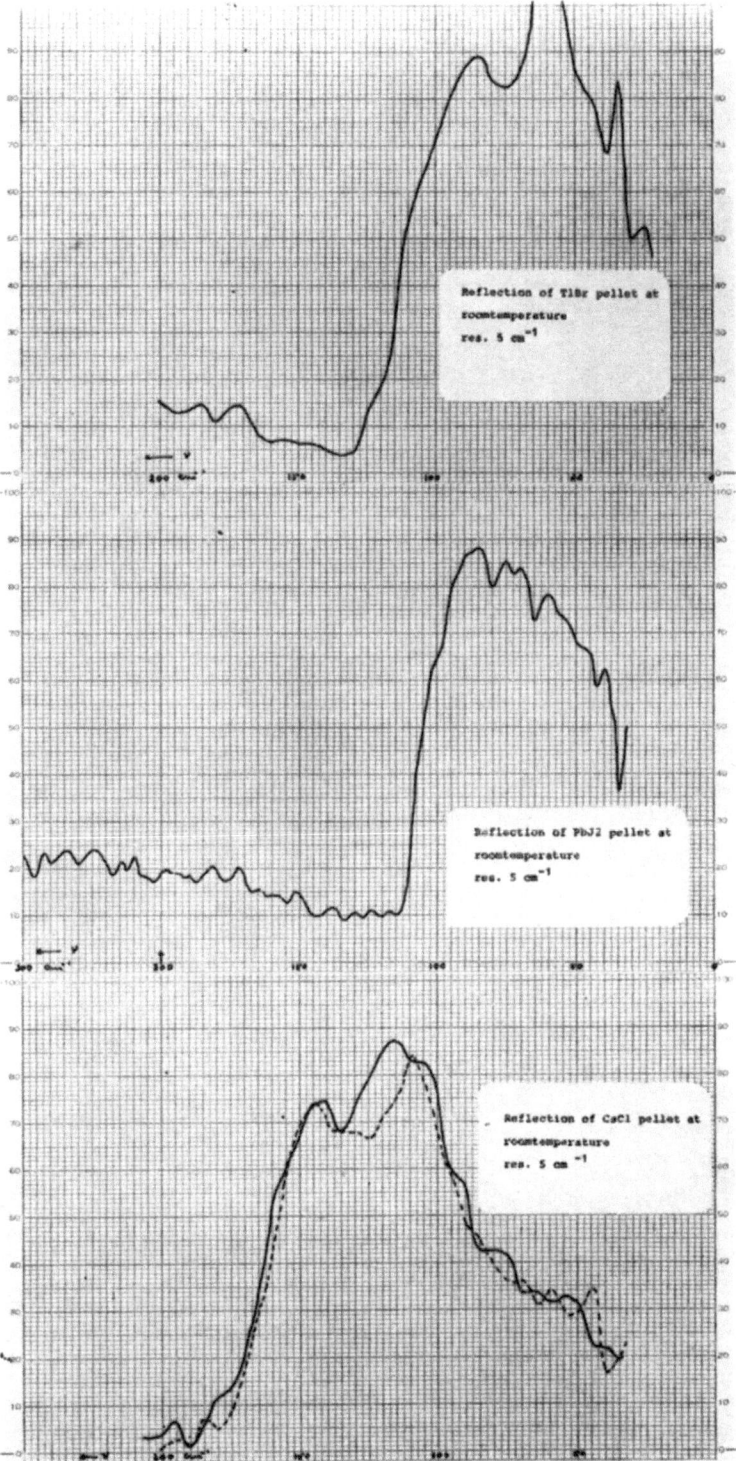

Fig. 7.   Reflection of some pellets for long-wavelength filtering.

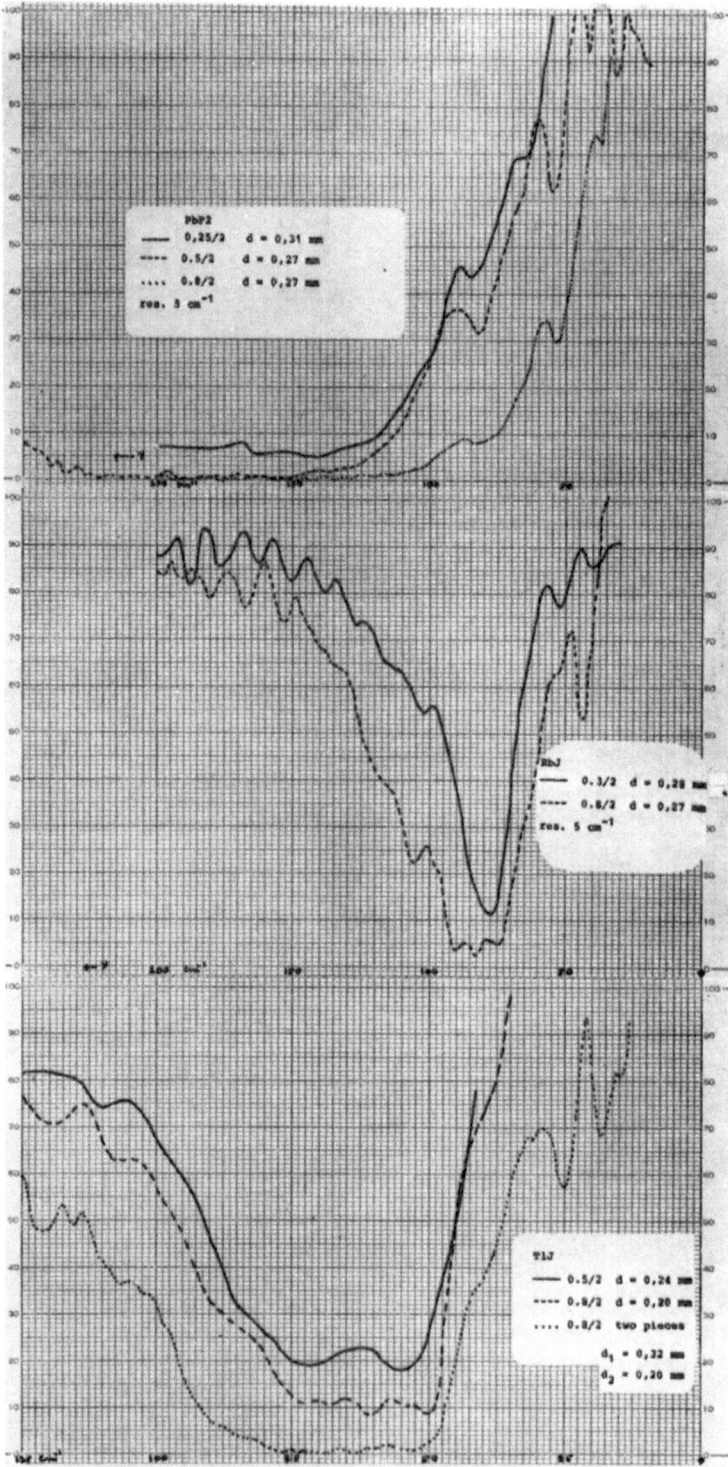

Fig. 8.   Transmission of powderfilters for various thicknesses and concentrations.

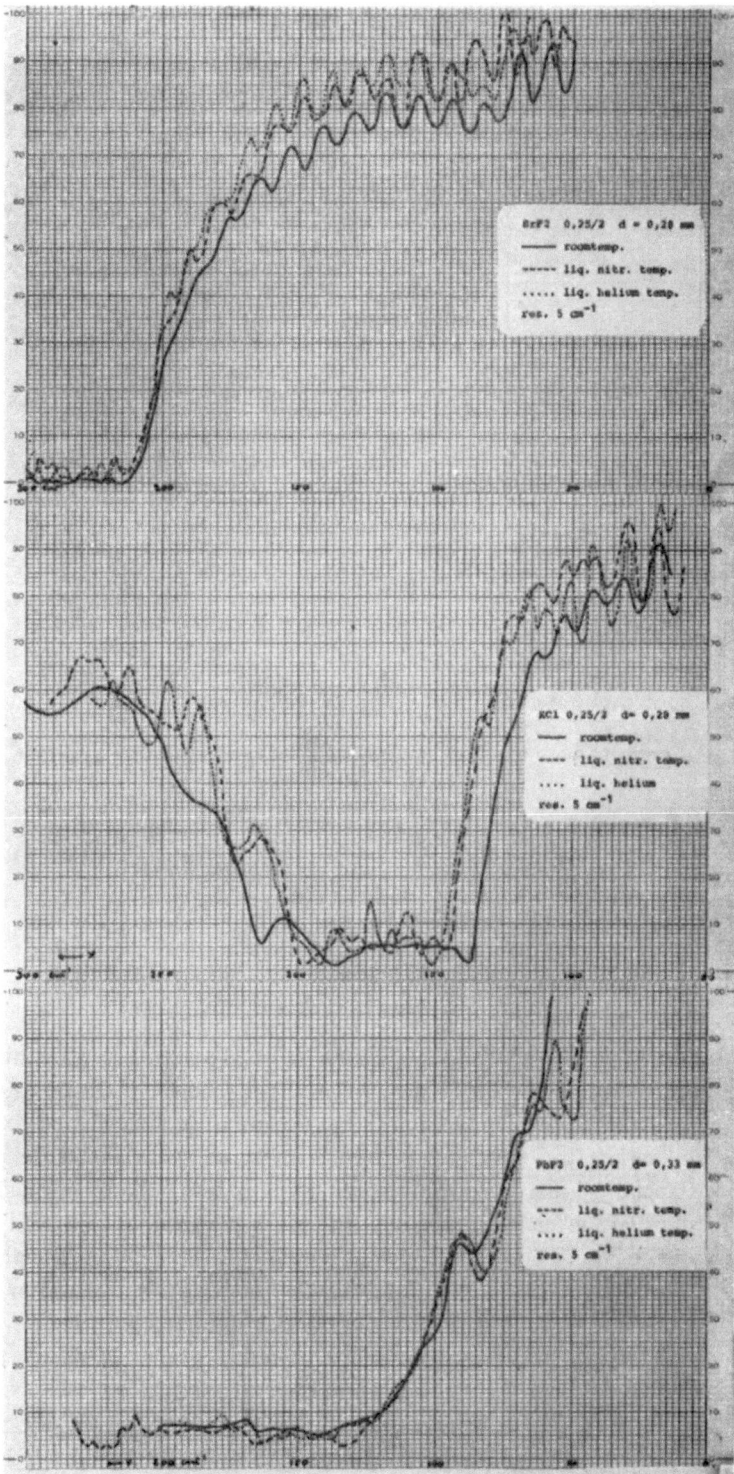

Fig. 9. Transmissions of powderfilters as a function of temperature.

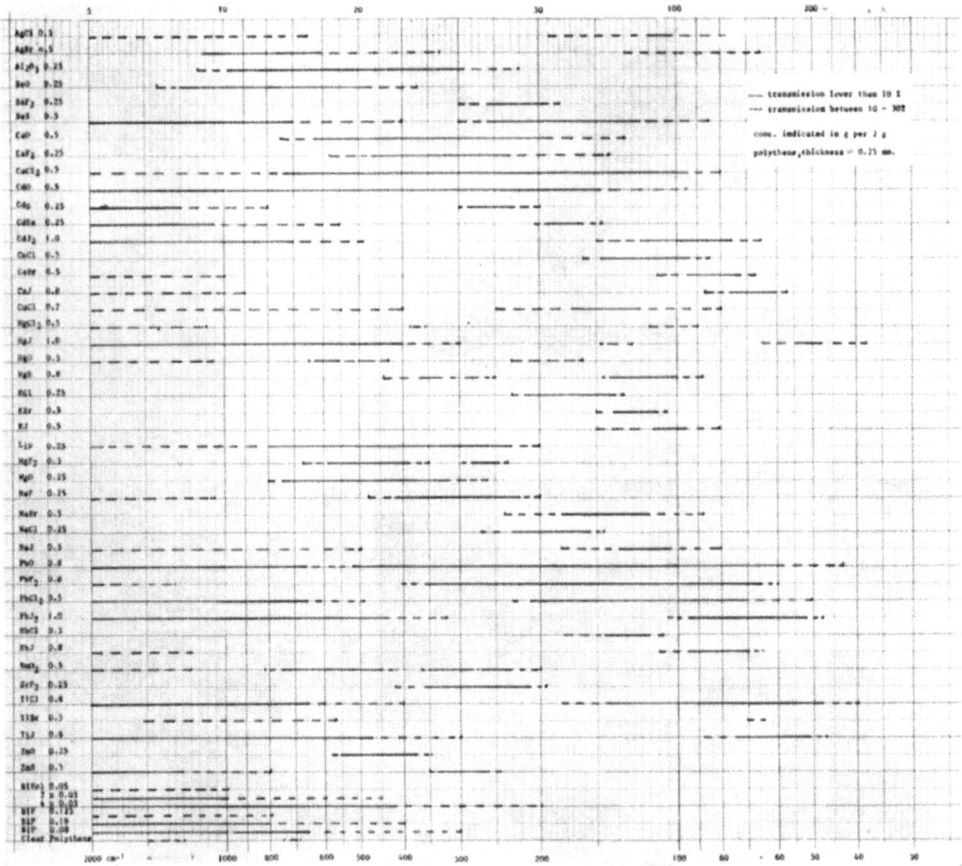

Fig. 10. Absorption chart for powderfilters.

4. Powder Filters

Based on the methods introduced by Yamada *et al.* [1] we made transmission filters by heating a mixture of polythene powder and the chosen material. After careful grinding and mixing the weighed quantities, the mixture was heated between two glass plates covered with a very thin oilfilm. The plates were therefore pressed between two electrically heated aluminium blocks. Cooling was accomplished by immersion in water and after some minutes the glass plates could be split very easily from the polythene. By applying a 0.25 mm wire as spacer between the glass plates a reasonably reproducible thickness of the filters has been obtained. For the lighter materials we chose usually concentrations of ca. 0.25 g mixed with 2 g polythene. For heavier materials higher concentrations gave better results as is shown in Figure 8 (PbF$_2$, RbJ, TlJ at different thicknesses and concentrations).

At lower temperatures the transmission curve changes, though less than in the case of the reststrahlen reflection. In Figure 9 results for KCl, PbF2, and SrF2 are shown.

This type of measurement will be continued specially because of the interesting results of Zwerdling and Theriault [4] which we could not reproduce hitherto.

For composing long wave pass filters for interferometric purposes we have made a chart giving the spectral limits for powder filters at transmissions of 10% and 30% (Figure 10). In this chart the composition of the filter is given as the amount of material added to 2 g polythene. In many cases a better result is obtained using still higher concentrations and therefore a more extensive survey is desirable.

Nevertheless the chart is very useful for the selection of powder materials on the basis of spectral transmission requirements. In addition an optimum concentration can be estimated.

5. Metal Mesh Filters

While playing with filters we also measured some commercially available sieves.* These are manufactured from metal foils of 30–50 μ thickness in which round holes are punched leaving a small cone rim. The transmission curve shows a remarkable high transmission and the bandwidth is relatively small (see Figure 11). Indicated is the diameter of the punched holes. Maximum transmission is obtained at a wavelength of two times the hole diameter. Shifting can be accomplished by immersion of the metal mesh in polythene powder, or in a powder filter. In future we hope that

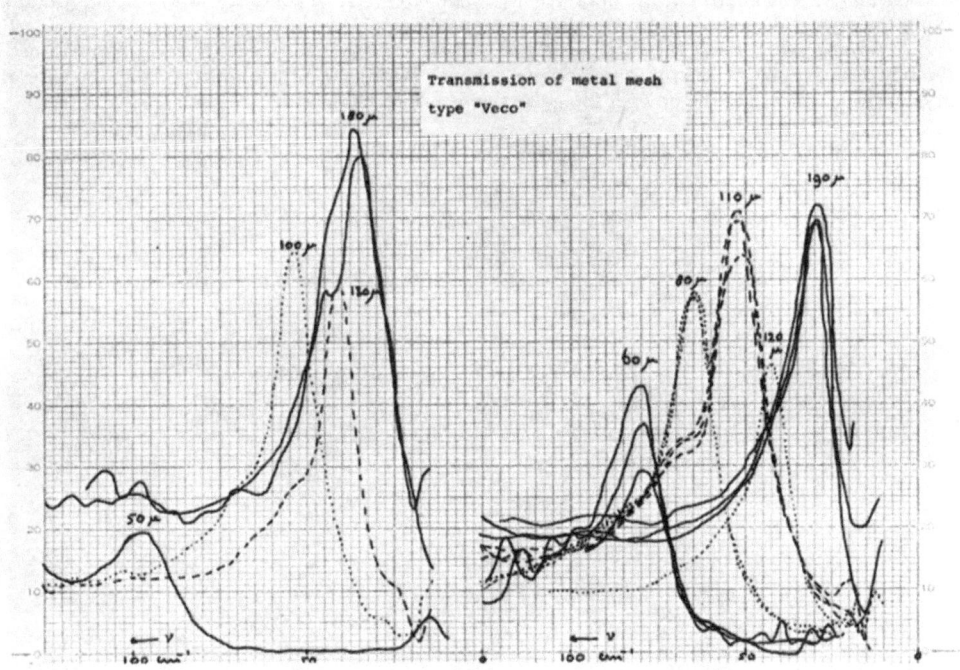

Fig. 11.   Transmission of various sieve-filters.

* Manufacturer: Veco – Eerbeek, Holland.

these techniques can be developed to obtain filters which are suitable for narrowband photometry in the far infrared.

Acknowledgement

For the work presented here we are very grateful to Dr H. J. van Linde of the computation centre for his assistance in the evaluation of the interferometer data. We also thank Prof. Dr J. Borgman for his interest and advice, J. Doornenbal of the workshop for all minor alterations in the apparatus, and D. Snel for his work in adapting the electronic circuitry of the stepping device.

References

[1] Yamada, Y., Mitsuishi, A., and Yoshinaga, H.: 1962, *J. Opt. Soc. Am.* **52**, 17.
[2] McCarthy, D. E.: 1963, *Appl. Opt.* **2**, 591 (part I); *Appl. Opt.* **4** (1965), 317 (part III).
[3] Hadni, A., Claudel, J., Morlot, G., and Strimer, P.: 1968, *Appl. Opt.* **7**, 161; and *Appl. Opt.* **7**, 1159.
[4] Zwerdling, S. and Theriault, J. P.: 1968, *Appl. Opt.* **7**, 209.
[5] Vergnat, P., Claudel, J., Hadni, A., Strimer, P., and Vermillard, F.: 1969, *J. Phys.* **30**, 723.

DISCUSSION

F. Melchiorri: The sharp cut off obtained by a single crystal in reflection may also be obtained using the crystal dispersed in polyethylene in transmission.

J. J. Wijnbergen: I think they are weaker and also the edge is not dependent on temperature in the powder case. We've seen that reflectivity depends strongly on the temperature in the case of a crystal alone. In case of the powder filter this seems not to be the case.

SUBMILLIMETRE AND MILLIMETRE MAPPING
USING INTERFERENCE FILTERS

P. A. ADE and J. A. BASTIN

Queen Mary College, London, U.K.

During the last ten years much consideration, especially at conference discussions, has centred on the relative advantages of transform interferometry as opposed to filters for astronomical measurements in the far-infrared wavelength range. The advantages of the use of ground based sites as opposed to balloon or rocket-borne platforms have also been discussed. Of the four possible combinations thus presented, three have recently been used successfully whilst the remaining possibility may well produce results in the very near future. Thus rocket and balloon borne filter instruments have been used by McNutt *et al.* [1], Harwit *et al.* [2], and Hoffmann *et al.* [3] to investigate the microwave background and map the galactic centre. Ade *et al.* [4] have used filters for solar mapping from ground-based sites. The use of transform interferometry has up to the present been less successful in respect of submillimetre astrophysical results. Nolt, using a ground-based site has detected an emission line centred at 11.7 cm^{-1} which could be of extraterrestrial origin whilst successful balloon- or rocket-borne transform interferometry is still awaited.

It is the purpose of this paper to show that the use of ground-based telescopes with interference filters centred on the various submillimetre windows represents a very satisfactory method of mapping sources and could represent an efficient means of point source detection in this wavelength range.

The great advantages of the use of ground-based sites is that for any given expenditure a far larger telescope can be used than could be rocket- or balloon-borne. If D is the linear dimension of the telescope the number of spatially resolvable elements in a given area of sky increases as D^2 and the intensity has the same dependence for point source observations. In addition, the techniques for telescope pointing from ground-based sites are well known.

We shall first consider the associated problems of atmospheric absorption and filter construction. The results of the application of the use of filters to solar scanning will be discussed and we shall conclude with a general discussion of the ground-based use of filters for mapping and point source detection.

1. Atmospheric Absorption

Pure rotational water vapour transitions give rise to the only important atmospheric absorption in this range. Relatively narrow-wavelength ranges of about 50 μm width between the more intense lines have sufficiently low absorption to allow appreciable atmospheric transmission to ground based sites. Table I lists these regions of absorption minima.

Manno and Ring (eds.), Infrared Detection Techniques for Space Research, 257–264. All Rights Reserved
Copyright © 1972 by D. Reidel Publishing Company, Dordrecht-Holland

TABLE I
Atmospheric absorption [5]

Wavelength (μm)	360	460	740	860
Minimum absorption dbs/km standard atm. experimental (standard atm. H$_2$O vapour 7.5 $\times$ 10^{-2} kg m^{-3})	50 $\pm$ 5	50 $\pm$ 5	12 $\pm$ 4	7.5 $\pm$ 1.5
Permissible precipitable water vapour for contour mapping kg m^{-2}	2.0	2.0	6.0	10

The choice of a suitable site is thus strongly controlled by the precipitable water vapour concentration, and for the windows centred at 360 μm and 460 μm either high altitude or high latitude sites are essential. The advantage of high altitude sites as a means of reducing the water vapour path length is obvious. The choice of sites of relatively high latitude is also made clear in Figure 1. Here sites have been chosen

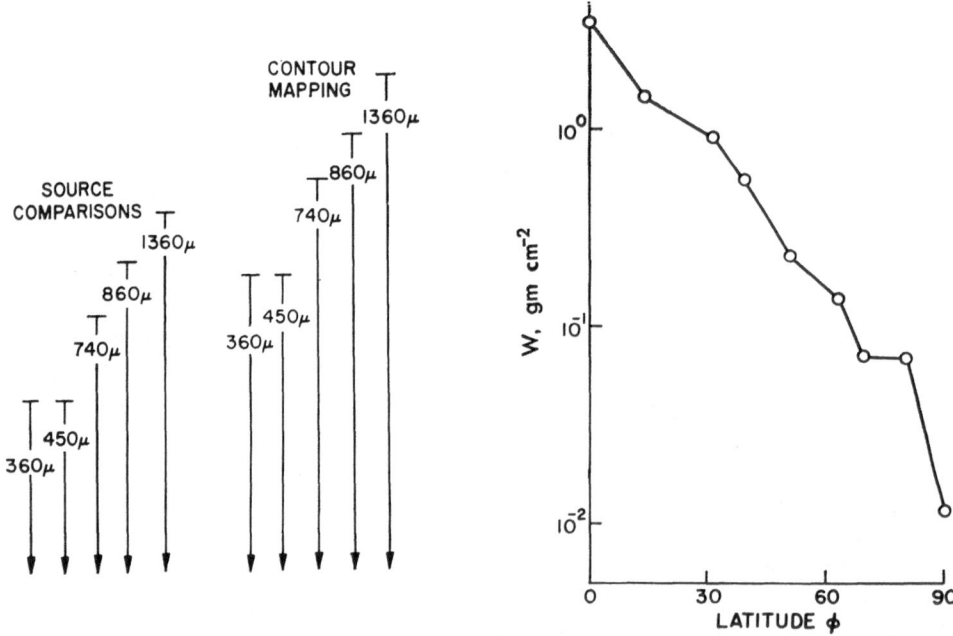

Fig. 1.   Precipitable water vapour for sites at different latitudes. In each case a dry site at a reasonably low altitude has been chosen. The right hand side of the figure refers to 3 db and 10 db criteria considered appropriate respectively for the maximum permissible atmospheric attenuation for source comparison and scanning observations (see [5]).

from the equator to the poles at roughly 10° intervals in latitude and at dry but relatively low sites. There is a very noticeable change, by a factor of about 200, in the precipitable water vapour content above the polar and equatorial sites, although of course the fraction of accessible sky for near polar sites is more limited.

As a general guide it may be concluded that a site is suitable for mapping measurements in all four windows when the dew point is below −15 °C. A more thorough discussion of the suitability of sites has been given by Gaitskell *et al.* [5]. Limitations resulting from atmospheric emission will be discussed in the final section.

2. Filters

Interference filters of the type first described by Renk and Genzel [6] represent a very convenient way of selecting the radiation transmitted in one or two closely proximate submillimetre atmospheric windows. Ideally the filter should be constructed so as to transmit all the radiation within such a window and cut out all radiation outside this wavelength range. Fortunately it is possible to vary the parameters of the filters used so that a reasonably good approximation to this criterion is obtained.

We now describe the characteristics of some filters developed at Queen Mary College and used during January 1971 for submillimetre solar mapping and we will also describe some improvements to their characteristics which have been made very recently. The transmission properties of these filters are shown in Figures 2 and 3.

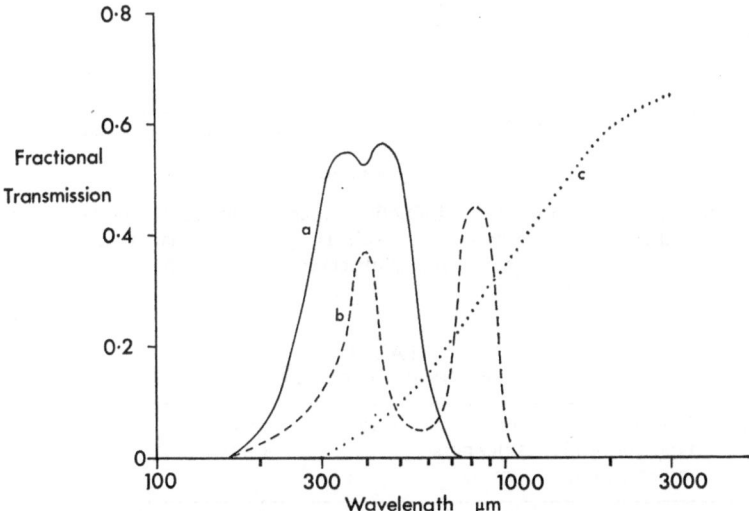

Fig. 2. Transmission characteristics of filters used for submillimetre and millimetre mapping. (a) 300–500 μm, 3-mesh filter: all elements of inductive type together with alkali halide filters, (b) 700–900 μm, 3-mesh filter: all elements of inductive type and 2 mm of carbon-impregnated polythene, (c) 6 mm carbon impregnated polythene. For further details of each of these filters see Table II.

Each filter consists of three parallel metal mesh grids constructed specifically to isolate wavelength regions defined by the limits 300–500 μm and 700–900 μm respectively. In each case ancillary filtering of alkaline halides [7] and carbon impregnated polythene [8] were used. The results were further enhanced in wavelength selectivity as a consequence of the short wavelength sensitivity cut-off of the InSb detector used in the Kinch-Rollin [9] mode. In the case of measurements centred at 1200 μm only carbon impregnated polythene was used.

The construction properties of the interference filters is shown in Table II. The first 700–900 μm filter shown in this table, whose wavelength transmission characteristics

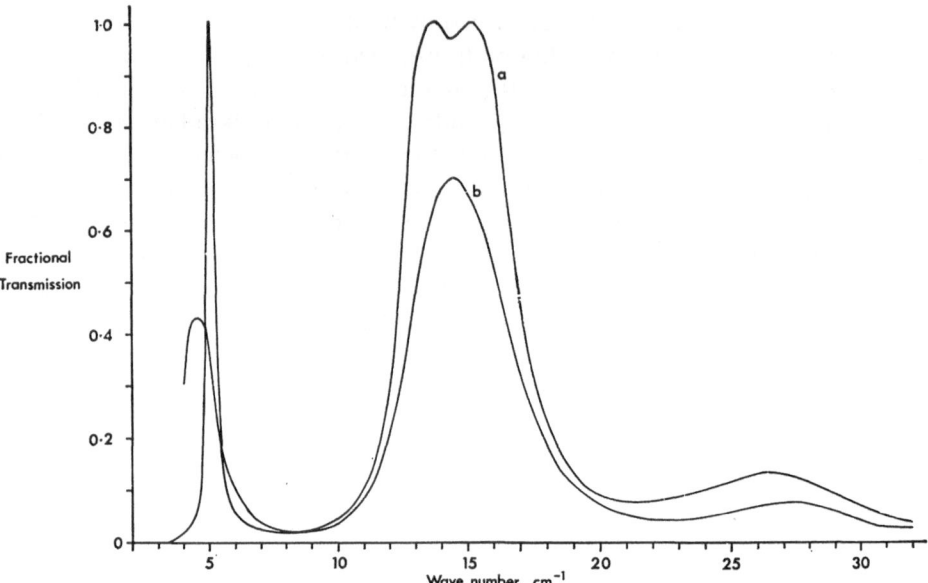

Fig. 3. Fractional transmission of an inductive-capacitive-inductive mesh. The details of the meshes used and their spacing is shown in Table II. (a) Transmission calculated from theory, (b) Experimentally observed transmission.

TABLE II
Geometrical properties of filters

Wavelength range μm	Grid 1 type and grid interval (μm)	Spacer thickness μm	Grid 2	Spacer thickness μm	Grid 3	Associated filtering
300–500	inductive 254	50	inductive 254	50	inductive 254	Alkali-halide
700–900	inductive 318	200	inductive 212	200	inductive 318	2 mm carbon impregnated polythene
700–900	inductive 318	200	capacitive 212	200	inductive 318	None
800–3000						6 mm carbon impregnated polythene

are illustrated in Figure 2 is seen to possess an unfortunate short wavelength transmission peak. This only affected the measurements to a negligible extent: partly because of the ancillary filtering used and partly because the solar energy transmitted at 800 μm was about an order of magnitude greater than that in the region centred at 400 μm. However, this somewhat unsatisfactory situation has now been corrected by the use of a central capacitative mesh within the filter. The properties of this improved filter are shown in Figure 3 and Table II.

In developing these filters we have been guided by theoretical methods originally developed by Marcuvitz [10] and Ulrich [11]. The use of an on-line computer facility enables the predicted wavelength transmission characteristics of any combination of

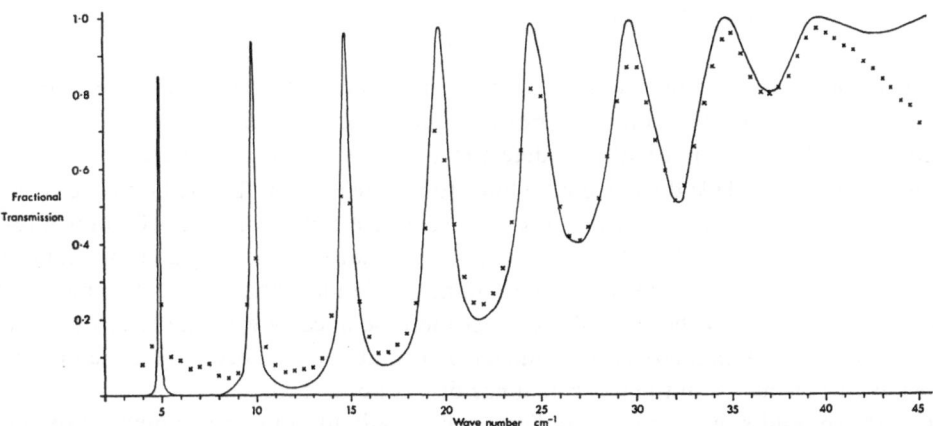

Fig. 4.   Transmission characteristics for a two grid etalon. The grids have a spacing of 750 μ and a grid constant of 212 μm. The continuous line shows the calculated transmission: crosses show observed values.

grids to be made within minutes, thus facilitating rapid development of new filters. In all cases the experimental transmission properties have been measured with the polarizing interferometer constructed and described by Martin and Puplett [12]. An example of the comparison between prediction and theory for a filter of two identical meshes is shown in Figure 4.

3. Observation

The above considerations of filter characteristics and atmospheric absorption show that it should be possible to use a site at about 3×10^3 m altitude at temperate latitudes to make useful mapping measurements in each of the windows shown in Table I. We now consider the angular resolution and detectivity which may be obtained using a conventional reflecting telescope at such a site. If the telescope area is A and the wavelength interval of the window is $\Delta\lambda$ throughout which the fractional atmospheric transmission is α then the energy received by the telescope in solid angle $d\Omega$ from an extended source of brightness temperature T is

$$E = B_\lambda(T)A\ \Delta\lambda\ \alpha\ d\Omega \tag{1}$$

where $B_\lambda(T)$ is the Planck function for a body at temperature T. We now assume that the optics of the telescope detector system are so arranged that the detector receives radiation from a small solid angle of sky of diameter just equal to the smallest interval which can be shown to be resolved on the basis of diffraction theory. This is clearly an optimum in the sense that, if we increase the acceptance angle of the detector, spatial detail is lost; whilst if we decrease this angle, the energy received by the detector decreases as well, without any appreciable increase in angular resolution. Writing this condition together with the low frequency approximation to $B_\lambda(T)$ Equation (1) becomes:

$$E = 1.83kcT\lambda^{-2}\ \Delta\lambda\ \alpha \tag{2}$$

where k and c are Boltzman's constant and the velocity of light (the dimensionless numerical factor 1.83 is a product of Bessel functions and powers of π).

Equation (2) shows that with the detector already described at this conference by Clegg and Huizinga [13] the detectable limit for a one second observation time corresponds to a temperature change of 5 K: we have here considered the 350 μm window and have set $\alpha = 0.1$ and $\lambda = 50$ μm. It is interesting that Equation (2) does not contain the telescope collecting dimensions. The number of independent spatial elements in a map of any region of the sky will however increase linearly with the telescope area. Thus for example with a telescope diameter of 1 m at 350 μm wavelength the number of resolvable elements on the solar or lunar disc is about 500.

The above results may now be applied to the possibility of solar or lunar mapping, the above data representing realistic estimates of equipment at present available for high altitude work. In the case of the Sun we would expect from an extrapolation of the work of Clark and Park [15] that the variations of brightness temperature associated with localized enhancements is of the order of 200–400 K. Unless there is appreciable loss of energy in the optical system or excessive atmospheric noise we see from Equation (2) that mapping of the enhancements is possible with adequate signal to noise and without loss in angular resolution. In the case of the Moon there are of course global temperature variations which are very much greater than the 5 K differential calculated above although the interesting detail corresponding for example to the contrast between highland and mare regions is only expected to be of this order.

Measurements made recently at the Pic-du-Midi Observatory do in fact bear out these predictions. The results have very recently been published [4] and therefore will not be described in detail. Briefly however these workers were not able to obtain sufficient signal-to-noise in their lunar scans to make worthwhile maps. Solar mapping with maximum angular resolution was however possible and the results are shown for three wavelengths in Figure 5 together with a Fraunhofer map for the same date. The enhanced regions are seen to correlate well with plage and sunspot areas and result from radiation originating in the lower solar chromospheric layers.

The above discussion has not considered the effect of noise due to fluctuations in sky emission and absorption. Park [14] has shown that this decreases rapidly with increasing frequency. For this reason Ade *et al.* [4] found it necessary to scan across the solar disc within about two seconds.

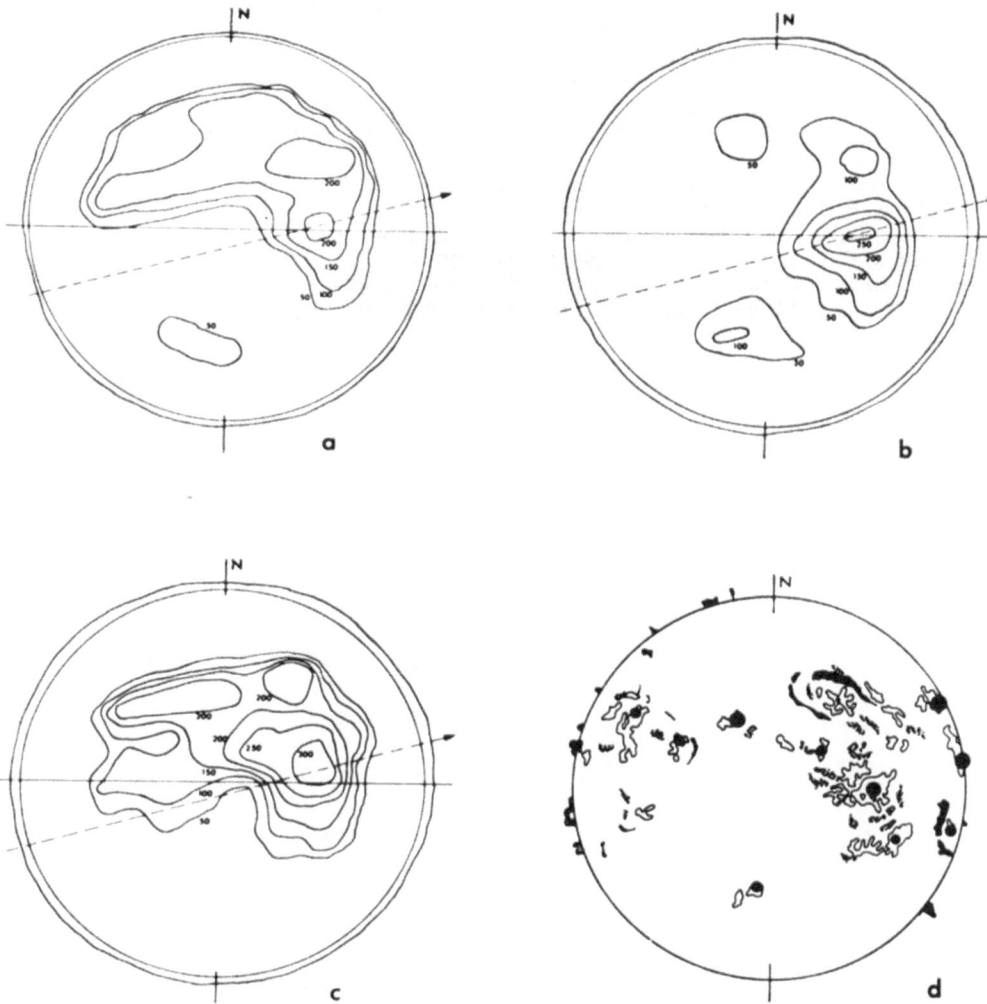

Fig. 5.   Solar far-infrared and visual features Feb. 8th, 1971. Maps (a), (b), and (c) refer respectively to scanning, in the direction of the arrow, at 400 μm, 800 μm, and 1200 μm [4]. The Fraunhofer map (d) shows sunspots ●, plages ⌇⌇, prominences ⬥. There is seen to be a strong correlation between the far infrared enhancements and the plage and sunspot visual features. This correlation has been previously noted in the case of millimetre enhancements [15, 16].

Acknowledgements

We wish to thank Professor D. H. Martin for providing the polarising interferometer and Mr E. Puplett for his assistance.

References

[1] McNutt, D. P., Shivanandan, K., and Zajac, B. J.: 1966, *Astron. J.* **71**, 1026.
[2] Harwit, M., Houck, J. R., and Fuhrmann, K.: 1969, *Appl. Opt.* **8**, 473.
[3] Hoffman, W. F., Frederick, C. L., and Emery, R. J.: 1971, *Astrophys. J.* **164**, L23.
[4] Ade, P. A. R., Beckman, J. E., and Clark, C. D.: 1971, *Nature Phys. Sci.* **231**, 55.
[5] Gaitskell, J. N., Newstead, R. A., and Bastin, J. A.: 1969, *Proc. Roy. Soc.* **A264**, 195.
[6] Renk, K. F. and Genzel, L.: 1962, *Appl. Opt.* **1**, 643.
[7] Yamada, Y., Mitsuismi, A., and Yoshinaga, H.: 1962, *J. Opt. Soc. Am.* **52**, 17.
[8] Blea, J. M., Parks, W. F., Ade, P. A. R., and Bell, R. J.: 1970, *J. Opt. Soc. Am.* **60**, 603.
[9] Kinch, M. A. and Rollin, B. V.: 1963, *Br. J. App. Phys.* **14**, 672.
[10] Marcuvitz, N.: 1951, *Waveguide Handbook*, M.I.T. Rad. Lab. Ser., McGraw-Hill.
[11] Ulrich, R.: 1967, *Infrared Phys.* **7**, 37.
[12] Martin, D. H. and Puplett, E.: 1969, *Infrared Phys.* **10**, 105.
[13] Clegg, P. E. and Huizinga, J.: 1971, this volume, p. 132.
[14] Park, W. M.: 1971, Ph.D. Thesis, University of London.
[15] Clark, C. D. and Park, W. M.: 1968, *Nature* **219**, 922.
[16] Beckman, J. E. and Clark, C. D.: 1971, *Solar Phys.* **16**, 87.
[17] Beery, J. G., Martin, T. Z., Nolt, I. G., and Wood, C. W.: 1971, *Nature* **230**, 36.

6. INTERFEROMETERS

INTERFEROMETRIC SPECTROMETRY FOR INFRARED ASTRONOMY

D. H. MARTIN

Dept. of Physics, Queen Mary College, London, U.K.

In studies of the reflectivities and absorptivities of laboratory samples at frequencies below about 400 cm^{-1}, the spectrometric technique which uses a two-beam interferometer has shown clear advantages over dispersive (grating) spectrometry, and the most sophisticated interferometers can in many respects better dispersive instruments at higher frequencies than this, up to $10\,000 \text{ cm}^{-1}$. In infrared astronomy, too, similar advantages can be gained in the right circumstances (as shown by the impressive spectra of planetary atmospheres recorded by P. and J. Connes [1] and by the satellite-based measurements of atmospheric emission described by Hanel [2]). However, these advantages may not be as readily achieved in astronomy and it is the special considerations arising in applying interferometers to astronomical spectrometry which I wish to examine in this paper.

First let me list the spectrometric advantages claimed for two-beam interferometry. Figure 1 illustrates a two-beam interferometer of the Michelson configuration and the

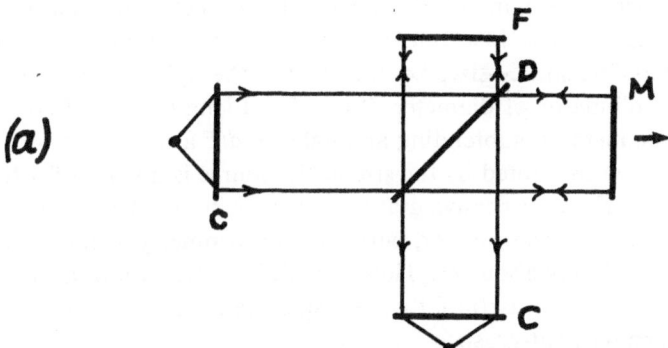

(a)

Fig. 1a. Michelson configuration for a two-beam interferometer. C, C source-collimator and detector-condenser; D beam-divider; F the fixed and M the moving mirror.

interferogram which is recorded by the detector as one mirror moves so as to vary the path-difference, d, between the two beams, up to a maximum value d_m. The Fourier transform of the interferogram gives a spectrum of the incident radiation with a resolution interval $\Delta\nu'$ equal to about $1/d_m$. This spectrometric technique thus has the complication of having to compute a Fourier transform, but it is claimed that one gains over dispersive spectrometry in the following ways:

(a) a greater light-grasp (and therefore a larger signal) at a given resolution [3];

Manno and Ring (eds.), Infrared Detection Techniques for Space Research, 267–280. All Rights Reserved
Copyright © 1972 by D. Reidel Publishing Company, Dordrecht-Holland

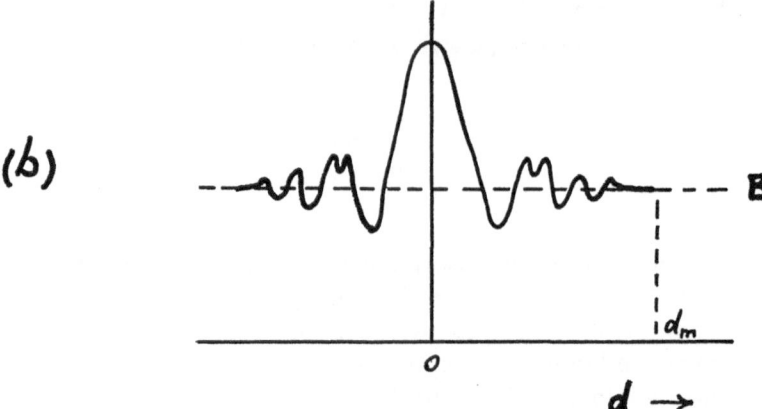

(b)

Fig. 1b. To illustrate an interferogram recorded by the detector as M moves to vary the path-difference, d, up to a maximum value d_m. The interferogram is symmetrical about the zero value of path-difference. The broken line labelled B is the 'background level' referred to in the text.

(b) a greater signal-averaging time (and therefore a lower noise-level) for a given recording time [4]; and

(c) a wider free spectral range without change of optical components.

These are, of course, advantages which must appear most attractive to astronomers but it is necessary to be clear about the circumstances in which these advantages accrue. I shall take each in turn.

(a) First consider the origin of advantage (a). The better the spectral resolution required of a spectrometric instrument, whether dispersive or interferometric, the smaller the permissible source-size, i.e. the smaller the light-grasp. Consider a spectrometer with a collimator of diameter D and focal length F (see Figure 2), and an effective source diameter d, subtending an angle $\delta = d/F$ at the collimator. The maximum resolving power as limited by the size of the source is different for the two types of spectrometer [3]. For a dispersive grating spectrometer the best $\Delta\nu/\nu$ is about equal to δ. For a two-beam interferometer, with circular symmetry around the beam axis, the best attainable $\Delta\nu/\nu$ is about δ^2. Thus, for a given resolution, an interferometric instrument permits a larger useful δ than a dispersive instrument, i.e. a larger source, which means a greater light-grasp.

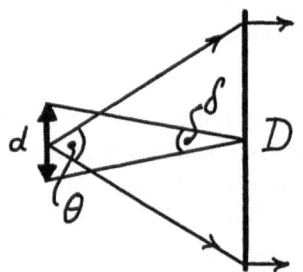

Fig. 2. To illustrate a collimator of diameter D with an effective-source of diameter d subtending an angle δ at the collimator.

However, it will be clear that a light-grasp advantage will be gained with an interferometer only if the sizes of the source and of available detectors are sufficient for the light-grasp to reach a greater value than that which would give the required resolution with a dispersive spectrometer. This might be less often the situation with the weak, small sources usual in astronomy than with the bright sources and large sources and samples of laboratory spectrometry. The same point can be put in a different way: the given source or available detector sets a limiting (maximum) value, δ_m, for δ, and only if the required resolution $\Delta\nu/\nu$ is smaller (better) than δ_m will an interferometer offer an advantage in light-grasp. The full advantage is gained only if the required resolution is δ_m^2.

To get quantitative guidance, consider Figure 3 which illustrates the dimensional parameters of a collector-spectrometer system and some of the relations between them.

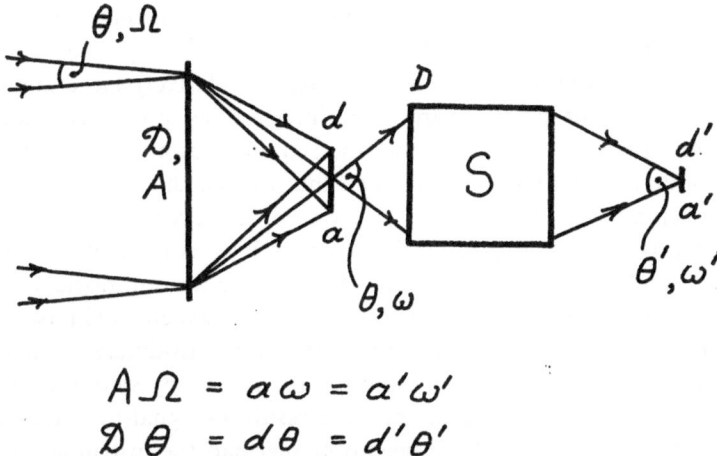

$$A\Omega = a\omega = a'\omega'$$
$$\mathcal{D}\theta = d\theta = d'\theta'$$

Fig. 3. To illustrate a collector of diameter $\mathcal{D}$ and area A which forms an effective-source of diameter d and area a from radiation which is incident within an angle Θ, solid angle Ω. The effective-source illuminates a spectrometer (dispersive or interferometric) S, which subtends an angle θ, solid angle ω, at the effective-source. The radiation is then focused on a detector with diameter d', area a'. The equalities in the figure are such that all the energy incident on A within Ω falls on a' within ω' (or vice-versa). The product $A\Omega$ is the 'light-grasp'.

First it should be noted that, except for astronomical objects of large angular diameter, say greater than 1°, the limit to the light-grasp is usually the difficulty of constructing a large collector rather thån obtaining a large detector. For example, a collector of 30 cm diameter and a source of 1° angular diameter, gives a light-grasp $A\Omega$ of about 0.4 cm² sterad; this can be matched by infrared detectors, most of which* have areas

* Though, as I have remarked above, most detectors in use have light-grasps of about 0.3 cm² sterad, the signal-output from a detector of a given type may depend only weakly on its area for a given flux of incident power and I am not at all sure that sufficient attention has been paid to the question of what the optimum detector-area is in the present context. This is, however, outside the scope of the present talk and the answer will involve details of construction (the thermal capacities of leads and solder, etc.).

of about 3 mm × 3 mm and acceptance angles of about 3 sterad, i.e. a light-grasp $a\omega$ of about 0.3 cm² sterad. To fill a similar light-grasp with a source of angular diameter 3′ would require a collector of diameter 600 cm. Now, assuming that it is the collector which sets the limit to the light-grasp, the relations in Figure 3 lead to the following form for the condition expressed above: the required resolution must be better (smaller) than $\delta_m = (\mathscr{D}/D)\theta$ if an interferometer is to give a light-grasp advantage, and as good as $\delta_m^2 = (\mathscr{D}/D)^2\theta^2$ for the full light-grasp advantage to be realized. If D is, say, 10 cm (a 'large' value), and $\mathscr{D}$ is 30 cm (also a 'large' value) then, for a source of angular diameter 3′, the required resolution which would make interferometry advantageous in terms of light-grasp is 1 in 300. That which would give the *full* advantage is 1 in 900. For a source of angular diameter 1° the corresponding figures are 1 in 16 and 1 in 256. Thus, for large sources the interferometric method gives a light-grasp advantage at quite modest resolutions, but for smaller sources the advantage is gained only at relatively high resolutions.

The spectra recorded by P. and J. Connes [1] were of planetary (small) sources but were at high resolution (1 in 10⁶); those recorded by Hanel [2] were of relatively low resolution but involved an atmospheric source of relatively large angular diameter. In both cases the light-grasp advantage of interferometry was realized. But, clearly, there are astronomical problems where the required resolution, or the resolution as limited by source-brightness, is lower than that for which interferometry would offer any light-grasp advantage. It should be stressed, however, that when the brightness of the source is an unknown quantity, so that it is not possible to anticipate the attainable resolution, less would be at risk in taking an optimistic assumption if an interferometer were used. This is because the data which might be taken with an interferometer running to an (optimistic) maximum path-difference d_m could be used to present a spectrum at a poorer resolution than $1/d_m$ (by smoothing) if the source-brightness proved to be inadequate and the resulting signal-to-noise ratio would be as good as that which would have been obtained had the instrument been set in the first place to run to the appropriate (smaller) value of maximum path-difference in the same total time. In contrast, setting the slits in a *dispersive* instrument to give a high resolution and then smoothing the resultant spectrum to give a lower resolution would give a poorer signal-to-noise ratio than would have been obtained had the slits been set to give the lower resolution in the first place.

Before going on to the second advantage claimed for interferometric spectrometry it should be remarked that the Fabry-Pérot interferometer used spectrometrically has the same light-grasp properties as the two-beam interferometer and the considerations above would apply equally well to such devices. Also I should mention the ingenious attempts which have been made to give dispersive devices the same light-grasps as interferometers (e.g. grating instruments having complicated grills rather than simple slits). These allow a restricted choice of operating conditions and have not been very widely used.

(b) I turn now to the multiplex advantage claimed for two-beam interferometric spectroscopy [4]. With a two-beam interferometric spectrometer the total intensity passing through the instrument falls on the detector throughout the recording time.

In contrast, with a scanning dispersive instrument, most of the throughput is dissipated on the jaws of the exit slit because each resolution interval is sampled sequentially. As a result, the detector noise for each resolution interval is effectively smoothed over the full recording time in the interferometric method whereas, with a scanning dispersive instrument, the smoothing time is simply the scan-time for one resolution interval. Consequently, the detector-noise level is less by a factor $N^{1/2}$, where N is the number resolution intervals recorded. This is the 'multiplex advantage' since it results in a significant improvement in the signal-to-noise ratio on the recorded spectrum provided the detector is the dominant source of noise. Other sources of noise can now be sufficiently reduced in laboratory spectrometry to leave detector noise dominant but there are special difficulties in astronomy where fluctuations of signal intensity pose a classical problem. When such fluctuations are the dominant noise, multiplexing can actually become a *dis*advantage (see below).

The signal fluctuations involved have several origins. For an extremely weak infrared source the photon shot-noise could be significant but detectors will have to be improved appreciably before such sources are open to astronomical studies (and when this happens both interferometric and dispersive methods will be limited by information statistics and there will be neither advantage nor disadvantage in multiplexing as far as this noise component is concerned since the fluctuations in different spectral components are uncorrelated). There is the possibility, which one cannot discount nowadays, of fluctuations of signal intensity in the time range involved which arise from intrinsic astrophysical phenomena in the source. Nevertheless, more likely to be encountered first are signal fluctuations of the kinds already familiar in other parts of the spectrum:

(1) Modulation by absorption or scattering in the atmosphere,
(2) Erratic pointing of the telescope on a small source.

In addition there is one which is characteristic of the long-wave infrared:

(3) Atmospheric emission and its fluctuations ('sky-noise').

The drastic effects which such signal-fluctuations can have in the interferometric method can most simply be recognized by reference to the general form of the interferogram in Figure 1b. At zero-path-difference all spectral components suffer constructive interference and this gives a central maximum having a magnitude equal to the incident signal intensity (reduced by the throughput efficiency of the instrument). At large values of the path-difference (such that $1/d$ is smaller than the widths of the narrowest features in the spectrum of the incident signal) the interferogram converges on a constant level equal to half the height of the central maximum. There is, therefore, a significant 'background level' (see Figure 1b), which could ideally be subtracted throughout the interferogram without changing the Fourier transform. The presence of the background is therefore irrelevant to the spectrometric technique in principle; in practice, however, it carries the fluctuations of the incident signal into the interferogram and these are misread, when calculating the Fourier transform, as if they were variations introduced by interferometric modulation, and give false structure in the spectrum. Since all spectral components contribute at full amplitude to this background level and fluctuate coherently due to (1) and (2) above (and similar remarks apply to the contribution of (3) to the incident signal), the effects on the computed spectrum of quite weak fluctuations can be great [5, 6]. In fact, since the resulting uncertainty or noise for a particular resolution interval in the computed spectrum is determined by the fluctuations of the total intensity rather than that of the particular spectral component alone, multiplexing is *dis*advantageous as far as signal-noise is concerned. Relative to a dispersive instrument, the signal-noise disadvantage of the interferometric method is a factor $N^{1/2}$ if the noise is white – the inverse of the detector-noise advantage.

There are, however, methods of reducing or eliminating these adverse effects of signal-fluctuations and one or other would probably need to be adopted in any infrared spectrometric astronomical experiment using a two-beam interferometer.*

    (i) Fast scanning.

    (ii) Phase or internal modulation.

    (iii) Use of two output and/or input ports.

The idea of the first method, fast scanning, was developed by Mertz. Signal-fluctuations of the kind involved here are of largest amplitude at low frequencies. Now, the more rapidly an interferogram in scanned, the higher the real-time frequencies of the Fourier components in the interferogram which correspond to the spectral frequencies of interest. The signal-fluctuations with which the true interferometric variations are confused are therefore smaller. For example, a spectrum over the spectral range $100 \, \mathrm{cm}^{-1}$ to $1000 \, \mathrm{cm}^{-1}$ gives Fourier components in the interferogram which have from 100 to 1000 oscillations in the length of the interferogram required to give a $1 \, \mathrm{cm}^{-1}$ resolution. If the interferogram were recorded in 100 s the real-time frequencies would be from 1 Hz to 10 Hz – a range in which signal-fluctuations would be marked. If, on the other hand, the recording time were 1 s the real-time frequencies would be 100 Hz to 1 kHz – for which the signal-fluctuations would be much less. The effective smoothing time for detector-noise could be maintained by accumulating 100 1 s interferograms instead of the single 100 s interferogram – ideally this would be an equivalent process for white detector-noise. It will be clear that a fast-scanning system, with accurate triggering and data handling for the accumulation of interferograms, requires sophisticated and expensive engineering and a fast detector. Such systems (e.g. Digilab) are commercially available for laboratory spectrometry but there have, as yet, been too few astronomical measurements using fast-scanning – or indeed, too few measurements of the frequency distributions of (1), (2), and (3) above – to make good estimates of the efficacy of fast-scanning for infrared astronomy. Experiments of the kind reported by Smyth [7] using a Digilab interferometer, are therefore of special interest.

The second method for reducing the effects of signal-fluctuations – 'phase' or 'internal' modulation – was also suggested by Mertz. It has been developed with marked success by Connes for the relatively near infrared and by Chamberlain and collaborators for the far infrared. In this method the 'fixed' mirror of the interferometer oscillates continuously so as to give a relatively high-frequency modulation

---

* The Fabry-Pérot multiple-beam interferometer does not have the multiplex advantage; it is a scanning instrument, as are the dispersive spectrometers considered here.

of the path-difference. The detector-system is tuned to this frequency and there is no chopper to give modulation of the signal in the usual way. Thus, in effect, the derivative of the interferogram is recorded, and since this contains all the spectral information a modification of the Fourier transformation programme will give the spectrum. Signal fluctuations carried by the background level are thus eliminated except for the components close to the modulation frequency which, with a fast detector, could be, say, 1 kHz. The theory of this method has been examined in detail by Chamberlain [6]. The construction of a suitable mirror oscillation system is not a trivial problem. Care must be taken to have an oscillatory movement of known and precisely constant form and amplitude. The amplitude which gives the best results must be sensibly chosen; clearly it must be less than the shortest wavelength to be recorded but too small a value would result in a reduction of the signal output relative to detector noise. In the near infrared, where the wavelength varies through a spectrum by perhaps a factor of 2 or 3, this may not be a serious problem but in the far infrared, where the variation of wavelength in a spectrum might be over a factor of 10 or more it is not possible to record with highest efficiency over the whole spectrum in one run. With careful design and understanding, excellent results in astronomical studies have been obtained by Connes [1] and recently by Chamberlain *et al.* [6] and they point clearly to the potential power of interferometry in infrared astronomy; the suppression of the fluctuations in the background level of the interferogram was crucial to this success.

There might well be simpler and more flexible ways of eliminating spurious contributions to spectra. Some were recognized in early analyses of the interferometric technique [4] but until recently at least, had not found effective exploitation in practice. These follow from the fact that there are essentially two exit ports in a two-beam interferometer. The interferometer divides the incident intensity into the two outgoing

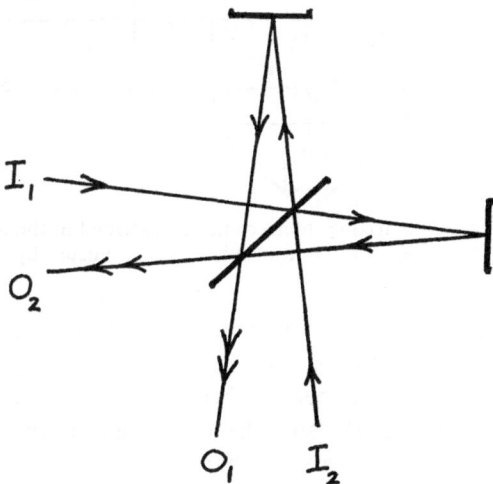

Fig. 4. Ray diagram for a two-beam interferometer which illustrates that an incident beam, I_1, contributes to two output beams, O_1 and O_2; also, that there is a second incident beam, I_2, which gives output beams at O_1 and O_2.

beams, the fraction going into one beam rather than the other varying with path-difference so that both produce an interferogram; the sum of the two interferograms is, clearly, ideally constant, that is to say the two interferograms are complementary. With a Michelson configuration (Figure 4) the second outgoing beam – usually unused – is that which goes back towards the source. If use could be made of the second beam it could serve to differentiate between those variations in an interferogram which are due to interference (which are complementary in the two beams) and those which are due to signal fluctuations on the background level (which are additive to both beams). To bring the second beam to a detector (preferably the same detector as used for the first beam) would clearly involve considerable optical complication. Moreover, a further difficulty arises from the fact that the second beam is produced by recombination of two partial beams which are inequivalent because one has been reflected twice at the beam divider and the other transmitted twice; as a result the two interferograms are not perfectly complementary should there be any absorption of transmitted beams in the beam divider.

A new kind of two-beam interferometer has recently been introduced [8] which allows the use of two complementary output beams with none of the complications referred to above. It is an interferometer in which the beam divider is a polarizer producing transmitted and reflected beams which are orthogonally plane polarized. The two output beams are differentiated, not by their direction of propagation, but by their plane of polarization.

Figure 5 illustrates the essential features of the method and Figure 6 a more effective way of realizing it which resembles the Michelson configuration. Referring to

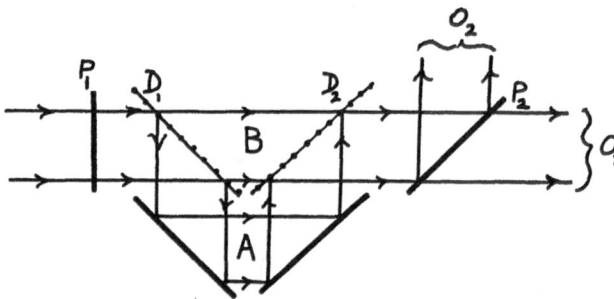

Fig. 5.   Ray diagram for the polarizing interferometer described in the text. There are two output beams, O_1 and O_2, which are respectively transmitted and reflected by the wire-grid polarizer P_2.

Figure 5, a collimated beam is plane-polarized at P_1 in the plane at 45° to the normal to the page. It is then divided by a flat wire-grid polarizer, D_1, into beam A polarized with its E-vector normal to the paper and beam B polarized at 90° to A. A and B are recombined at the wire-grid D_2 and the combined beam finally passes through polarizer P_2 which has its axis parallel to that of P_1 or at 90° to that direction. With a monochromatic source, the beam is elliptically polarized after recombination at D_2 with an ellipticity varying periodically with increasing path-difference between beam A and beam B. After P_2 the beam is plane-polarized with an amplitude which varies

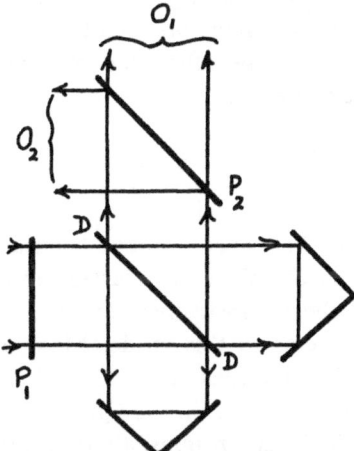

Fig. 6.   Ray diagram for a polarizing interferometer showing the two output beams, O_1 and O_2, which are respectively transmitted and reflected by the wire-grid polarizer P_2. The corner reflectors (one fixed and one moving) have their axes at 45° to the planes of polarization of their incident beams and rotate the planes through 90°.

periodically with path-difference in the same way as in a Michelson interferometer (including the same intensity, assuming an unpolarized source). That is [8].

$$I_p = \frac{I_0}{2}(1 + \cos \varDelta)$$

and

$$I_t = \frac{I_0}{2}(1 - \cos \varDelta)$$

where $\varDelta = (2\pi/\lambda)x$ (where x is the path difference) and I_0 is the intensity of the plane-polarized beam incident on D_1. The case I_p is for parallel P_1 and P_2, and I_t is for crossed P_1 and P_2.

This complementary nature of the interferograms for the two orthogonal orientations of the polarizer (see Figure 7) makes it possible to eliminate the high mean-level

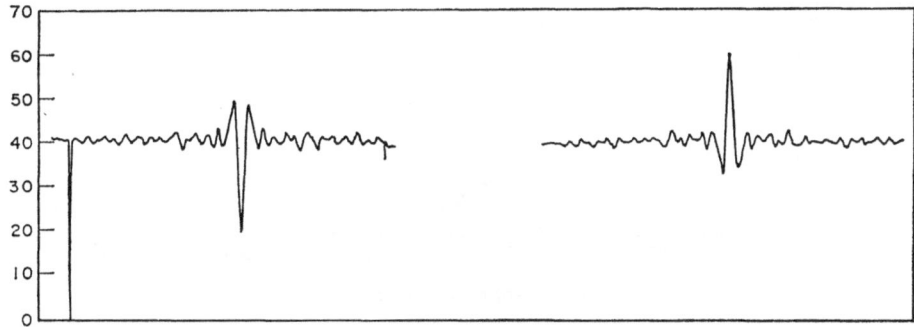

Fig. 7.   Far infrared interferograms obtained with crossed and parallel polarizers, P_1 and P_2.

of an orthodox interferogram by alternating the orientation of P_1 instead of chopping in the usual way. Alternatively two detectors could be used in opposition, one receiving the beam transmitted by, and the other that reflected by, P_2. In either case the output for a monochromatic source is:

$$I = I_p - I_t = I_0 \cos \Delta$$

which oscillates about the true zero-level.

Figure 6 illustrates the interferometer we have been using in this way. Corner reflectors are used as rotators of the plane of polarization to give a configuration close to that of a conventional Michelson instrument so that it is not difficult to convert an existing Michelson interferometer to this mode. Most of our recent measurements have in fact been made with a modified R.I.I.C. F.S. 720. The polarizer P_1 is not placed in the collimated beam as drawn in the figure but is close to the source and is in the form of a 'chopper' which, as it rotates, alternates the orientation of P_1 between the parallel and perpendicular settings with respect to P_2. As noted above, the resulting interferogram oscillates about zero, rather than about a mean-level at half the height of the central maximum. The depth of modulation is as great as that given by a dielectric beam-divider at peak efficiency because the latter sends half the useful signal back towards the source which compensates for the loss at P_1 with an unpolarized source. A further increase of x2 would be given if two detectors were used in opposition in the manner described above.

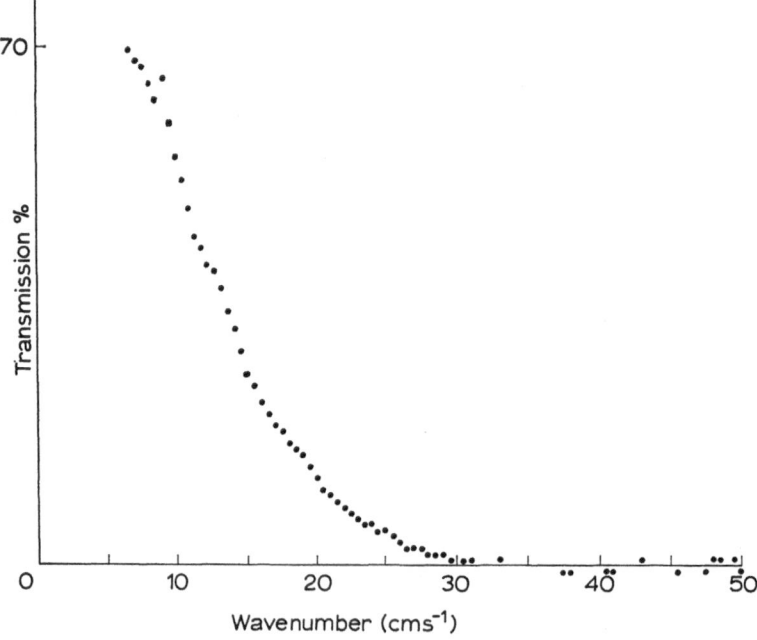

Fig. 8.   Transmission through a sample of lunar dust, 1 mm thick.

The following figures illustrate laboratory spectra obtained with a polarizing interferometer; they were taken with a cooled InSb detector and a wire-wound beam divider.

Figure 8 shows the strong variation with frequency near $\lambda \sim 1$ mm of the transmission through a 1 mm sample of lunar dust. The measurement was made on a sample being investigated by Dr J. A. Bastin and his colleagues at QMC, in connection with the NASA Apollo project. The thermal diffusivity of the Moon's surface is due in large part to radiative heat transfer. It has been found necessary to study a wide range of sample thicknesses to separate the scattering and other contributions to the transmission spectrum of such a finely powdered material as this.

Figure 9 shows part of a high resolution spectrum of water vapour. The full spectrum

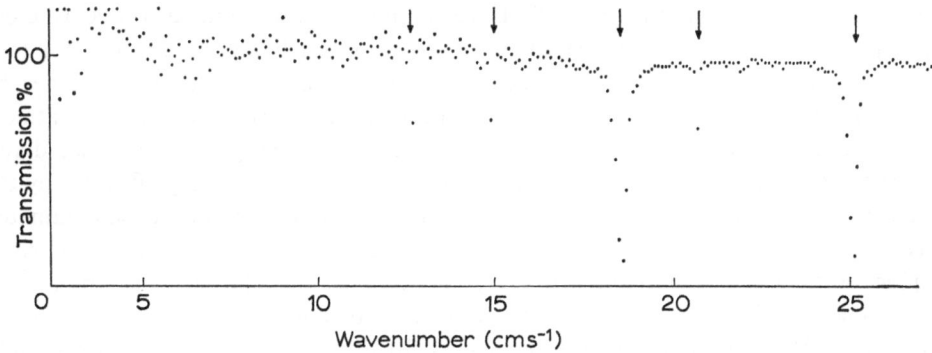

Fig. 9.   Transmission spectrum of water vapour.

ran up to 70 cm⁻¹ (where the detector loses its responsivity). The recording time for this spectrum was about 14 min but the detector time-constant was such (0.1 s) that this could have been reduced to about 3 min without change in signal-to-noise ratio, were our tape-punch fast enough. Some of the structure which resembles noise is due to weak absorption lines.

Beckman [9] has described at this meeting a polarizing interferometer which will be used shortly in a balloon-borne experiment to measure the cosmic background radiation in the submillimetre spectrum.

The method has advantages other than that dealt with here (see below) and is very simple to construct and operate; its main present deficiency is that available grid beam-dividers are limited to operation at frequencies below about 150 cm⁻¹, but extension to higher frequencies can be anticipated.

I have referred above to the methods for reducing or eliminating the noise due to signal fluctuations which are carried by the background level of the interferogram. They do not remove all effects of signal variations. The height of the interferogram *above* or *below* the background level will itself be proportional to the incident signal strength. This height at most points in a normal interferogram is, however, much smaller than the background level itself and noise associated with it has therefore a much smaller effect on the computed spectrum. It might be a limiting factor in the performance of an instrument once the background level has been suppressed however. To reduce the effects of this the total incident signal could be monitored to provide a (fluctuating) scaling factor for the interferogram but this could lead to spurious contributions to a computed spectrum unless the fluctuating component in the incident signal had the same spectral composition as the total signal; as far as effects (1) and (3) above are concerned, that would be a dangerous assumption to make. This is a problem which is difficult, perhaps impossible, to solve in a general way (except by using a method in which each spectral component is separated in a time much smaller than the fluctuations and the intensity of each one is recorded simultaneously, i.e. a dispersive instrument with a photographic plate or other multi-element detector). Where the fluctuating components arise from a source geometrically different from that of main concern something might be done. Atmospheric emission signals, for example (case (3) above), come from an extended source, whereas the astronomical source of interest will often be more nearly a point source. It is then possible to compare the signal from the astronomical source with that from a nearby point in the sky by rapid alternation using an oscillating mirror and, since the atmospheric fluctuations probably involve patches of the sky having larger angular diameters than the astronomical source, they would not be modulated significantly by the comparison process and their contribution to the interferogram would be eliminated (except, of course, for that component in the fluctuations close to the comparison frequency). Such a comparison process could replace the chopper in a polarizing interferometer using two detectors but could not, of course, replace the oscillating mirror in a phase-modulation interferometer. In any case, the comparison switch-system would need to be very carefully designed if a high signal-to-noise ratio were to be obtained in the interferogram.

The structure of a two-beam interferometer in principle allows the comparison to be made without a switching system because it has, in principle, two input beams as well as two output beams (see Figure 4). One input channel could be used continuously for the direct beam from the astronomical source while the other would be used for the reference beam. For a normal Michelson interferometer there is considerable optical complication in using both input ports in this way but the polarizing interferometer described above also has two input beams, differentiated by plane of polarization rather than by direction of propagation, and it is much easier to make use of them, as illustrated in Figure 10. If the atmospheric signals entering the two input ports suffer identical fluctuations, the atmospheric contributions are automatically cancelled when the difference of the two output beams is recorded to give an interferogram. This possibility has yet to be put to use in an astronomical experiment.

This is perhaps a point at which I should comment on the noise signals which can come from components in the optical system, for example, thermal emission from mirrors or choppers. These would intrude much in the same way that atmospheric emission does. Several speakers at this meeting have commented on the fact that the sensitivities of cooled infrared detectors are now such that photon-shot noise from optical components, which appear in the aperture of the detector but do not have zero

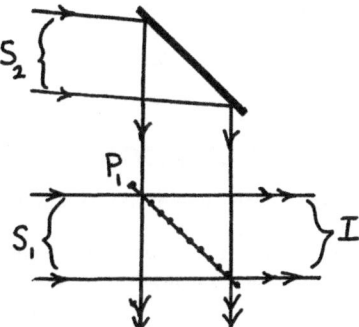

Fig. 10. To illustrate the two input beams, S_1 and S_2, which can be passed through a polarizing interferometer by transmitting the first through, and reflecting the second from, the polarizer P_1. The two input beams make a small angle to each other, the first deriving from an astronomical source and the second from a nearby point in the sky.

emissivity, might be sufficiently important for it to be necessary to cool them to the temperature of liquid nitrogen or even that of liquid helium. I would suggest that fluctuations in the thermal emission due to variations in temperature might well be greater than photon shot-noise.

(c) Finally I turn very briefly to the third of the advantages claimed for interferometric spectrometry over dispersive methods, that is the ease with which a wide spectral range can be covered in a single pass. This is based on the fact that, whereas the overlapping orders of a grating dispersive instrument lead to a need for several good low-pass filters with sharp cut-off characteristics if a range greater than an octave is to be covered and for mechanisms to change them during a scan, there is no such necessity with an interferometer. While this is so, it is not quite the case that there is never a need, with an interferometer, to change optical components and to cover different spectral ranges in separate runs. A beam-divider's efficiency varies with frequency; at low frequencies this is because of interference effects within the thickness of the divider and at higher frequencies because of absorption bands in the material of the divider. The wire-grid polarizing beam-divider has a good efficiency at very low frequencies (and should displace the lamellar-grating devices previously used for very low frequency studies) and maintains this high efficiency down to wavelengths comparable with the grid's spacing. With a polarizing interferometer it is possible at present to work from essentially zero frequency up to, say, 100 cm^{-1} with high and nearly constant beam-divider efficiency. As finer grids become available this limit should increase appreciably.

References

[1] Connes, J. and Connes, P.: 1966, *J. Opt. Soc. Am.* **56**, 896.
[2] Hanel, R. A.: 1972, this volume, p. 84.
[3] Jacquinot, P.: 1960, *Rep. Prog. Phys.* **23**, 267.
[4] Fellgett, P.: 1958, *J. Phys. Rad.* **19**, 187.
[5] Martin, D. H.: 1967, in D. H. Martin (ed.) *Spectroscopic Techniques for Far Infrared, Sub-millimetre and Millimetre Waves*, North Holland, Amsterdam, Ch. 1.

[6] Chamberlain, J.: 1971, *Infrared Phys.* **11**, 25.
[7] Smyth, M. J.: 1972, this volume, p. 284.
[8] Martin, D. H. and Puplett, E.: 1969, *Infrared Phys.* **10**, 105.
[9] Beckman, J. E.: 1972, this volume, p. 63.

DISCUSSION

R. A. Hanel: I agree that the Michelson interferometer has definite limitations if we go to very small sources. It will reach its maximum application in planetary work rather than in stellar work. One might mention another property of the instrument which is sometimes an advantage and this is the wavenumber precision, because you use your laser as a scale and all other wavenumbers are related to that frequency.

FAR INFRARED BROAD BAND INTERFEROMETRY

J. P. BALUTEAU, N. EPCHTEIN, J. GAY, and J. P. VERDET

Observatoire de Meudon, Meudon, France

The infrared group in Meudon Observatory has developed a broad band interferometer providing spectra with a resolution up to 0.1 cm^{-1} in the spectral range 25 μ to 1 mm. This range can be studied from a gondola that allows us a mean atmospheric transmission greater than 99%.

The interferometer makes use of the cat's eye system whose advantages are well known (Figure 1).

The cat's eye eliminates the effects of non strictly parallel displacements of the mirrors. It also refocuses the pupils limiting the beams resulting into a reduced size of the device even though the large étendue of the beam: the whole Sun is imaged with a 40 cm telescope.

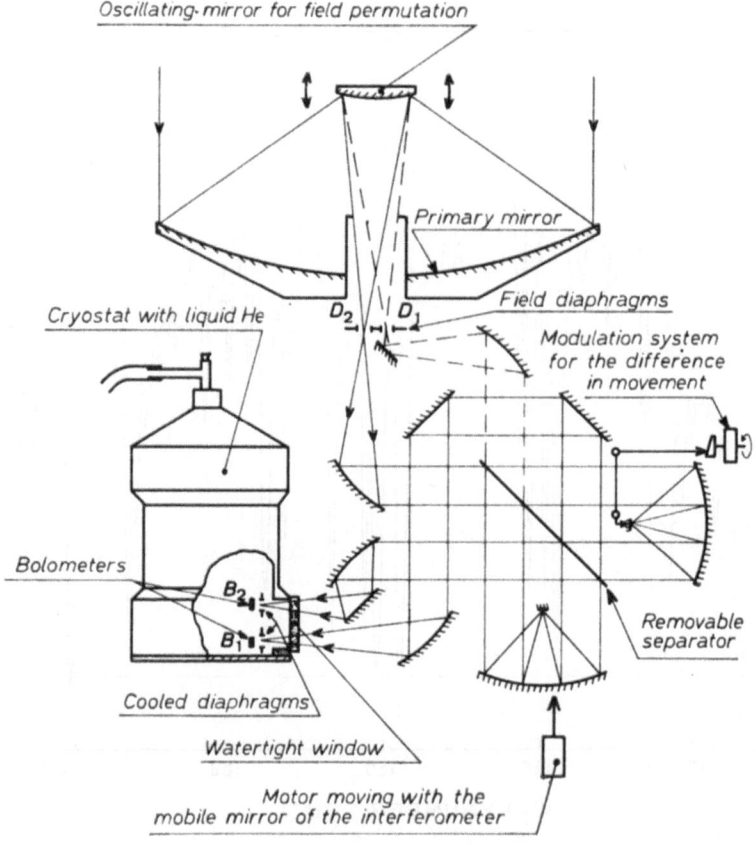

Fig. 1. Schematic of the interferometer.

Manno and Ring (eds.), Infrared Detection Techniques for Space Research, 281–283. All Rights Reserved
Copyright © 1972 by D. Reidel Publishing Company, Dordrecht-Holland

Another feature of this interferometer is the internal modulation of the path difference. This is achieved by vibrating the secondary mirror of the cat's eye.

This has been done in order to suppress the fluctuations of the mean level induced by the imperfection of the pointing of the gondola.

We are also able to separate the input and output beams. This allows us two entries. (The secondary mirror of the telescope oscillates and focuses the image of the source on two different pupils, this technique eliminates the background radiation from the sky.)

Since we have two separate output beams we may use two detectors so we may double the observation time or the spectral range, for example one detector with a quartz window (50 μ to 300 μ) another with an ICs window (20 μ to 50 μ).

Compared to a classical system we have a signal-to-noise ratio improvement by a factor $2\sqrt{2}$. That means a gain by a factor 8 in the available observational time and this is a very important point for balloon-borne experiments where the observation duration is rather short (about 4 to 6 h).

However since the two beams do not have the same attenuation coefficient the

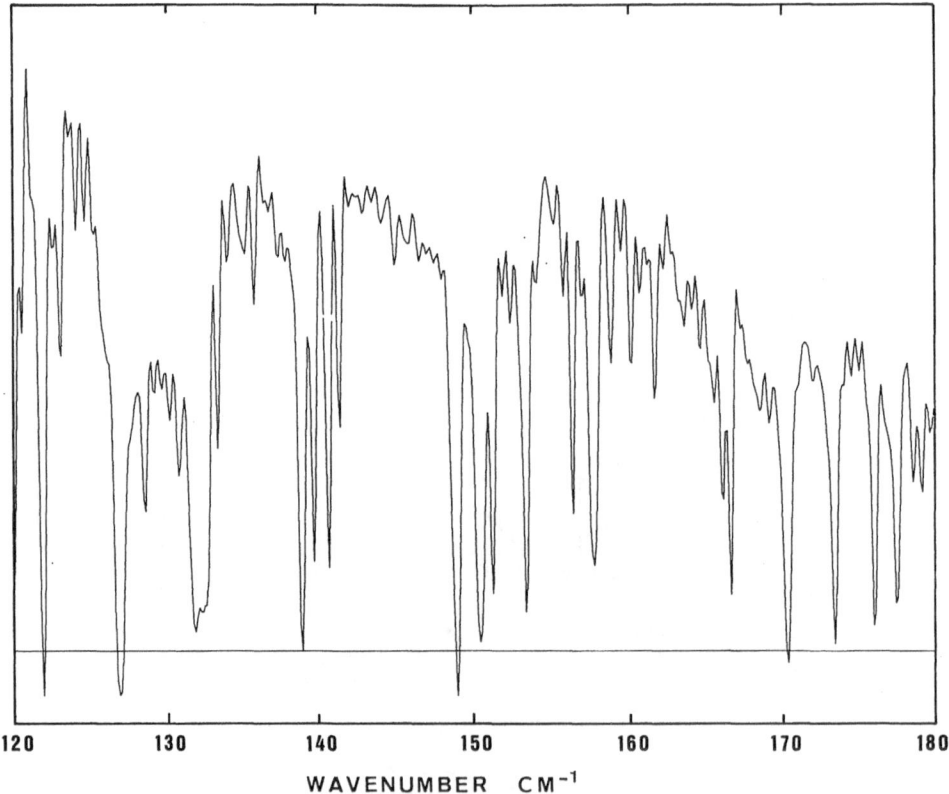

Fig. 2.   Laboratory water vapour spectrum between 100 and 200 cm^{-1} (pressure 170 mm Hg, spectral resolution: 0.25 cm^{-1}).

absorption in the beamsplitter creates a dissymmetry in the interferograms. This is of no importance for high signal to noise ratio experiments but because of the beam-splitter absorption, hence emission, up to now limits the study of faint objects.

We are now studying carefully this phenomenon and hope to get rid of it.

Three such spectrometers have been realized. The first one is a prototype used in the astronomical observatory at Gornergrat (3100 m) near Zermatt (Switzerland). It studies spectral range 2 mm–300 μ. It is now equipped with a pair of bolometers which are as identical as possible (10%). Their NEP are 3.5 and 4.5×10^{-14} W Hz$^{1/2}$, they enabled the detection of Jupiter in 900 μ range (T = 145 ± 14 K) and the sub-millimetric radiation of Orion Nebula ($\sim 10^4$ FU at 1 mm). We are now studying these quite recent results obtained during last winter. The second one is suited to solar observations and it has brought the most successful results. The experiments have been realized on the solar gondola 'Astrolabe' developed by CNES. The detectors are two pneumatic ones (ONERA) (Jobin Yvon). These experiments have allowed us to measure the Sun brightness temperature in the range 50 μ to 500 μ. The third one, also intended for balloon-borne experiments, has been built to study the planet Jupiter and the Orion Nebula. Figure 2, shows a water vapour laboratory spectrum obtained with this interferometer.

In conclusion we have realized a spectrometer well suited to balloon-borne solar experiments because of its reduced size, weight, double beam technique and modulation of the path difference, and we are now trying to study fainter objects.

DISCUSSION

R. A. Hanel: A general comment on sky noise. We hear that people are generally limited by sky noise and try to devise means to overcome it. Have you any information on how severe the sky noise is?

N. Epchtein: We've obtained several sky spectra and we are now working on them. We feel how-ever that the sky noise at 20 μ is not so important.

R. A. Hanel: I tend to agree and I believe that many times people have seen an unexplained phenomenon in their measurements, and they blame sky noise for it.

RAPID SCANNING MICHELSON INTERFEROMETER

M. J. SMYTH

University of Edinburgh, Edinburgh, U.K.

Our experience at Edinburgh, in using a commercial rapid-scan Michelson inter-ferometer for obtaining infrared spectra of stars at moderate to high resolution, suggests that such an instrument would have immediate application to infrared space research, especially from aircraft. The instrument is the Model 297 interferometer spectrometer manufactured by Block Engineering, Inc. [1]. It is compact (50 cm long), portable (22 kg), and has given no trouble during several visits to distant observatories. The attainable resolution, 0.5 cm^{-1}, is quite high enough for space applications, which will be limited by collecting optics that are small by ground-based standards. Three beam-splitters cover the wavelength range 0.6 to 25 μm.

Using an earlier version of the rapid-scan interferometer, [2] and [3] obtained numerous spectra of stars, and also spectra of the Moon from an aircraft at 12.5 km. Several of the new Model 296 are now in use by astronomers, and results have been published by Smyth *et al.* [4]. The rapid-scan technique can be regarded as established, and indeed small instruments of this type have operated successfully aboard space vehicles [5].

While the stepping interferometer (e.g. Connes *et al.* [6]) is preferable for resolution of 0.1 cm^{-1} and higher, such an instrument, weighing several tons, is clearly unsuited to space research; nor is such high resolution likely to be required. The IRIS continu-ous-scan interferometer has operated with conspicuous success aboard the Nimbus-3 satellite [7]. Where point sources are concerned, Mertz [8] has pointed out that, unless signal-dependent noise can be eliminated, the multiplex advantage of the Fourier spectrometer disappears, and even turns into a disadvantage, compared with spectrum scanning. His rapid-scan version of the interferometer therefore employs a mirror moving at a speed that produces fringe frequencies of the order of 1 KHz, so that low-frequency signal-dependent noise (stellar scintillation, tracking errors) may be filtered out electronically.

Scintillation and tracking noise having been overcome, a major technical problem is the requisite uniform rectilinear motion of the mirror. In the Model 297, the (plane) moving mirror is carried by a very stiff linear airbearing; thus the instrument requires a small compressed-air supply. The reflecting back surface of the moving mirror forms the moving element of a second small (reference) Michelson inter-ferometer fed by a helium-neon laser; the resulting sinusoidal fringe signal is frequency-stabilized, in our case at 2 KHz, by servo feedback to the electromagnetic mirror drive. Accelerometer feedback allows the drive to operate up to 45° from the horizontal. Since the laser reference wavelength is 0.6328 μm, a spectrum lying between 1–5 μm, say, will produce interferogram frequencies between 1266 and 253 Hz. A 'white' light, also feeding the reference interferometer, generates a narrow interferogram whose peak serves to define zero path-difference. The laser (sinusoidal) and 'white' (peak) signals are combined into a composite digital reference signal, which begins with a bipolar trigger-pulse marking zero path-difference and the beginning of the active scan, followed by 2^n ($10 \leqslant n \leqslant 15$) bipolar clock-pulses marking precisely equal increments of path-difference.

Because the interferogram is scanned in a few seconds, the signal/noise ratio in a single interferogram of a celestial source is necessarily low. Successive interferograms are therefore digitized and 'co-added' in an electronic computer, building up dynamic range and signal/noise ratio. The composite reference signal is used to ensure that corresponding samples of the interferograms are averaged. To guard against slight thermal drift between signal and reference channels, we terminate co-adding after about 20 min, and perform Fourier transformation and phase correction [9]. Figure 1

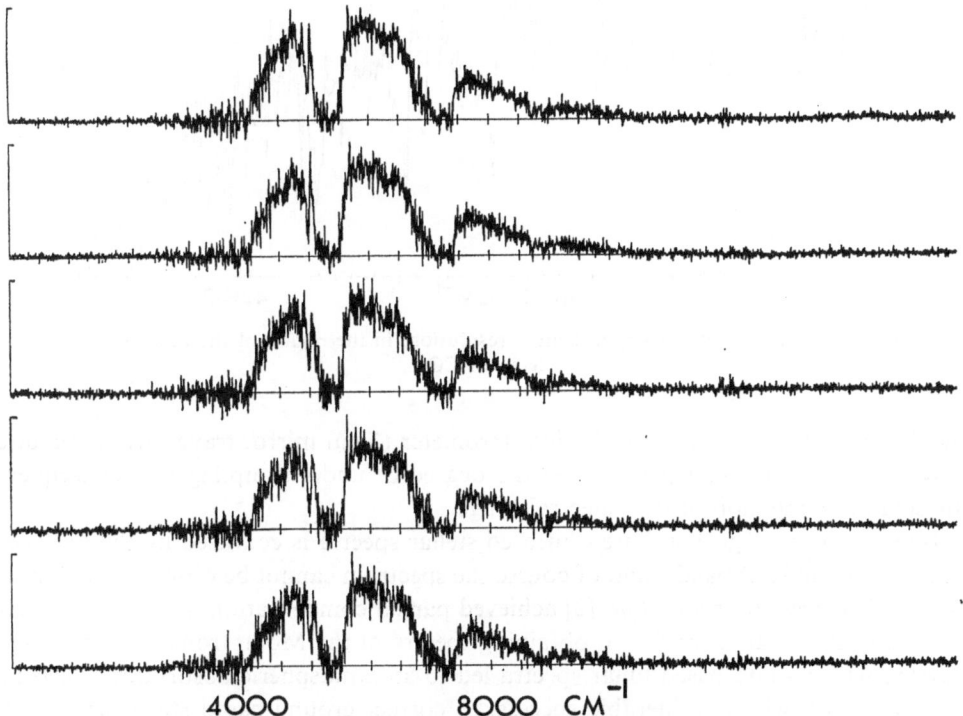

4000   8000 CM^{-1}

Fig. 1.   Spectra of α Scorpii, 1–2.5 μm; each is obtained from the average of
several hundred interferogram scans.

shows successive spectra of α Scorpii, each obtained from co-adding several hundred interferograms. This vividly illustrates the deleterious effect of atmospheric H_2O absorption bands in ground-based infrared spectroscopy! Final spectra are obtained by averaging as many of these intermediate spectra as necessary. It should be noted that the spectra of Hanel et al. [7] are obtained from single relatively slow scans.

It is our practice to record star- and reference-signals on an analogue magnetic tape recorder, for subsequent playback into our home computer. There would however be considerable advantages in co-adding the interferograms in a mini-computer attached directly to the interferometer: signal quality could be monitored immediately, the quantity of data to be recorded would be greatly reduced, and sample spectra could be obtained during an observing run. Use of a small computer in an aircraft should not pose serious problems.

The resolution of our present spectra is limited by a co-adding capacity of 4K samples ($1K = 2^{10} = 1024$). These give 4 cm^{-1} resolution with a free spectral range of 8000 cm^{-1}, or 2 cm^{-1} resolution with a free spectral range of 4000 cm^{-1} if we sample at half the laser frequency. A small portion of a 2 cm^{-1} spectrum of α Scorpii, in the region of the CO first overtone band, is shown in Figure 2. To obtain

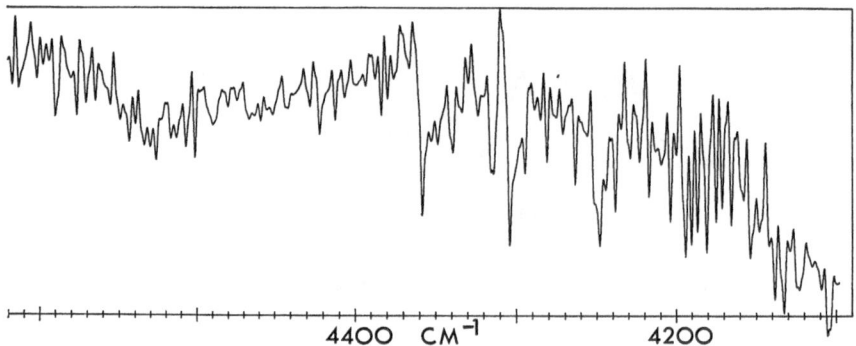

Fig. 2. Spectrum of α Scorpii, 2 cm^{-1} resolution, in the region of the first overtone band of CO.

the full 0.5 cm^{-1} resolution of the interferometer (1 cm mirror travel) would require 16K or 32K samples, depending on the degree of under-sampling and consequent aliasing and reduction of free spectral range.

Interpretation of ground-based infrared stellar spectra is confused by the presence of atmospheric H_2O bands, and of course the spectrum cannot be observed at all near the band centres. Johnson et al. [2] achieved partial compensation by flying the early interferometer in an aircraft and obtaining spectra of the Moon from 12.5 km. Comparison with ground-based lunar spectra led to an atmospheric absorption spectrum that was used, with considerable success, to correct ground-based stellar spectra. It will be useful similarly to fly the Model 297, in order to apply corrections to higher-resolution ground-based spectra. With sufficiently long flight times it may be possible to obtain direct absorption-free spectra of the brighter infrared sources. It seems clear that the commercial rapid-scan Fourier spectrometer, such as that of which we have experience at Edinburgh, is adequate to obtain moderate-resolution spectra from aircraft.

It will be a challenging problem to obtain stellar spectra in the 10 μm region, where thermal background dominates. The two inputs of a Michelson interferometer can be used to subtract background, but little success appears to have been achieved so far. The sky background, but not the optics background, will be much reduced by aircraft observations. It is interesting to note that a version of the Model 297 with optics cooled to 77 K has been constructed [10].

References

[1] Curbelo, R. and Foskett, C.: 1971, *Aspen International Conference on Fourier Spectroscopy, 1970*, p. 221.

[2] Johnson, H. L., Coleman, I., Mitchell, R. I., and Steinmetz, D. L.: 1968, *Commun. Lunar Planetary Lab.*, No. 113, 83.

[3] Johnson, H. L. and Méndez, M. E.: 1970, *Astron. J.* **75**, 785.

[4] Smyth, M. J., Cork, G. M. W., Harris, J., and Wallace, T.: 1971, *Nature Phys. Sci.* **231**, 104.

[5] Hohnstreiter, G. F., Sheahen, T. P., and Howell, W.: 1971, *Aspen International Conference on Fourier Spectroscopy, 1970*, p. 243.

[6] Connes, J., Delouis, H., Connes, P., Guelachvili, G., Maillard, J.-P., and Michel, G.: 1970, *Nouvelle Revue d'Optique Appliquée* **1**, 3.

[7] Hanel, R. A., Schlachman, B., Clark, F. D., Prokesh, C. H., Taylor, J. B., Wilson, W. M., and Chaney, L.: 1970, *Appl. Opt.* **9**, 1767.

[8] Mertz, L.: 1965, *Transformations in Optics*, Wiley, New York.

[9] Mertz, L.: 1967, *Infrared Phys.* **7**, 17.

[10] Engel, J., Wijntjes, G., and Potter, A.: 1971, *Aspen International Conference on Fourier Spectroscopy, 1970*, p. 289.

DISCUSSION

R. A. Hanel: There is a very simple method to get rid of all the drifts. This is to use the front surface of the mirror, which you use in the main interferometer, for your fringe control. One does not really pay any penalty if one images the primary on the interferometer as one should, because the center is taken up by the shadow of the secondary anyway. Also as far as fast scanning is concerned, the technique Mertz uses in adding up the spectra is not used by us. In the Nimbus case, one interferogram generates one spectrum.

M. J. Smyth: What is the lowest fringe frequency?

R. A. Hanel: About 8 Hz in our case. Also, I should say that we have used the differential method on Venus. It is virtually impossible to balance the interferometer over the whole frequency range. One can do it at one wave number, but not over the whole range, because all the optical components are not that well matched. But it is really not necessary to do so either, because you use a star in one input and then you switch over to the other input and you cancel out the small residuals.

P. Léna: If you make one spectrum per interferogram and then you add the spectra afterwards, how do you find out the position of the central fringe and with what accuracy?

R. A. Hanel: In the case of Nimbus we do not add up spectra, but when adding up spectra, you don't need to know the central fringe because you are adding up in the spectral domain already, if you use the power transform.

FIELD COMPENSATED INTERFEROMETER
FOR STELLAR SPECTROMETRY

A. GIRARD

ONERA, Chatillon, France

The interferometer has dimensions $18 \times 18 \times 80$ mm, and is composed of a fixed assembly of four optical elements. Thus there are no internal moving parts.

It is essentially a Michelson interferometer in which one of the mirrors is mechanically slightly inclined with respect to the axis, according to the well-known setting 'franges de coin d'air'. It exists, therefore, a linear variation of the path difference in the plane of localization of the fringes. The image of the star of unknown spectrum is focused on a surface element of this plane corresponding to a fraction of the interfringe.

The compensation systems employed make the path difference largely independent of the inclination of the radiation with respect to the axis. In these conditions, the analysis of the Fourier transform is made by variation of the path difference obtained by moving with uniform speed the whole interferometer perpendicularly to the fringes 'de coin d'air'. The path difference obtained is 0.6 mm.

References

[1] Girard, A.: 1970, 'Nouveau dispositifs de spectrometrie stellaire', Proceedings du Colloque sur la Spectroscopie par Transformation de Fourier, Aspen.
[2] Laurent, J. and Portat, M.: 1971, 'Spectrometrie stellaire entre 1,8 et 6 μ', Proceedings du Colloque sur l'astronomie infrarouge, Liège, Juin.

DISCUSSION

J. Ring: Dr Girard's system is elegant and undoubtedly useful. It is perhaps worth mentioning two other field-compensated Michelson interferometer systems being developed in my group by J. Schofield.

The first is a large instrument of the type described by Mertz. It has glass optics of ~ 10 cm diameter and is designed for spectrometry of nebulae in the range 1–2.5 μm, with a resolving power of 10^4 at 1 μm. The moving retroreflector is on an air bearing and its position is controlled using laser fringes. The variable wedge is servo-controlled to synchronize with the moving carriage. The resolution-luminosity product is 1.5 m²sr.

The second is an improved version. Only one motion is required and the field-compensation is achromatic. This system has a similar aperture to the first but has a higher resolving power (10^5 at 0.5 μm) and a resolution-luminosity product of 13.6 m²sr. It covers a wavelength range 0.5–2.5 μm.

A. Girard: By what means did you obtain achromaticity and what is the spectrum then?

J. Ring: All we do is to use the formulae which you derived; we apply them separately for two wavelengths and we adjust the dispersive powers of the glasses, in such a way as to make the field compensation exact for two wavelengths. The instrument we built is compensated for wavelengths of about 4500 Å and about 6000 Å and of course it has departures between these limits.

A SISAM INTERFEROMETER AND A SIMPLE MICHELSON-INTERFEROMETER WITH SPHERICAL MIRRORS FOR SPACE APPLICATION

H.-J. BOLLE, M. BOTTEMA,* W. VÖLKER, and A. ZICKLER

Meteorologisches Institut der Universität München, München F.R.G.

Abstract. Descriptions of a SISAM-interferometer for space application and of the performance of a Michelson type Fourier transform interferometer are given. The benefits and disadvantages of both instruments in different applications are discussed. The usable wavelength range for the SISAM ranges from 1 to 9 μ (extensible up to 100 μ). The Michelson interferometer is designed for the far infrared beyond 25 μ. The obtained spectral resolution is 0.2 cm^{-1} and 1 cm^{-1} respectively. Atmospheric effects on high resolution measurements from balloon or aircraft are discussed.

1. Introduction

Four advantages make interferometers attractive for space research:
 (i) high throughput or large étendue;
 (ii) signal-to-noise gain by multiplex operation;
 (iii) small size and compact construction;
 (iv) no order filter problems.
The multiplex advantage becomes always then important when detector and background noise would otherwise limit the resolution. In order to make full use of this advantage it is necessary that the spectral interval of interest is at least two orders of magnitude larger than the desired resolution.

The throughput gain with respect to grating spectrometers is $2\pi f/1$, with $f=$ focal length and $1=$ slit height of the spectrometer. The effective $f/1$ ratio is seldom smaller than 30 for conventional spectrometers so that the throughput gain is in the order of 200, if full use is made of the permitted field of view.

The size reduction is considerable if one regards that a 1.8 m Ebert monochromator is not superior to a 40×40 cm^2 interferometer. No volume is wasted as in the case of spectrometers where only a small fraction of the volume is used for mechanical and optical components.

The order filter problem becomes especially serious for work in the far infrared because for most sources the intensity increases with increasing wave number.

Let us now consider two problems of atmospheric physics which partially have their parallels in astrophysics as well, and which by themselves are regarded as very important research areas for space application.

The first problem is the derivation of physical state parameters of atmospheres with known or at least partially known composition. The second one is the optical determination of atmospheric constituents and the measurement of the concentration of

---

* Ball Brothers Research Corporation, Boulder, Colo., U.S.A.

Manno and Ring (eds.), Infrared Detection Techniques for Space Research, 289–305. All Rights Reserved
Copyright © 1972 by D. Reidel Publishing Company, Dordrecht-Holland

certain gases in the upper atmosphere of planets. To be more specific: One important question is what the far infrared spectrum of Jupiter looks like, because it would then be possible to determine the H_2/He ratio and the temperature structure of its atmosphere. Another problem is the water vapour distribution in the Earth's mesosphere.

The Jupiter problem can be attacked by emission spectroscopy. We will see that we will need moderate spectral resolution for this purpose, but the spectral interval of interest is broad. It ranges from about 100 to 400 wave numbers – a very difficult range for grating spectrometers, and just a problem where a Michelson-Fourier transform spectrometer shows its advantages.

If we now turn to the other problem, we shall see that what we really need is a high resolution, in the order of 0.2 cm^{-1}, but only a narrow spectral interval of about 1 cm^{-1} is of interest. In fact what would be necessary are only two very narrow filters of 0.2 cm^{-1} halfwidth, about 1 cm^{-1} apart from each other. On the other hand the time for one measurement is very limited, in the order of 0.5 s. These are conflicting specifications, if we think of using a Michelson interferometer for this task. In order to achieve such a resolution we need a mirror travel of 5 cm and $5/\Delta x$ sample points, $\Delta x = 1/2\Delta\nu$ being the sample interval which depends on the spectral bandwidth $\Delta\nu = (\nu_2 - \nu_1)$ of interest. If we make $\Delta\nu$ small by applying a narrow band filter we lose the multiplex advantage and have to integrate each single sample point in order to arrive at the same S/N ratio, if we make $\Delta\nu$ large we need many data points which takes the same time. Therefore for this problem we arrived at the conclusion that a SISAM interferometer would be a more suitable tool. We shall now turn to the instrumental aspects.

2. Description of the SISAM Interferometer

The SISAM (Spectromètre Interférentiel à Sélection par l'Amplitude de Modulation) is a direct readout spectrometer, not a Fourier transform spectrometer, and has first been described by Connes [1]. It is basically a Michelson interferometer in which the plane mirrors are replaced by gratings (Figure 1). These gratings act as narrow-band pass filters. Only radiation of a spectral interval defined by the field of view and the angular dispersion of the gratings reaches the detector. The arrangement uses a compensator plate which is separated from the beamsplitter. The power reaching the detector (Figure 2) is a function of the path difference between the two plane waves arriving from both gratings at the detector. The interesting term is the cosine of $2\pi\nu x$ where ν is the wave number and x the path difference. Now the compensating plate is turned with a certain speed which induces a change of the optical path in the one arm of the interferometer over several periods:

$$\nu x = ft,$$

where f is the frequency of the modulation. The modulated part of the power becomes then

$$P = 2\varepsilon B(\nu) \cos(2\pi ft) \, d\nu.$$

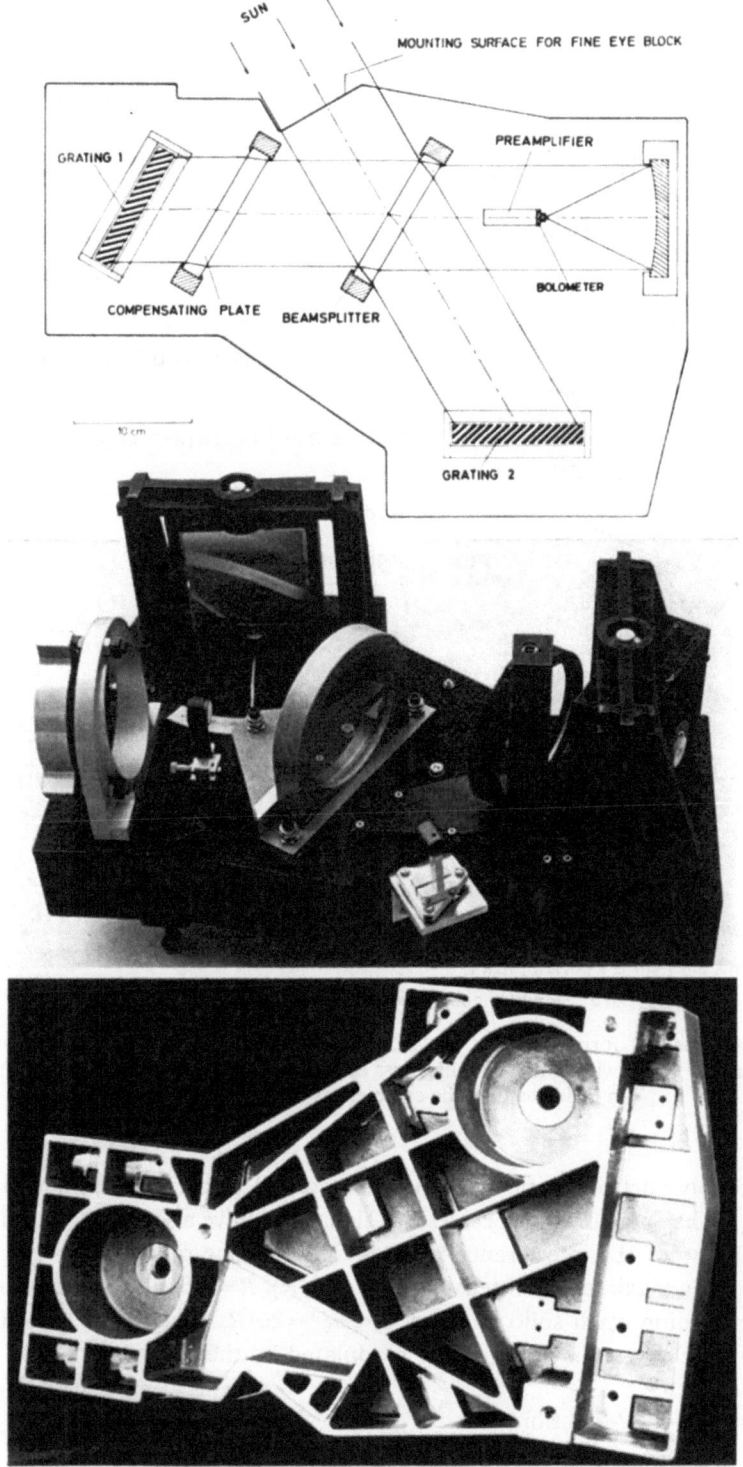

Fig. 1.

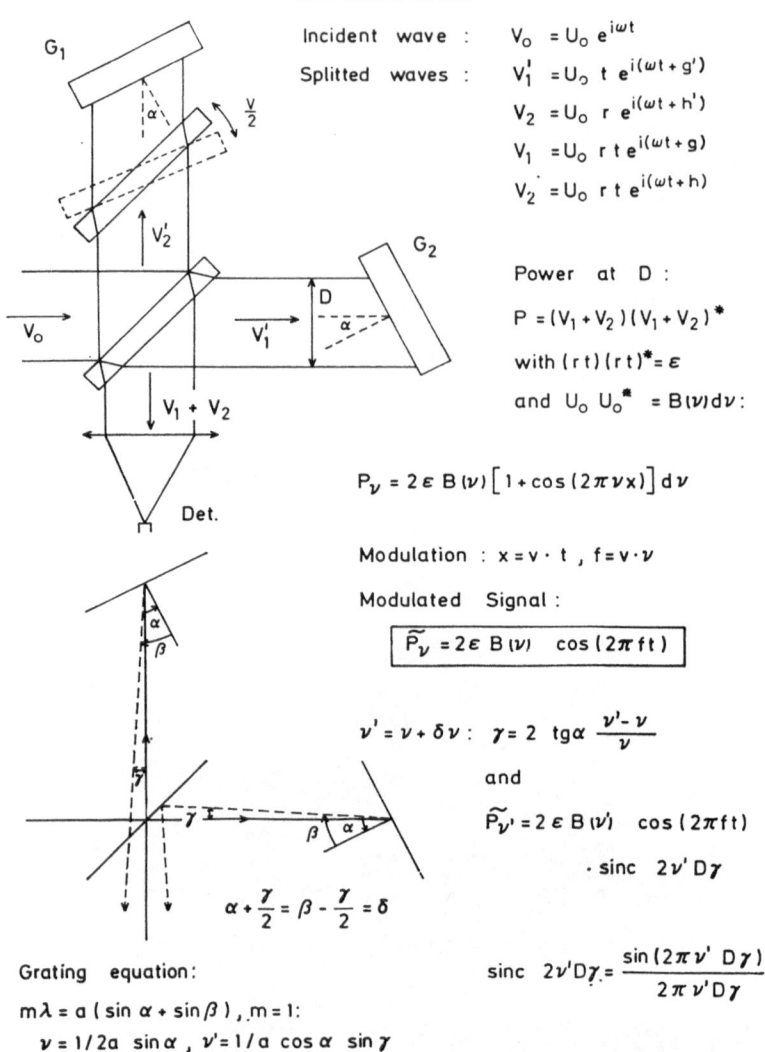

Incident wave : $V_0 = U_0 e^{i\omega t}$

Splitted waves : $V_1' = U_0 t e^{i(\omega t + g')}$

$V_2 = U_0 r e^{i(\omega t + h')}$

$V_1 = U_0 r t e^{i(\omega t + g)}$

$V_2' = U_0 r t e^{i(\omega t + h)}$

Power at D :

$P = (V_1 + V_2)(V_1 + V_2)^*$

with $(rt)(rt)^* = \varepsilon$

and $U_0 U_0^* = B(\nu) d\nu$:

$$P_\nu = 2\varepsilon B(\nu)\left[1 + \cos(2\pi\nu x)\right] d\nu$$

Modulation : $x = v \cdot t$, $f = v \cdot \nu$

Modulated Signal :

$$\boxed{\widetilde{P}_\nu = 2\varepsilon B(\nu)\ \cos(2\pi f t)}$$

$\nu' = \nu + \delta\nu$: $\gamma = 2\ tg\alpha\ \dfrac{\nu' - \nu}{\nu}$

and

$\widetilde{P}_{\nu'} = 2\varepsilon B(\nu')\ \cos(2\pi f t)$

$\cdot\ \mathrm{sinc}\ 2\nu' D\gamma$

$\alpha + \dfrac{\gamma}{2} = \beta - \dfrac{\gamma}{2} = \delta$

Grating equation:

$m\lambda = a(\sin\alpha + \sin\beta)$, $m = 1$:

$\nu = 1/2a\ \sin\alpha$, $\nu' = 1/a\ \cos\alpha\ \sin\gamma$

$\mathrm{sinc}\ 2\nu' D\gamma = \dfrac{\sin(2\pi\nu'\ D\gamma)}{2\pi\nu' D\gamma}$

Fig. 2.   Mathematical formulation of the SISAM function. The modulation velocity v can be interpreted as the speed with which the thickness of a fixed compensator plate would have to swell in order to give the same effect as a rotating plate.

This is the only part of the power which is recorded by an ac-amplifier tuned to the frequency f. The detector that receives this signal will have a size that is not larger than the central disc of the interference pattern. The angular radius of the disc which is assumed to be a maximum of intensity is $\zeta = \sqrt{(2/R)}$. $R = \nu/\delta\nu$ is the spectral resolution. The corresponding small solid angle is $\Omega_1 \cong \pi\zeta^2 = 2\pi/R$. If we allow for off axis rays, the interference pattern is additionally modulated by the

$$\mathrm{sinc}\ \frac{R}{4\pi}\ \Omega\text{-function}$$

which has its first minimum at $R\Omega_0/4\pi = 1$, $\Omega_0 = 4\pi/R$. Beyond this point the sign of the sinc function is reversed (a feature which can be accounted for in the SISAM by apodizing it with a rhomboidal aperture). For $\Omega_1 = 2\pi/R$ the sinc function has the value of 0.64 which represents only a minor distortion so that Ω_1 can be regarded as the maximum solid angle corresponding to the resolution R.

Now, also radiation of other wave numbers than ν is present in the incident beam. Due to the diffraction of the gratings this radiation arrives at the interferometer exit under non-zero angles with respect to the optical axis. So it undergoes the same sinc function modulation as off-axis rays with the wave number ν arriving from an extended source. This causes a rapid cut off for $|\nu' - \nu| > 0$ and we obtain only a narrow spectral band at the detector mainly limited by the resolution $R = mN$ of the grating (m = order, N = number of rulings).

The instrument which was constructed according to this theoretical consideration has the shape shown in Figure 1. All components are mounted on a ripped magnesium structure. The gratings have a ruled area of 102×102 mm^2 with 150 rulings per mm, the theoretical resolution being $R = 15\,300$. They are mounted on circular discs of 100 mm in diameter which are connected by a steel band in order to guarantee precise equal rotation. The beamsplitter is made of BaF$_2$ bloomed with a multilayer stack of Ge-ZnS (Figure 3). The compensating disc is also a BaF$_2$ plate. It is rotated

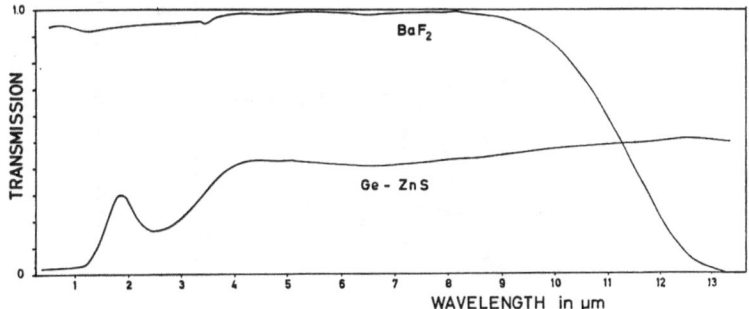

Fig. 3.   Optical characteristics of the beamsplitter materials.

by an angle of ± 0.3 deg by a linear cam (double Archimedes spiral) driven by a synchronous motor. The detector is an immersed Barnes bolometer which defines the FOV to be 19 arcmin. This is 60% of the diameter of the Sun, the main source for occulation measurements in the infrared. Equipped with the present components the instrument can be used up to 8.5 μ. By changing the components – gratings, beamsplitter, and compensating plate – it can be fitted to other spectral intervals. For use in the 1.9 μ water vapour band, gratings with 300 rulings per mm were used in second order together with a beamsplitter and compensating plate made of glass. The beamsplitter coating was simply evaporated aluminum. Some spectra are shown in Figure 4. They show a resolution of about 0.2 cm^{-1} while the theoretical resolution is 0.15 cm^{-1}.

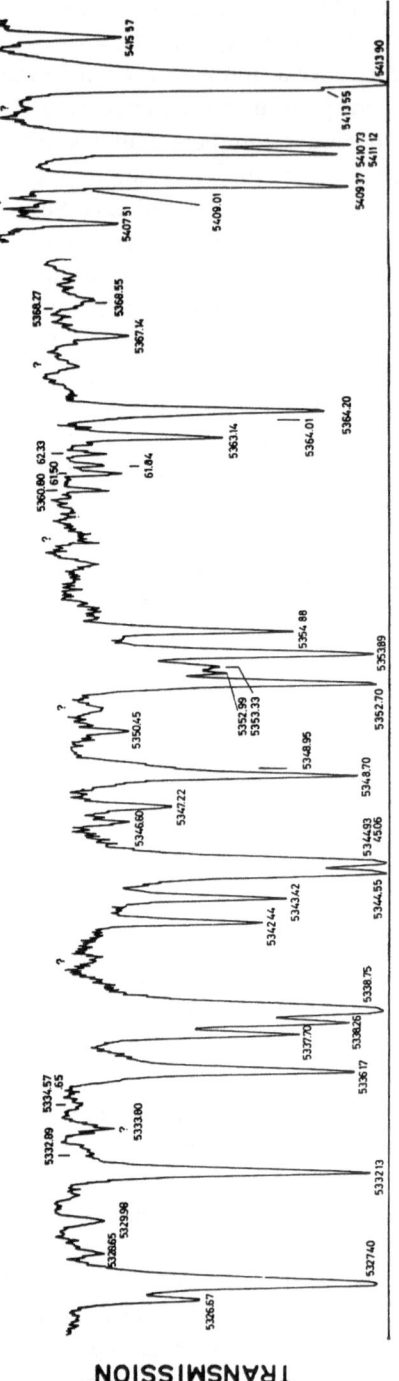

Fig. 4. Spectra obtained with the SISAM in the 1.9 μ region.

3. Design Criteria for a Michelson Interferometer with Spherical Mirrors

Michelson interferometry is almost entirely based on the assumption of plane reflecting mirrors. We may, however, for space application, assume that we have to collect as many of the available energy as possible with a telescope. In this case an image of the object, which may be a star or a planet, is formed in the focal plane of this telescope. In order to direct the radiation into a conventional Michelson interferometer, a relay mirror has to be used and at the exit of the interferometer again a collimator which focuses the radiation upon the detector. Since simplicity is mandatory in some space applications it seems to be worthwhile to consider a Fourier spectrometer with spherical mirrors which can directly be attached to the telescope, having the same speed. Such an optical arrangement seems to be especially advantageous for rocket instrumentation.

One of the two spherical mirrors has to be moved in order to obtain the Fourier transform of the spectrum. In contrast to the conventional instrument the image of the movable mirror is displaced with regard to the image of the fixed mirror by a distance equal to the path difference (Figure 5). This defocusing effect places an upper limit on

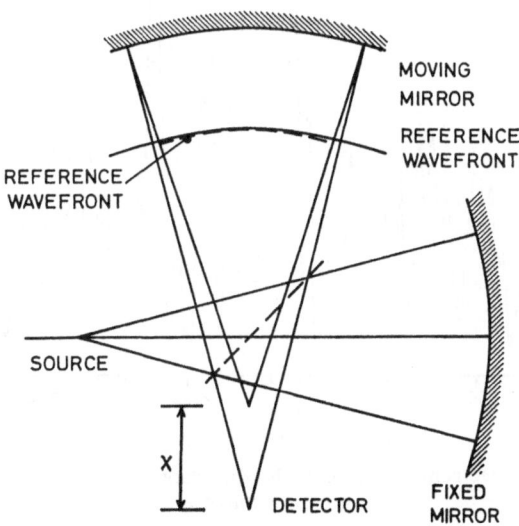

Fig. 5.   Principle of Michelson Fourier transform interferometer with spherical mirrors.

the usable path difference and therefore on the resolution. However, in applications where no extreme resolution is demanded, this is not essential. The theory of such a simple interferometer has been developed by Bottema [2, 3]. The main result of these considerations is that the path difference which still permits 64% modulation for a point source is given by

$$x_{64\%} = x_m = \pm(1/\nu)(\rho/a)^2$$

where ρ = curvature of the mirrors, a = mirror radius. The spectral resolution $\nu/\delta\nu$ corresponding to this path difference is

$$R = \nu/\delta\nu = 2(\rho/a)^2.$$

With an $f/6$ beam ($\rho/a = 2f/a = 12$) a resolution of $R = 288$ is thus obtained, and with a $f/8$ system $R = 512$. The always small size of stellar objects (except the Sun and the Moon) introduce negligible errors. The interferometer with curved mirrors is therefore theoretically capable to resolve features of 1.75 cm^{-1} at 500 cm^{-1} and of 0.35 cm^{-1} at 100 cm^{-1}.

An experimental test was performed with such an instrument. The experimental parameters were $\rho = 300$ mm, $a = 25$ mm. The beamsplitter was a 6.25 μ mylar film. A mercury lamp was used as source, a Golay cell with a SiO_2 window as detector and a 0.1 mm sheet of black polyethylene as additional filter. The entrance aperture and the detector diameter were 3.5 mm. The path difference was limited to ± 3 mm. A resolution of about 2 cm^{-1} was obtained (Figure 6). The theoretical resolution of a

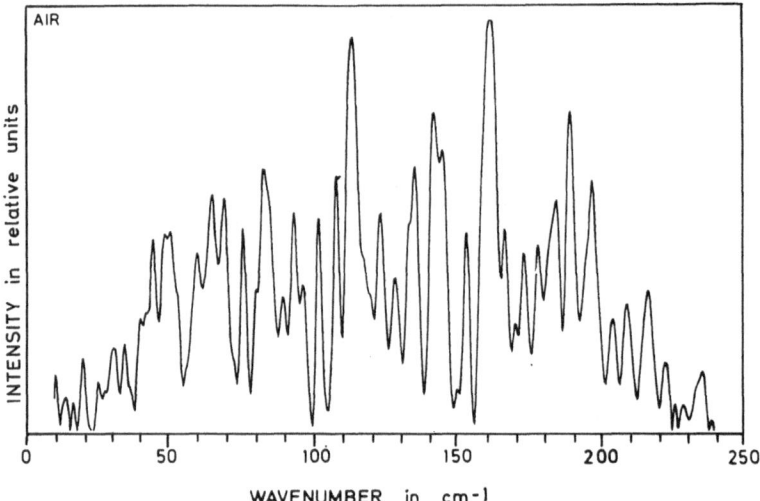

Fig. 6.   Water vapour spectrum obtained with Michelson-Fourier transform spectrometer with spherical mirrors.

Michelson interferometer with plane mirrors would have been

$$\delta\nu = 1/2x_m = 1/6 \text{ mm}^{-1} = 1.7 \text{ cm}^{-1},$$

for the same mirror travel.

4. Requirements for Occultation Experiments

Occultation experiments can be carried out from orbiting spacecrafts or fly-by missions on planetary atmospheres.

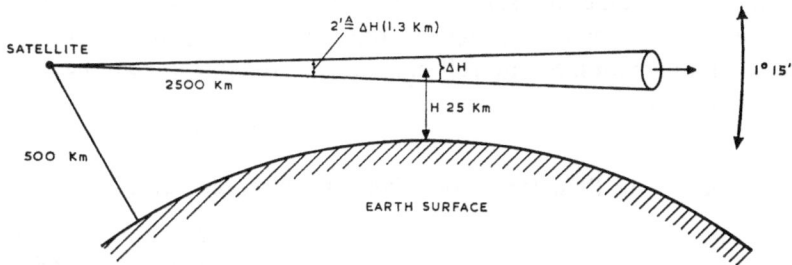

Fig. 7. Schematic of occultation experiment on planetary atmosphere.

Let us assume a satellite in a circular Earth orbit at 500 km height above the surface (Figure 7). When the satellite is emerging from the Earth's shadow and the instrument is pointing towards the Sun, the optical path Sun – instrument is tangential to the Earth surface, moving upward as the satellite progresses in its orbit. If we regard the water vapour absorption in the spectral region 3814–3825 cm^{-1}, its magnitude decreases as the minimum distance between the optical path and the Earth surface increases. However, we see that with a spectral resolution of 0.2 cm^{-1} the absorptions remain measurable up to a minimum distance of 80 km (Figure 8). This is no more true for a spectral width of 1 cm^{-1}. The solid angle Ω_1 that corresponds to the resolution of 0.2 cm^{-1} at 3800 cm^{-1} is $2\pi/19\,000 = 3.3 \times 10^{-4}$, the field of view being 35 arcmin, which is about the diameter of the Sun. For optimum vertical resolution, however, a narrow field of view is desired. A height difference of 10 km at the minimum distance of the optical path from the Earth surface corresponds to a field of view

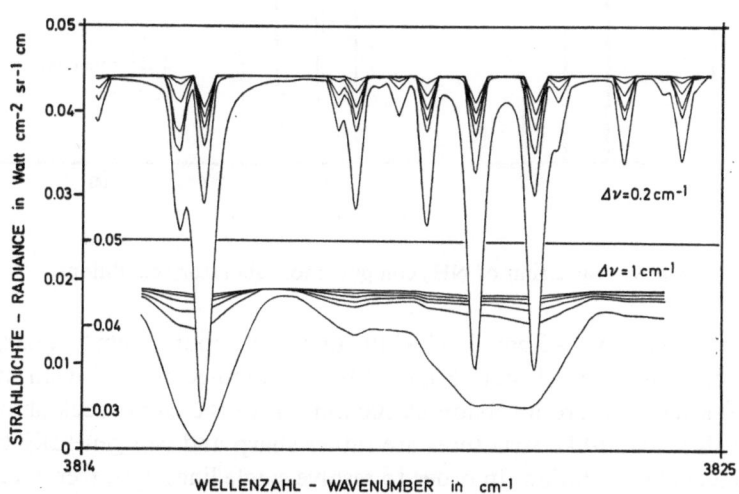

Fig. 8. Expected radiance for tangential transmission of solar radiation through the Earth's atmosphere in the 3814–3825 cm^{-1} region. The minimum heights of the optical path above ground are 31.2 km, 42.7 km, 50.5 km, 61 km, 70.5 km, and 80 km respectively. The water vapour mixing ratio in the upper atmosphere is 10^{-5}.

of 14 arcmin corresponding to a solid angle of 0.5×10^{-4} and a spectral resolution of 125 000 requiring measurements at higher order with larger gratings which can only be achieved with difficulties. So, no full but still considerable use can be made of the throughput advantage.

5. Requirements for Planetary Work Examplified on the Jovian Atmosphere

As has been pointed out by R. Hanel and S. D. Smith during this symposium, the thermal structure of planetary atmospheres can be deduced from radiometric emission measurements in narrow spectral bands. For Jupiter it is possible to obtain as well the temperature structure as the H_2/He mixing ratio from measurements, e.g. in the spectral region between about 100 and 400 cm^{-1} where the maximum of the planet's emission occurs. The temperature sounding of the upper atmosphere can be made in the NH_3-band which has the structure shown in Figure 9. There are clusters of lines

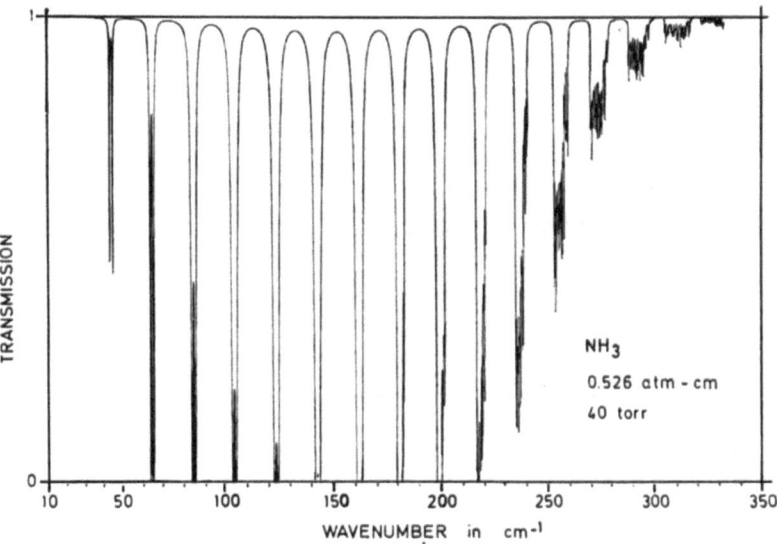

Fig. 9.   Spectrum of NH_3 computed for laboratory conditions.

in spectral intervals of 2–5 cm^{-1} of width that are about 19 cm^{-1} apart. A closer inspection of such a line cluster (Figure 10) as computed for a Jovian atmospheric model with a temperature inversion at 100 km above the cloud deck shows that the features of these band-like structures are rather sharp and can generally be discriminated with 2 cm^{-1} resolution. In order to resolve single lines, however, a resolution of 0.15 cm^{-1} would be necessary. The radiance in between the NH_3-bands depends on the H_2 amount present in the atmosphere and the H_2/He mixing ratio. A Fourier spectrometer with spherical mirrors as described can easily do the job without losing any of the desired information.

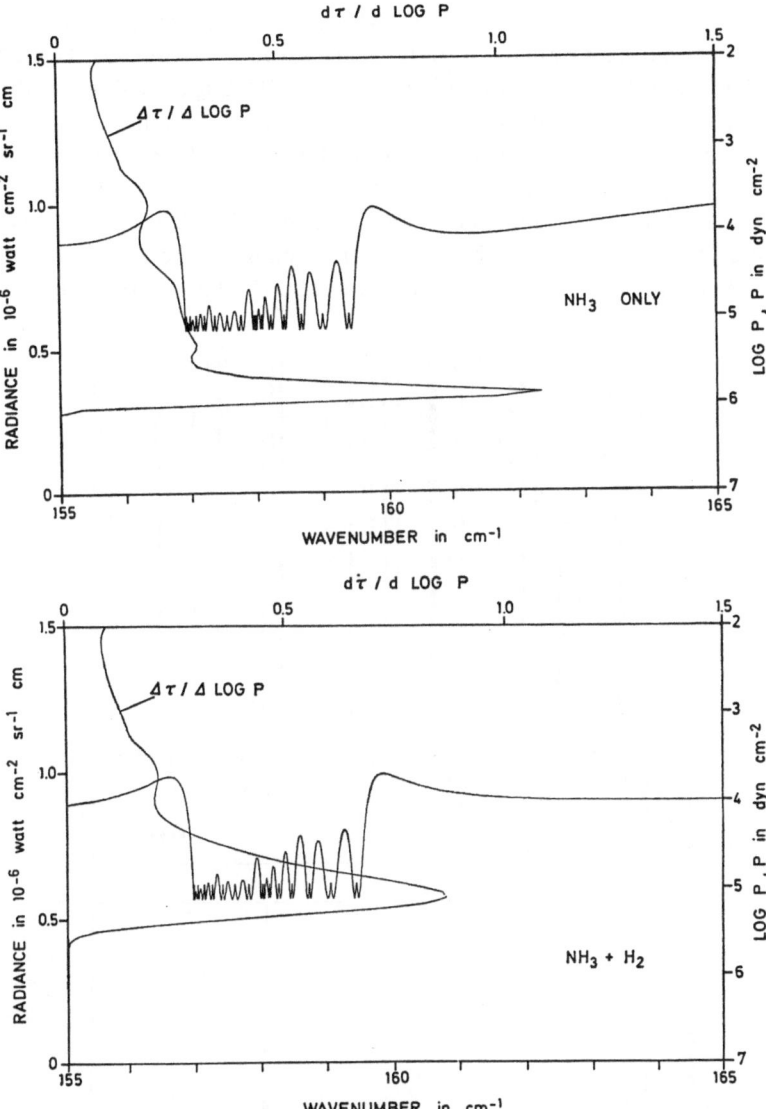

Fig. 10. Computed Jupiter emission in the 155–165 cm⁻¹ region for NH₃ only (top) and NH₃ plus H₂ (bottom). The assumed H₂/Hₑ ratio is 5. The spectrum as well as the weighting functions $\Delta\tau/\Delta \log P$ are shown.

6. Remark on Atmospheric Effects

Very often the question is asked, whether or not experiments as proposed in the preceding section on Jupiter can be made from aircraft or balloons. In order to discuss this question it has first to be regarded that these measurements have to be made in specific spectral intervals where the emission of the planet is high and the reflected solar radiation negligible. According to Planck's law, the maximum of the emission occurs for the Earth and the outer planets at wave numbers smaller than 1000 cm^{-1}. Below 750 cm^{-1} our atmosphere strongly absorbs and emits in CO_2, H_2O, N_2O, and O lines or bands, perhaps also other constituents as $(H_2O)_2$ contribute to this absorption and radiation. The question of using an aircraft or a balloon for the measurements must already very carefully be regarded when looking at the water vapour spectrum only. Figures 11 and 12 show the water vapour transmission and emission spectra computed for an atmospheric model containing 899 μg cm^{-2} H_2O above 12.5 km and 101 μg cm^{-2} H_2O above 31.2 km which can be regarded as an upper limit. The used distribution is given in Table I. The O I-line at 63 μm is not included

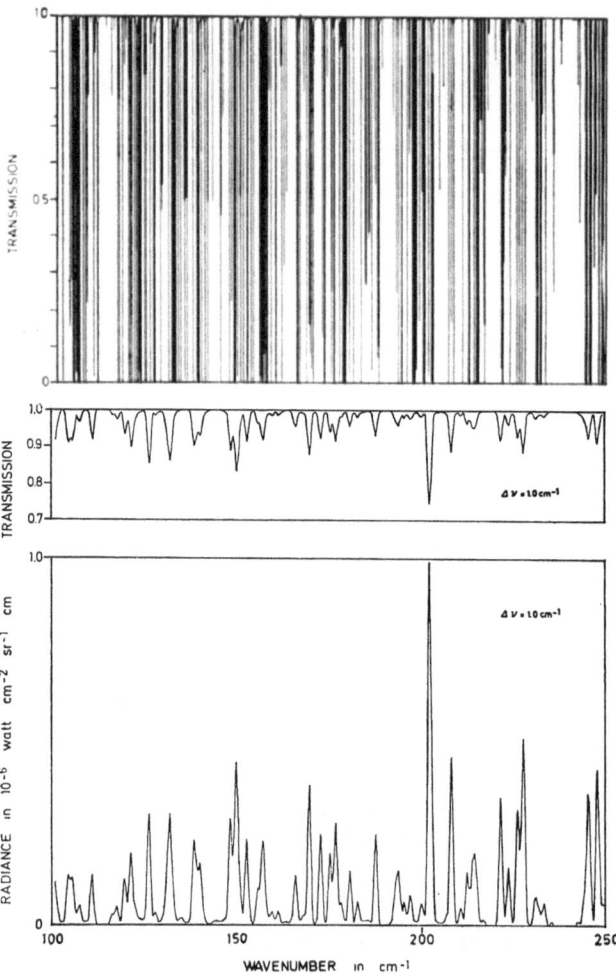

Fig. 11.   Water vapour 100–250 cm^{-1} transmission spectrum (upper two graphs, with different resolution) and emission spectrum (bottom) at 31.2 km altitude. The observation angle is 45°. The upper atmospheric water vapour mixing ratio is 10^{-5}. The transmission is plotted for two spectral resolutions, 10^{-3} cm^{-1} and 1.0 cm^{-1}, the emission for $\Delta\nu = 1.0$ cm^{-1} only.

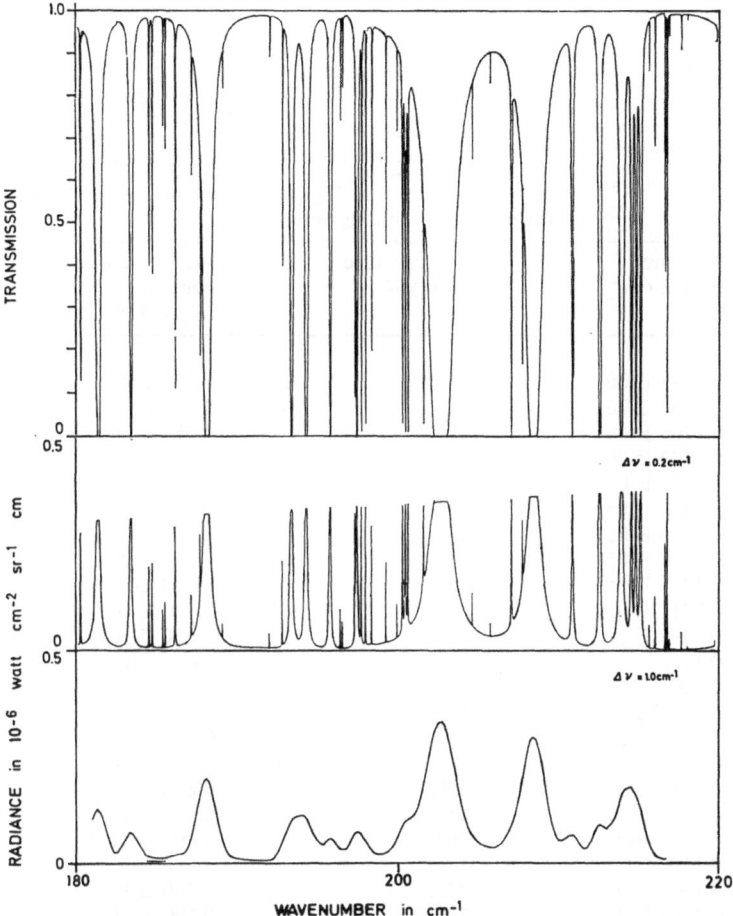

Fig. 12. Water vapour 180–220 cm^{-1} transmission spectrum (top, 10^{-3} cm^{-1} resolution) and emission spectrum (lower two graphs, 0.2 and 1.0 cm^{-1} resolution) at 12.525 km altitude. Other data as in Figure 11.

in these spectra. For high resolution spectroscopy which requires longer recording time, also changes in the upper atmospheric water vapour content during the measuring time must be regarded. Figure 13 demonstrates the transmission and emission for an atmospheric model with 20% of the amount of water given in Table I (mixing ratio 2×10^{-6}).

The influence of the water vapour emission can be eliminated or at least be reduced experimentally by using instruments which switch the point source under observation against the sky. The absorption, however, can only very approximately be accounted for in most cases. Therefore, if high radiometric accuracy is required, as in planetary studies, the influence of the Earth's atmosphere should be excluded as far as possible.

TABLE I

Model atmosphere used in Figures 11 and 12
(Stratospheric mixing ratio 10^{-5})

Pressure in dyn cm^{-2}	Temperature in deg K	Water vapour density in g cm^{-3}
0.1786 E-06	216.7	0.260 E-08
0.1502 E-06	216.7	0.175 E-08
0.1204 E-06	216.7	0.118 E-08
0.1003 E-06	216.7	0.950 E-09
0.6995 E-05	216.7	0.490 E-09
0.4981 E-05	217.3	0.210 E-09
0.4000 E-05	218.7	0.120 E-09
0.3521 E-05	219.5	0.112 E-09
0.2930 E-05	220.7	0.100 E-09
0.2511 E-05	221.7	0.790 E-10
0.1998 E-05	223.1	0.620 E-10
0.1499 E-05	225.0	0.460 E-10
0.1001 E-05	227.7	0.310 E-10
0.7482 E-04	231.5	0.230 E-10
0.4983 E-04	239.3	0.150 E-10
0.4087 E-04	243.2	0.120 E-10
0.3012 E-04	249.4	0.840 E-11
0.2011 E-04	257.8	0.540 E-11
0.1507 E-04	263.9	0.400 E-11
0.1102 E-04	270.7	0.280 E-11
0.1002 E-04	270.7	0.260 E-11
0.7497 E-03	270.7	0.190 E-11
0.5900 E-03	270.7	0.150 E-11
0.4012 E-03	264.6	0.110 E-11
0.3101 E-03	260.7	0.830 E-12
0.1969 E-03	253.8	0.540 E-12
0.1507 E-03	247.1	0.420 E-12
0.1041 E-03	236.7	0.310 E-12
0.7445 E-02	227.5	0.280 E-12
0.5114 E-02	217.7	0.160 E-12
0.4050 E-02	211.9	0.130 E-12
0.3011 E-02	204.7	0.100 E-12
0.2105 E-02	196.2	0.700 E-13
0.1502 E-02	188.7	0.500 E-13
0.1037 E-02	180.7	0.400 E-13

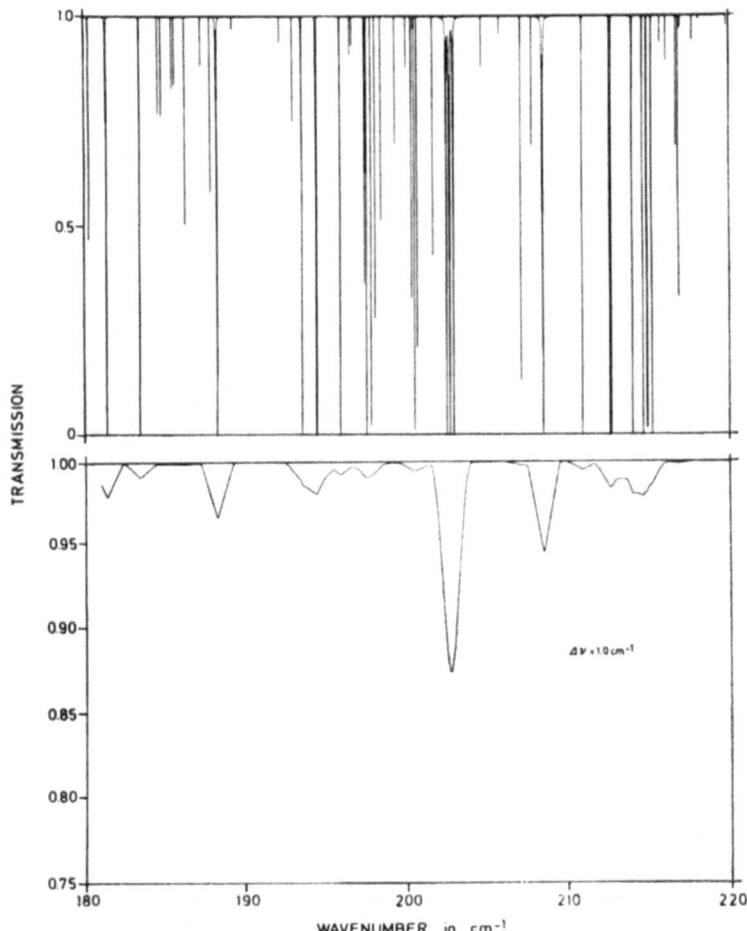

Fig. 13a–b. Water vapour 180–220 cm^{-1} transmission spectrum ((a), 10^{-3} cm^{-1} and 1.0 cm^{-1} resolution) and emission spectrum ((b), 10^{-3} cm^{-1} and 1.0 cm^{-1} resolution) at 31.2 km altitude for 2×10^{-6} mixing ratio. Observation angle 45°.

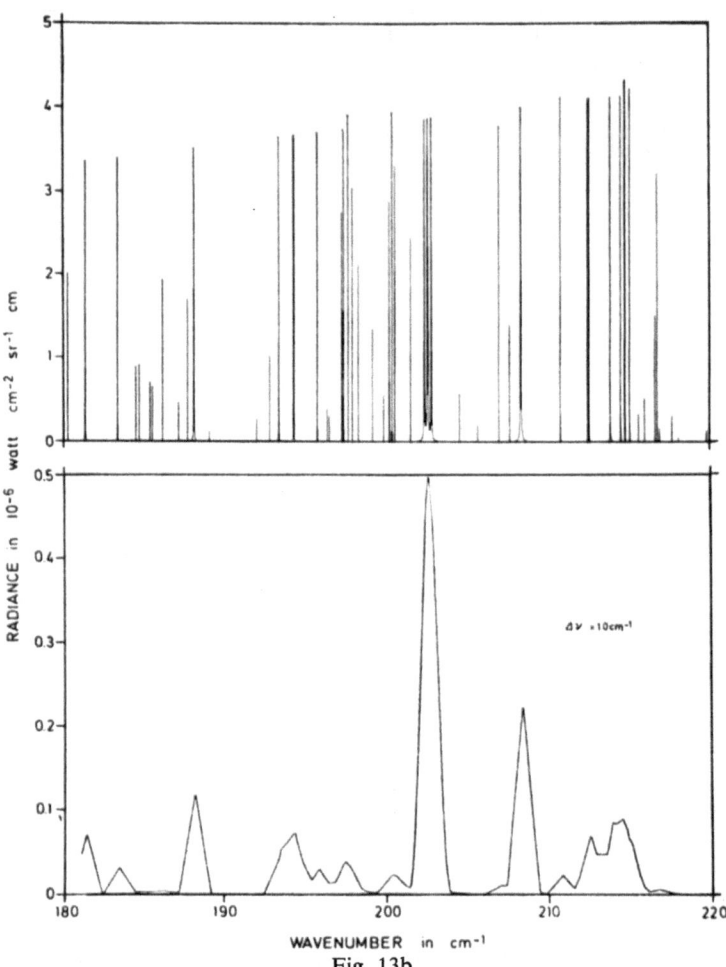

Fig. 13b.

Note added in proof: One of the authors has further investigated the theory of Michelson interferometers with spherical mirrors, and found that their performance can greatly be improved if both mirrors are moved simultaneously in opposite direction (M. Bottema, to be published in *Appl. Opt.*).

References

[1] Connes, P.: 1957, 'Principe et réalisation d'un nouveau type de spectromètre interférentiel', Thèse, Paris; *Rev. Opt.* **38** (1959), 157; *Rev. Opt.* **38** (1959), 416; *Rev. Opt.* **39** (1960), 402.
[2] Bottema, M.: 1970, 'A Simple Interferometer for Low-Resolution Far Infrared Fourier Spectroscopy', in H.-J. Bolle and R. Beffert (eds.), *Experimente für meteorologische Satelliten oder eine Weltraumstation*. BMBW Forschungsbericht W 70-70.
[3] Bottema, M. and Bolle, H.-J.: 1971, 'An Interferometer with Spherical Mirrors for Fourier Spectroscopy', in G. A. Vanasse, A. T. Stair, Jr., and D. J. Baker (eds.), *Aspen International Conference on Fourier Spectroscopy, 1970*, AFCRL-71-0019, Special Report No. 114.

DISCUSSION

J. Ring: I am puzzled by your choice of instruments. You said you chose the Sisam rather than the Michelson because of the time that was allowed, but surely with the multiplex advantage, even if you had to use the restricted sampling theorem to take fewer samples, you must gain in a given time with the Michelson. A second point: I don't see what advantage you gain with the spherical Michelson, other than slightly greater energy, to compensate for the disadvantage of this asymmetry.

H.-J. Bolle: We don't have any disadvantage in using the spherical mirrors because we get the spectral resolution of 1, 2, or 3 wave numbers which we require and we can directly connect this interferometer to a telescope without using relay mirrors. With regard to the first point, if we use a restricted spectral interval with the Michelson interferometer we are losing the multiplex advantages at the same time. In a very narrow interval there is not much energy but nevertheless we have to sample over a large number of points. Essentially we need only two points, one in the center of the line and one in the window next to the line, so we can make the measurement in a shorter time with the Sisam.

J. Ring: Let's come back to the spherical mirrors. It seems that by using those you have built in an uncontrolled apodization of your interferogram, because it is automatically apodized as it goes out of focus and you can't choose the apodization.

R. A. Hanel: Well, I think that for the low resolution for the particular application it was a convenient engineering compromise.

LIQUID HELIUM-COOLED GRATING MONOCHROMATOR

M. HERSE

Service d'Aeronomie du CNRS, Verrières-le-Buisson, France

At wavelengths greater than $2\,\mu$ background and instrumental thermal radiation reaching an infrared detector presents a problem because its energy is often superior to that from the target source (Figure 1). An infrared detection system may be con-

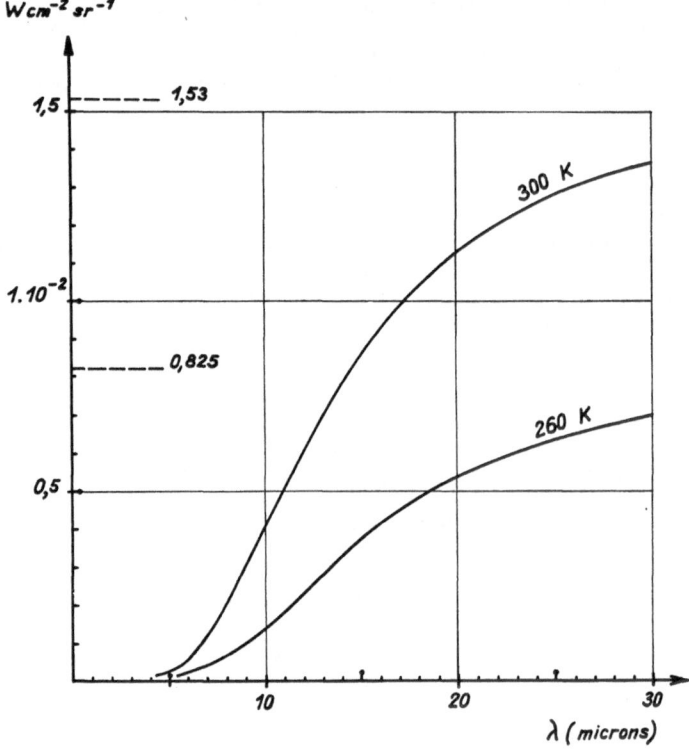

Fig. 1.   Integrated black-body energy from $\lambda=0$ to λ.

sidered to comprise an optical system defining the spectral interval, either by filters (direct) or by Michelson interferometry (indirect), a spatial interval defining system (telescope, baffles, etc.), and one or more detectors and their associated electronics. The highest performance generally requires liquid helium-cooled detectors with the immediate surroundings of the detector also maintained at low temperatures. On the other hand, the necessity of a vacuum cryostat usually imposes a warm window in the optical path, at least for ground-based and balloon-borne experiments.

The spectral interval is usually defined by ambient temperature instruments (grating spectrometer or Michelson interferometer).

Manno and Ring (eds.), Infrared Detection Techniques for Space Research, 306–312. All Rights Reserved

Although the emissivities of the mirrors is usually very low, the total thermal emission of these warm elements is far from negligible and subject to the same fluctuations as the environment. Finally, in the case of ground-based experiments, the sky emission is quite strong. The combined effects of these energy sources produces three principal effects on the detection system: first, this thermal energy flux incident on the detector raises its temperature and thus decreases its responsivity. Secondly, its temporal fluctuations will have a component which will pass through the electronic bandpass filter. Finally, this flux itself is associated with a photon noise power. For ultra-sensitive detection systems having large spectral bandwidths, this noise can become comparable to the intrinsic detector noise itself (Figure 2). The noise reaches a maximum value near the peak of the Planck curve (Figure 3).

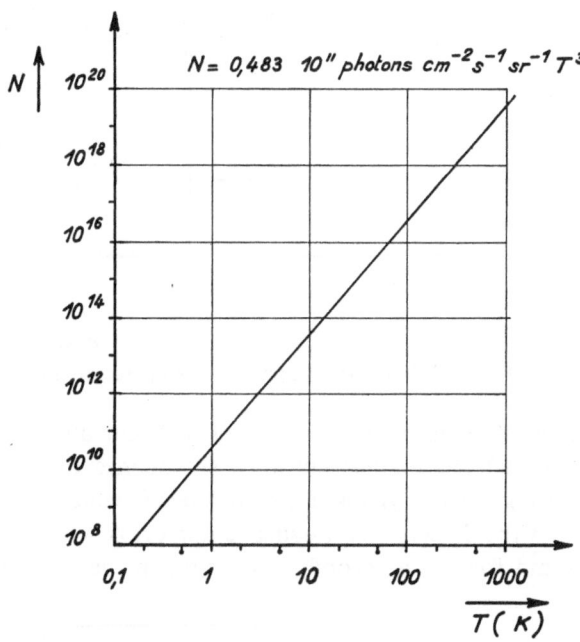

Fig. 2.   Integrated number of photons emitted by a black body.

For the eclipse of 12 November, 1966 we developed a novel apparatus. The experiment was designed to measure solar infrared limb darkening using the Moon as an occulting body in order to obtain the required high spatial resolution (1 arcsec). The most interesting wavelength from the astrophysical point of view is around 25 μ. A wavelength of 23 μ was chosen in order to take advantage of an atmospheric window because to achieve the necessary stability our apparatus needed to be ground-based at high altitude (4100 m). The only commercially available filter was a 10 μ large-band filter. We therefore decided to construct a grating monochromator.

To reduce the thermal background, the monochromator was incorporated into the cryostat, necessary in any case for cooling the detector. The mounting is shown

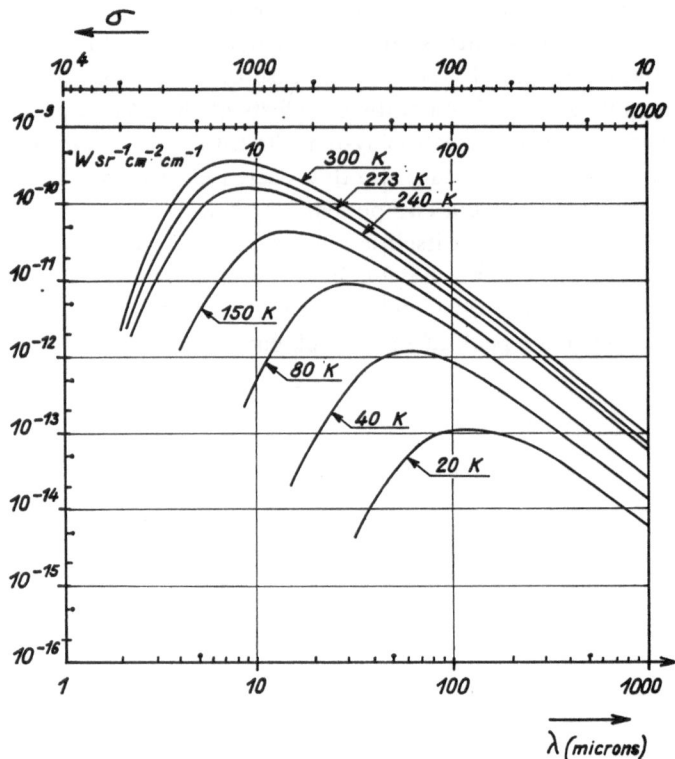

Fig. 3.   Photon noise of a black body as a function of temperature and wavelength.

schematically in Figure 4. Placed at the focus of a 25 cm diameter, 1 metre focal length telescope, a cooled rectangular entrance slit defines the solar region to be measured. A collimating mirror yields a parallel beam which is dispersed by the grating and focused on the detector by a concave mirror. A second mirror interrupts the dispersed beam and focuses a second wavelength in the second order (the first

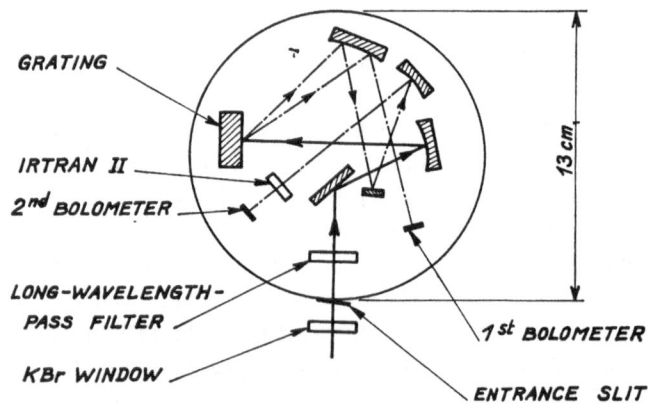

Fig. 4.   Optical layout.

order wavelength is blocked by a sheet of Irtran II) upon a second bolometer. Located just behind the entrance slit, a 12.5 μ long-wavelength pass filter prevents admission of higher order wavelengths. Spectral profiles obtained with a Grubb-Parsons monochromator are given in Figure 5.

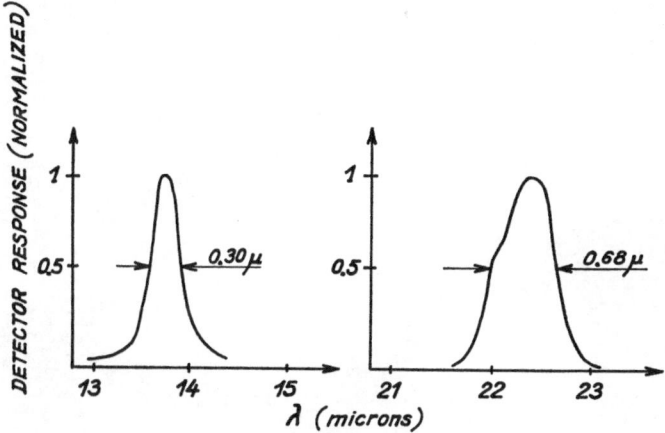

Fig. 5. Bandpass characteristics.

Both detectors were Texas Instruments bolometers, 2×5 mm optimized for 3.6 K operation (boiling point of helium at the altitude of experiment site). Figure 6 shows a photograph of the monochromator mounted on the removable optical platform of the cryostat (described in a separate paper by G. Chanin). All mirrors were of nickel-plated duraluminum mounted on pure aluminum supports fixed to the optical platform of the cryostat. The Bausch and Lomb grating was of 40 lines/cm and of 30×30 mm surface. A movie camera made a continuous record of the solar image upon the entrance slit through the cryostat window of KBr.

To further reduce environmental thermal radiation, a 3 M Velvet Coating painted light shield surrounded the monochromator and was mechanically and thermally anchored to the bottom of the helium reservoir. Adjustments were easily carried out in the visible using sodium light in a higher order. A three-wavelength monochromator has also been constructed using the same principles but incorporating the three detectors in a single mount (see paper on detectors by G. Chanin).

This basic design was carried further by J. P. Baluteau, G. Chanin, and J. M. Fitremann using mechanical systems to produce continuous or step-by-step rotation of the grating. In the Baluteau version, an electric motor within the cryostat vacuum envelope produces a continuous rotation of the grating using a cardioid cam yielding a linear wavelength scan. This system has been used at the astronomical station at Gornergratt (Switzerland) in the 18–35 μ spectral range. A modification of the optics permitted covering the 7–15 μ range (Figure 7). This system was also used at the Coudé focus of the 152 cm telescope of the Observatoire de Haute Provence as a control on a lunar thermoluminescence experiment of M. L. Chanin. The modulation was ob-

Fig. 6. Photograph of the bottom of the cryostat.

tained by chopping between the target and a 300 K or a 77 K black body. This procedure enables the computation of the target temperature if no significant parasitic radiation enters the system after modulation. Spectral resolving power measured with the Grubb-Parsons monochromator was of the order of 40. Angular resolution was 10 arcsec.

Figure 8 shows several lunar scans (two minutes per scan). Figure 9 shows the output corrected for telescope parasitic light for a crater 60 h after passage of the morning terminator and Mare Serinitatus on the morning therminator.

The spectral resolution of this class of systems depends upon the following parameters: size of entrance slit imaged upon the detector plane, grating dispersion and

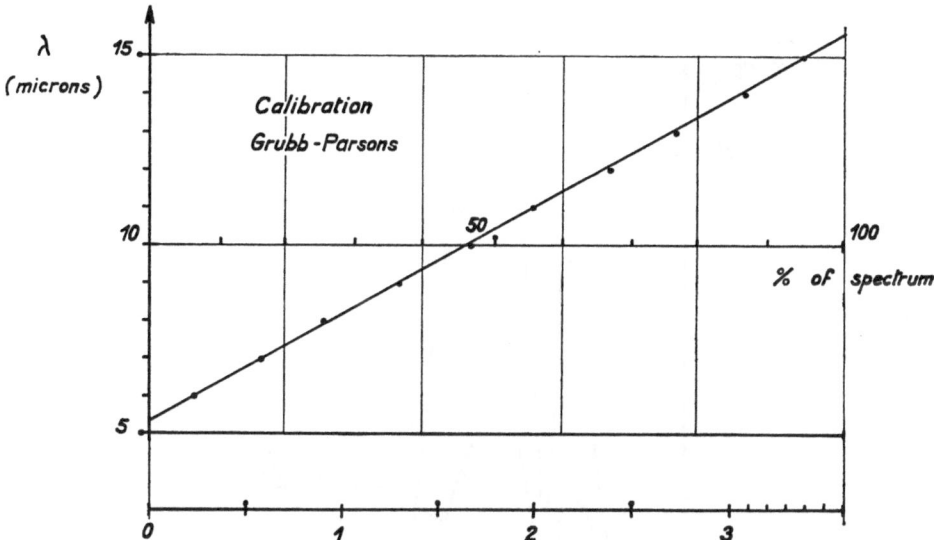

Fig. 7. Wavelength on the bolometer as a function of the rotation of the motor.

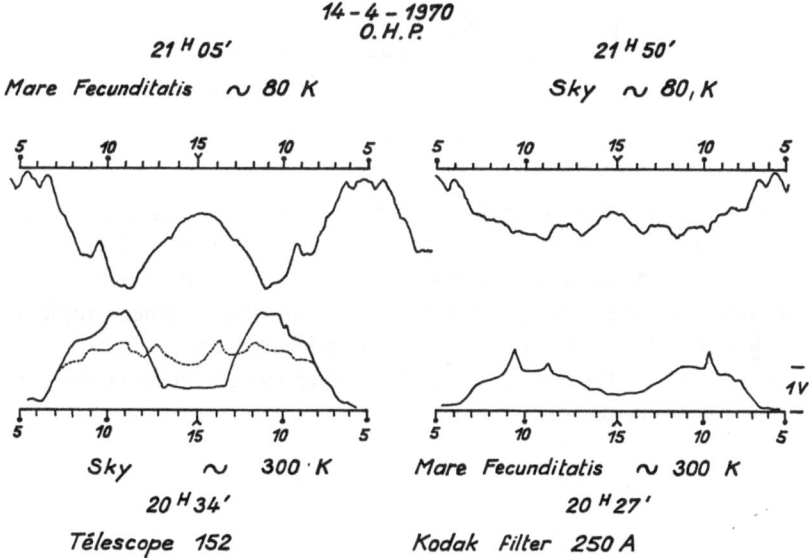

Fig. 8. Recordings of the Moon and sky modulation with respect to black-body temperature of 80 K and 300 K. 152 Telescope-Observatoire de Haute Provence, France.

detector dimensions. For a 13 cm optical platform a resolving power of 40 is easily obtained. At very long wavelengths, the number of grating rulings actually used may be the limiting factor. At these medium resolutions, the noise contribution from background radiation is usually negligible. For all these systems, detector resistance (and responsivity) is practically unchanged from its blanked-off value. The infrared trans-

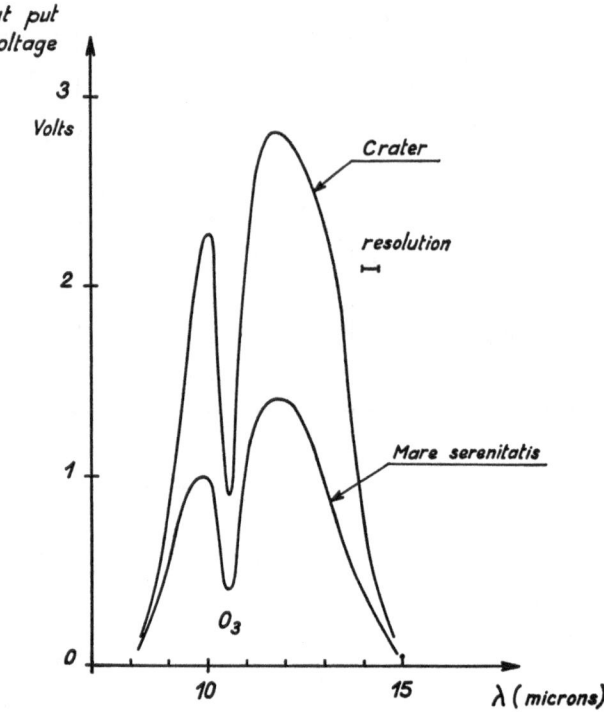

Fig. 9.   Spectra of the Moon.

mission is quite good because of the high mirror reflectivity and excellent grating efficiency. The same mounting with only a change of grating and order separating filter can cover the entire spectral range from visible to millimetre wavelengths. Medium resolution spectroscopy is therefore possible with a rather simple, compact, and portable apparatus: suitable for both ground-based and balloon-borne experiments. The cooling of the entire spectral defining system permits the detector to operate at its optimum performance level.

ONE- AND TWO-DIMENSIONAL COMPUTATIONAL PROBLEMS ASSOCIATED WITH INTERFEROMETRY

YVON BIRAUD

Observatoire de Meudon, Meudon, France

1. Introduction

It is almost unnecessary to point out that computational problems, especially in interferometry spectroscopy, are becoming of greater and greater importance. The computer now is really a part of the experiment and its influence is felt in a variety of applications. At first it is often used for data acquisition, then for the pretreatment, and in the end for the scientific exploitation of the data. But we think that it is more important to point out the very rapid improvements of the computers' technology and especially of the algorithms they use. These two aspects allow us not only to solve present problems but also to look at these problems from a different and new point of view. For example some deconvolution and new spectroscopy techniques are possible only because of powerful or well-suited computing devices.

2. The Fourier Transform Algorithms (FT-Algorithms)

The evolution of these algorithms is one of the best examples of the progress realized in the reduction of computing time and storage so that they are not now limiting the resolution in Fourier spectroscopy. The limitation is due to technological imperfections.

Nevertheless we shall mention the old, classical FT-algorithm which calculated N^2 sine or cosine values and performed N^2 additions and multiplications (in the case of both N input and output samples). I think that this algorithm has still to be used for small numbers of samples because of its flexibility: one may choose the input and output sampling intervals and also the spectral domain to be computed. One can take advantage of this last facility by computing only the few last points of the Fourier transform and verify that they have negligible magnitudes. This is a very good way to check that the sampling interval for the input function is sufficiently small and is close to the Shannon-Nyquist value.

The last advantage is very important. The classical FT-algorithm is a true real-time algorithm. This means that you need not wait until you get the last input point to perform the FT. You may calculate, just after a new point is available, its influence on the whole output function. Hence, you may, for example, visualize the FT on a display and see the resolution increase with time. This algorithm has been wired in a special FT hardware [1] and is very useful when you want to test your equipment and also in spatial experiments when real-time control is important.

But the main disadvantage of this algorithm is that it is very time-consuming. It requires N^2 operations and the computing time becomes rapidly prohibitive [2].

The great success of the Cooley-Tukey algorithm which is now very well-known and widely used, is due to the reduced number of operations involved, $N \log_2 N$, which allows us to transform very large arrays in a moderate duration of time. I will simply mention some improvements to the initial algorithm. The so-called 'Decimation in Time' algorithm saves computation time (by a factor 2.8) as well as storage requirements (by a factor 4) for real-odd interferograms. The principle described in [2] is to sort the data into two subsets, take their FFT and make a linear combination of them. This operation is repeated $\log_2 N$ times ($2N$ is the number of input/output samples) (Figure 1). Then, taking advantage of the parity of both the input and output functions more time can be saved. The computing time on an IBM 360/75 of the FT

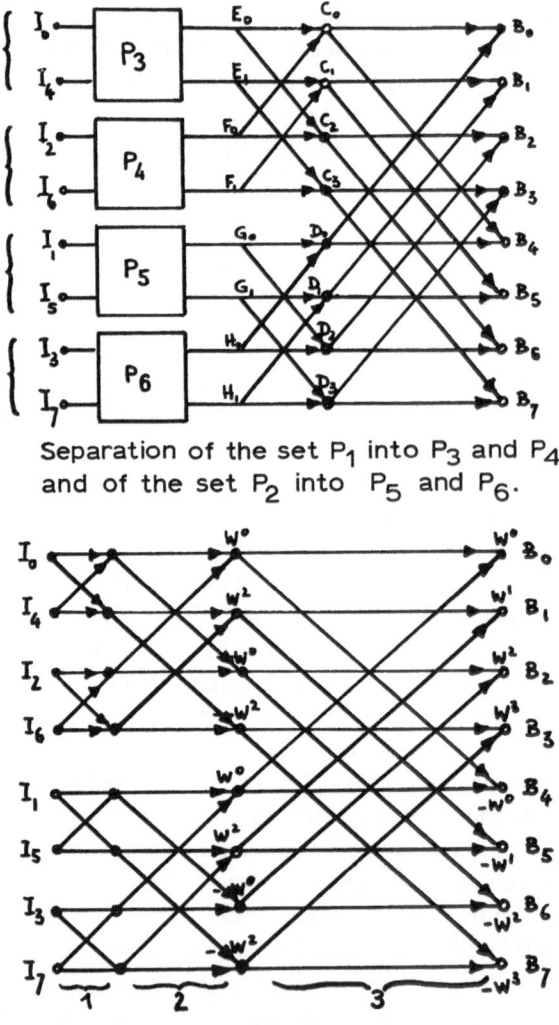

Separation of the set P_1 into P_3 and P_4 and of the set P_2 into P_5 and P_6.

Fig. 1.   Diagram of 'Decimation in Time' technique for an 8 points FFT.

of a real-odd function with one million samples is at present about 9 min. This allows a very fast treatment of a 2-dim. array of dimension 1000×1000 (Figure 2).

Let us now consider the limitations of this algorithm. First it is not a real-time one. You need the totality of the input samples to perform the transform. But the calculation is so fast that, in some cases, you may transform the part of the function you know before the arrival of the next input point.

Another disadvantage is that you calculate only the Shannon points on the FT (that is to say the samples taken at abscissa multiples of the Shannon interval: $1/2\,N$). But these samples are often not convenient (for example as when you want to plot the curve or study the fine shape of an absorption line). Then one is obliged to program the Shannon interpolation formula and this is quite difficult [2]. Because of the truncation of the sinc function which is the weighting function of this interpolation one must make an error. All that one can calculate is a coarse upper limit of this error and doing that obliges one to make certain assumptions about the shape of the spectrum. Some counter-examples prove [3] that one may make infinite errors when applying the Shannon interpolation to sampled and non strictly band-limited functions.

Another improvement [4] to the FFT allows one to compute only a few points of

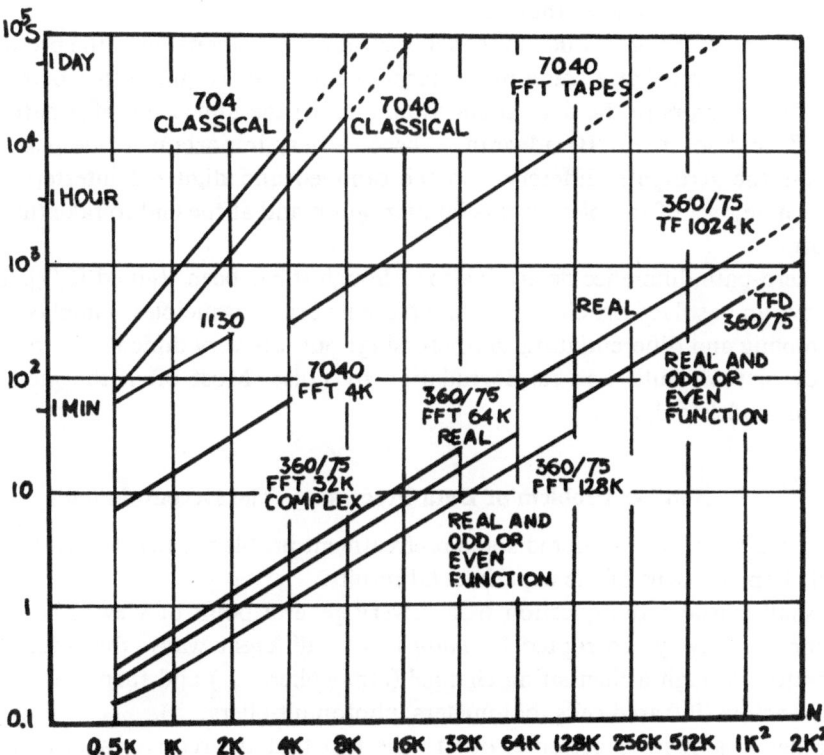

Fig. 2. Diagram of Computation Times for a FFT versus the number N of input/output samples. (1 k = 1024) Dotted curves are extrapolation.

the FT when there must be a great number of samples on the input function. Consequently as in the classical FT algorithm, you may choose your spectral domain.

3. Computers for Fourier Spectroscopy

The advantages of Fourier spectroscopy are very well known – large throughput, multiplex advantage, high resolution – and are too obvious to be discussed. But it is not a direct measurement of the spectrum which is obtained by Fourier transforming the interferogram. And the spectroscopist wants to get the spectrum as rapidly as possible. This seems to be no longer a problem.

Fourier analysis has become more and more important in the last few years and it is amazing to see the number of Fourier analyzers or computers that have been studied and are often available on the market. A very complete review of them has been made by Connes and Michel [1] and Hoffman [5]. They consider special-purpose devices (both digital and analog) as well as the computers implementing the FFT and are connected with multi-purpose large computers. They discuss the ambiguous notion of 'real-time' and pseudo real-time. The point I want to emphasize here is that the Fourier analyzers can be chosen only when the experiment is very well defined and in routine exploitation. When flexibility is desired or dictated by the improvements or changes in the spectrometer they fail.

Another disadvantage is due to the fact that they are not capable of the pretreatment of the data. I think that a small multi-purpose computer is quite often better suited. It can correct errors in the data acquisition, discard the unneeded information, filter the input, and so on.... Its advantages become very important if one is obliged to determine the zero-path difference on the sampled and digitized interferogram, to generate a new set of samples on this interferogram and at the end to take the Fourier transform.

This zero-path difference determination, though it has been studied [6, 7], is, as far as I know, an unsolved problem. It is based upon various principles: matched filtering, oversampling and differentiating, zero crossing; but it is very difficult [8] to study the accuracy of the results and its degradation by noise. Most often the methods are tested on models only.

4. Another Problem of Data Processing: The Deconvolution

In Section 3 we have considered some pretreatment problems. Let us now look at the possible improvement of the experimental results.

We shall consider the question from a very general point of view. The measured quantities are always corrupted by *noise* in two different ways: the information is transmitted through a fluctuating channel (atmosphere...) and then is measured by noisy detectors (infrared cells, bolometers, photomultipliers...).

Another loss of information is due to the fact that the receiver always *filters* the quantity to be measured, altering those signal frequencies that are transmitted and failing to transmit those that are greater than the cut-off frequency [9].

In the absence of noise, this deterioration of the information is theoretically of no importance because if the measured quantity is either time or band limited, a purely mathematical approach allows one to restore the information to its true value.

But in the presence of noise the deconvolution problem becomes very difficult and few results have been published about the restoration of noisy data.

If one cannot recover the information beyond the cut-off frequency then it is best to correct the attenuation due to the filter. This is not, in our opinion, a deconvolution. We may call it an apodization. Many different techniques have been studied [10, 11] in order to restore the 'image' in a certain sense: maximization of the contrast,

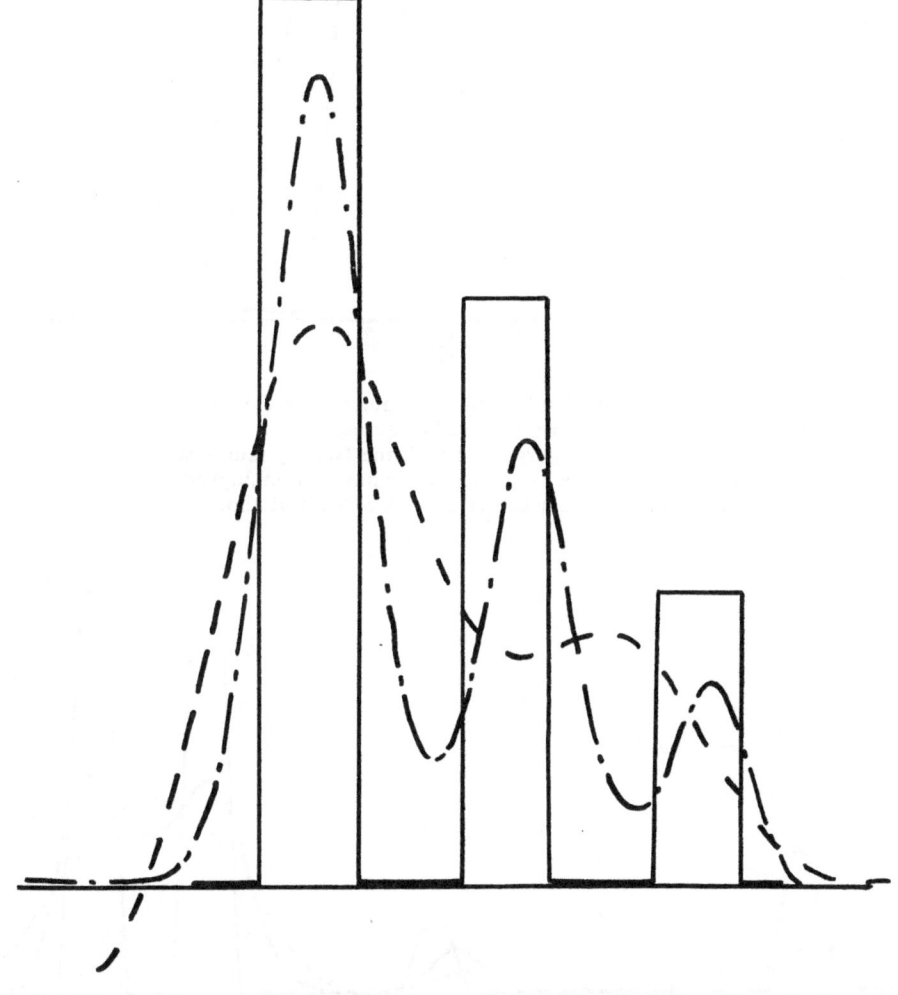

Fig. 3.   An example of positive deconvolution: Result of Biraud's method.

——: The true profile to be restored.

− − −: The result after correction of the broadening due to the transfer function of the instrument.

·—·—: The positive restoration; the resolving power has been multiplied by 3.

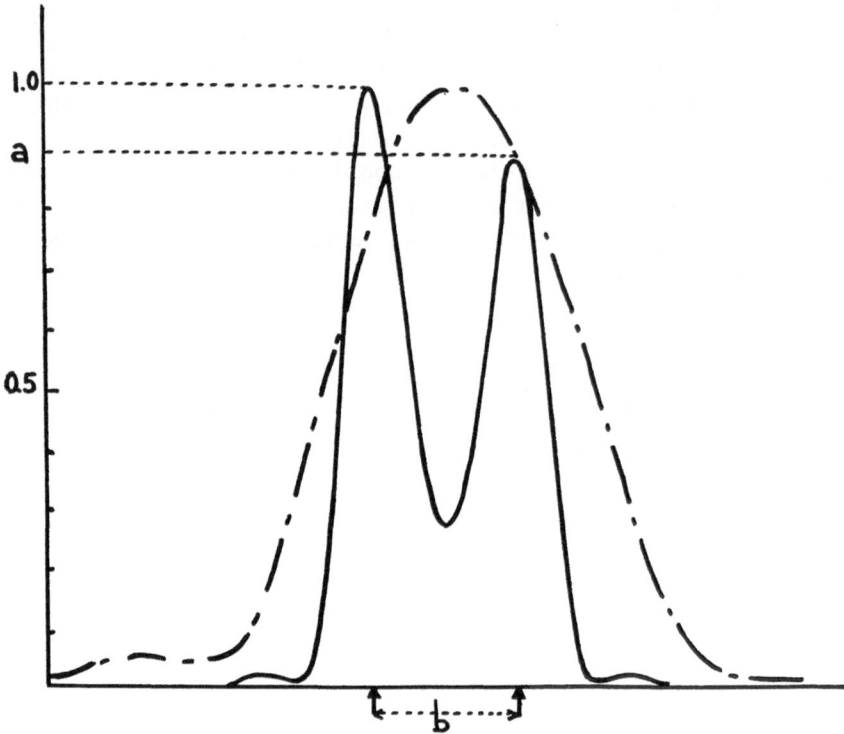

Fig. 4. Biraud's deconvolution applied to radio-astronomy. The E-W brightness distribution of Cygnus A radio-source.
—·—: The measurement performed with the Nançay radio-telescope.
———: The positive deconvolution. The resolving power has been multiplied by 2.75. The parameters a and b are correct (error 1%). S/N ratio: about 80.

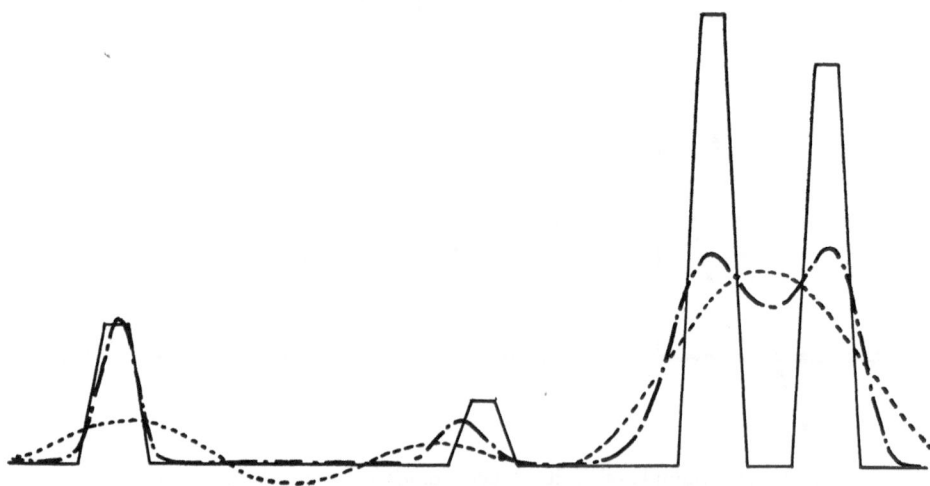

Fig. 5. Frieden's deconvolution technique.
- - -: Inverse filtering.
—·—·—: Positive restoration.

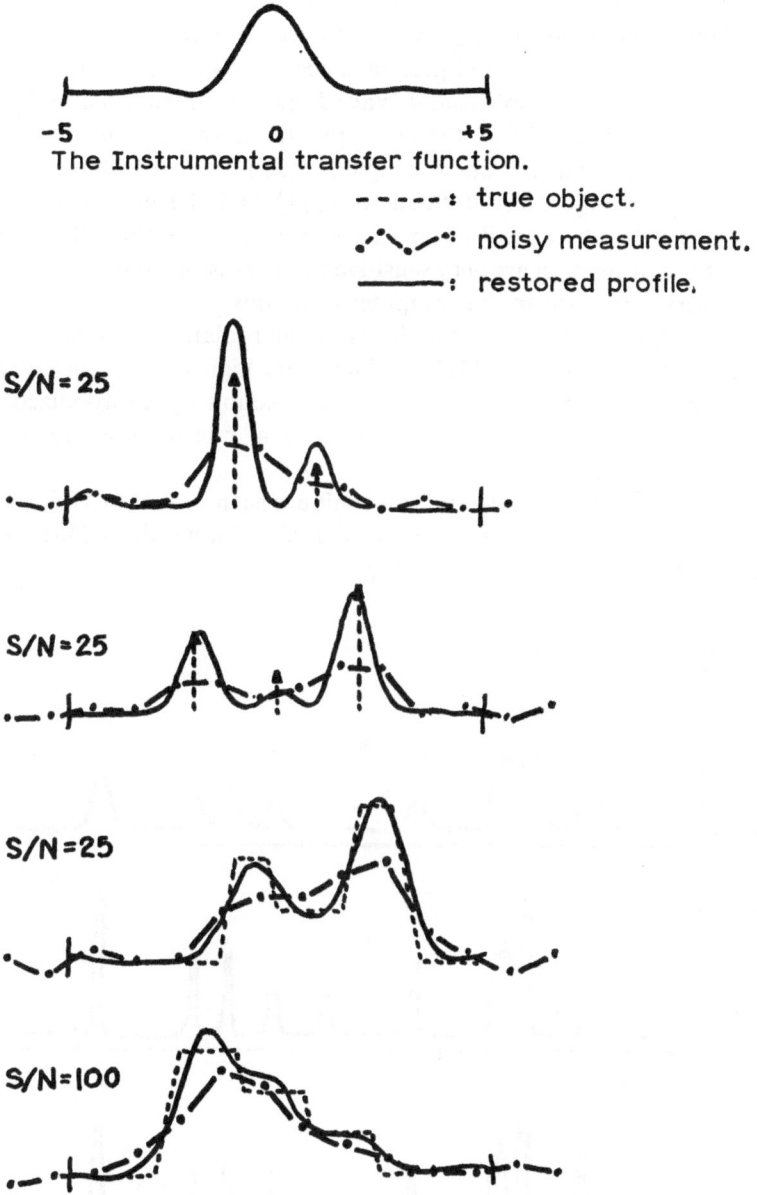

Fig. 6. Hershel's deconvolution technique. Deconvolution of Dirac impulse functions and of 'staircase' functions.

separation of a satellite from a strong line; but these improvements are always performed at the cost of other image characteristics which are altered.

However if one wishes to recover the information beyond the cut-off one must add some a priori information concerning the function to be restored.

Many analytic continuation techniques have been studied [12, 13, 14, 15]. Some of them make use of the Prolate Spheroidal Wave Functions introduced by Slepian and Pollak [16]. But they are very difficult to use in practice because of the noise and of the lack of precision in their calculation; they quite often provide oscillating restitutions. All the preceding techniques are difficult to apply and the published results concern only noise-free data because it has been shown [17, 18, 19] that all the techniques using an analytic continuation are very sensitive to the noise as well as to the necessary truncations of series involved in the computer programs.

Some other methods make use of the Shannon interpolation formula; one of them is a noise-free case [20], the other [21] considers very high signal-to-noise ratio data, and replaces the instrumental impulse response function by an apodized function, everywhere positive and also narrower than the real one in order to increase the resolving power.

Let us now mention the 'model-fitting' techniques such as that of Dolby [30]. They may be of some interest if one has an almost complete knowledge of the spectrum to be restored, but they may also be very dangerous.

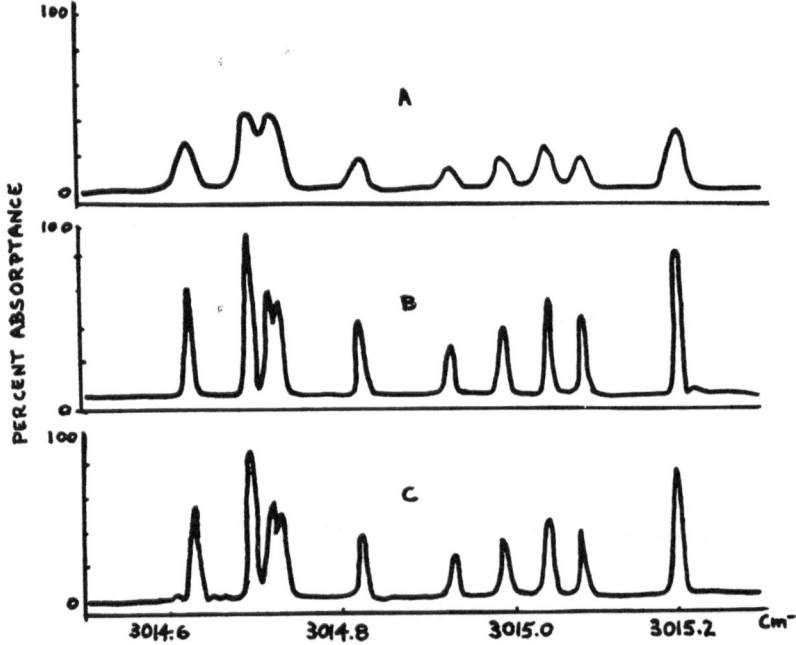

Fig. 7.   Jansson's deconvolution of a portion of a band of CH_4. Curve A: the measured spectrum. Curve B: the deconvolution with the measured response function. Curve C: the deconvolution with the response function approximated by a Gaussian function of the same half-width.

Restoration processes using the very weak constraint of positiveness of the source [9, 22, 23] have shown that one can achieve super-resolution even in the case of noisy data. Many later studies have tried to take advantage of the positiveness and even of the lower and upper bounds of the predicted 'object' [24] to [27]. Some of the preceding techniques are applicable to two-dimensional problems. (Figures 3, 4, 5, and 6.)

Finally, but not pretending to be exhaustive, we will mention a very interesting improvement of the Van Cittert iteration method used first to determine the response function of infrared spectrometers and also to enhance the spectra [28, 29]. (Figure 7.)

5. Walsh Functions, Hadamard Matrices

Joseph L. Walsh published his work on these functions to which he gave his name in 1922 [31]. The Walsh functions whose values are restricted to $+1$'s and -1's became of great interest when binary computers appeared. Although Fourier transformation has been extensively studied and although very fast algorithms have been found, it nevertheless is true that sine and cosine functions are not at all fitted to the computers' calculation [32].

Before looking at the applications of these very interesting mathematical tools, let us study their most important properties.

One way to obtain Walsh functions is to use Rademacher's functions [34] defined on the interval $(0, 1)$ where we have:

$$X_0(\theta) = 1$$

$$X_n(\theta) = \text{sign} (\sin (2^n \pi \theta)), \text{ Rademacher function of order } n.$$

Their properties are:

Null mean value $\quad \bar{X}_n = 0 \quad \forall_n$

Orthogonality $\quad \langle X_n, X_m \rangle = \delta_{m,n}.$

However they do not form a complete set.

Let us now define the Walsh functions. If we write n in its binary form:

$$n = \sum_{i=0}^{\infty} n_i 2^i$$

then

$$\text{wal} (n, \theta) = \prod_{i=0}^{\infty} (X_{i+1}(\theta))^{n_i}$$

and

$$\text{wal} (m, \theta) \times \text{wal} (n, \theta) = \prod_{i=0}^{\infty} (\text{wal} (2^i, \theta)^{m_i \dot{+} n_i}$$

with

$$-\tfrac{1}{2} \leqslant \theta < +\tfrac{1}{2} \quad \text{(See Figures 8 and 9.)}$$

and where $\dot{+}$ denotes the modulo-two addition.

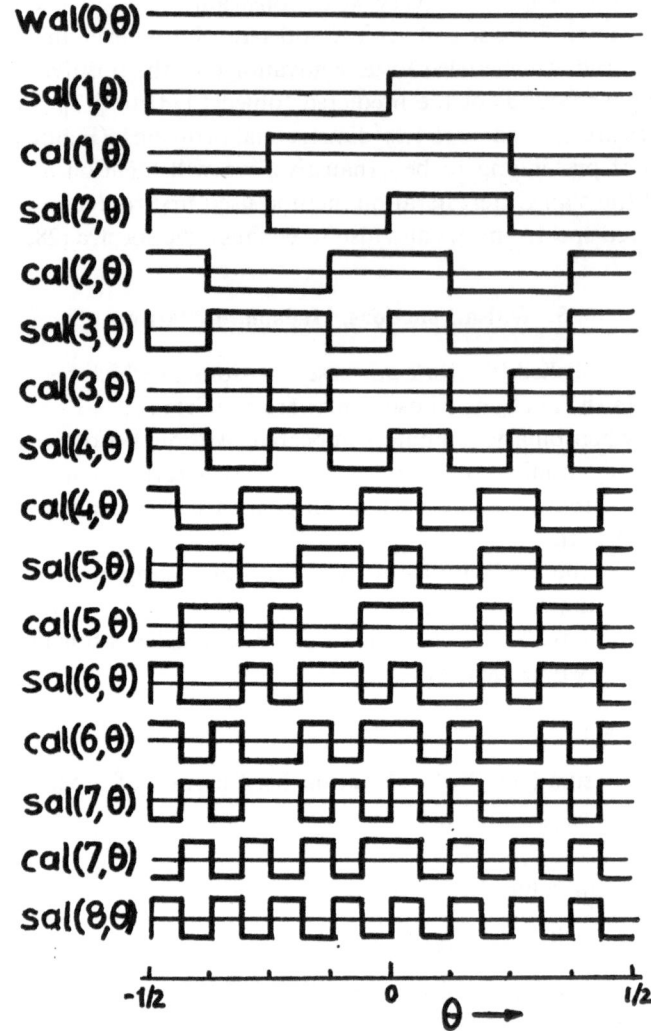

Fig. 8.   Walsh functions wal $(0, \theta)$, cal (i, θ), and sal (i, θ).

As a definition we also can use an iterative formula:

$$\text{wal } (2j + p, \theta) = (-1)^{[j/2]+p}\{\text{wal } (j, 2(\theta + \tfrac{1}{4})) + (-1)^{j+p} \text{ wal } (j, 2(\theta - \tfrac{1}{4}))\}$$

$$\text{wal } (0, \theta) = 1 \quad \text{if } \theta \in (-\tfrac{1}{2}, +\tfrac{1}{2}]$$

with

$p = 0$ or 1

$j = 0, 1, 2, \ldots$

$[j/2]$ is the greatest integer not exceeding $j/2$.

Now let us summarize their properties:

$$\text{wal } (h, \theta) \times \text{wal } (k, \theta) = \text{wal } (h \oplus k, \theta),$$

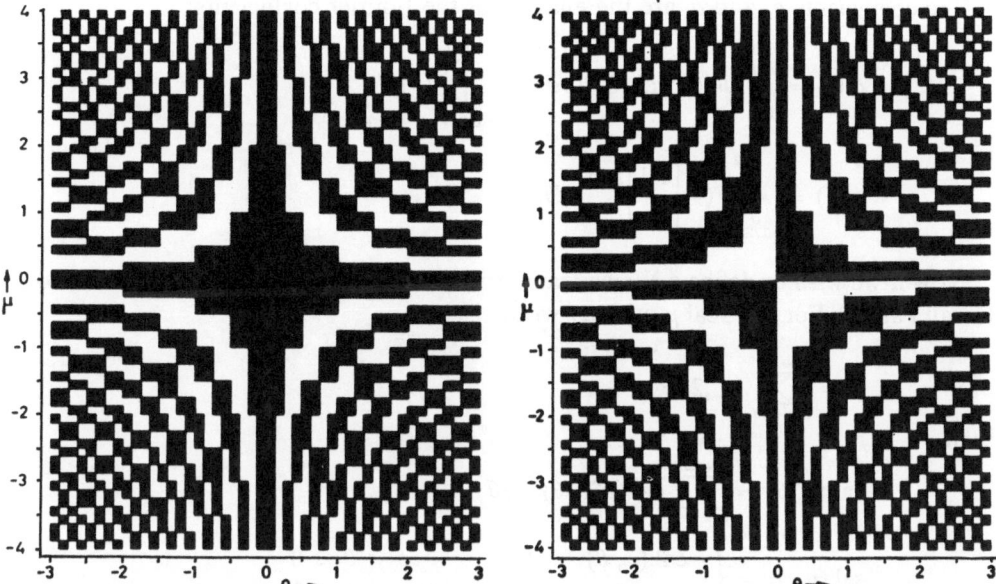

Fig. 9.   (left) The functions cal (μ, θ) in the interval $-3 < \theta < +3$, $-4 < \mu < +4$. A function, e.g. cal $(1.5, \theta)$, is obtained by drawing a line at $\mu = 1.5$ parallel to the θ-axis. cal $(1.5, \theta)$ is $+1$ where this line runs through a black area and -1 where it runs through a white area. At borders between black and white areas use the value holding for the absolutely larger μ. The function cal $(\mu, 1.5)$ is obtained by drawing a line at $\theta = 1.5$ parallel to the μ-axis and proceeding accordingly.
(right) The functions sal (μ, θ) in the interval $-3 < \theta < +3$, $-4 < \mu < +4$. The values $+1$ and -1 of the functions are obtained by drawing lines as explained in the caption of Figure 12. At borders between black and white areas use the value holding for the absolutely smaller μ or θ. There are no functions sal $(0, \theta)$ or sal $(\mu, 0)$.

where $\oplus$ is the modulo-two addition.

$$\text{wal } (j, \theta)^2 = \text{wal } (j \oplus j, \theta) = \text{wal } (0, \theta) = 1$$
$$\text{wal } (j, \theta) \times \text{wal } (0, \theta) = \text{wal } (j, \theta).$$

The preceding properties confer to the Walsh functions the character of a commutative Abelian group. Since they are orthogonal and complete we can define, exactly like a Fourier series expansion, a Walsh series expansion of any function $F(\theta)$ defined on $[-\frac{1}{2}, +\frac{1}{2}]$,

$$F(\theta) = a(0) \text{ wal } (0, \theta) + \sum_{i=1}^{\infty} [a_c(i) \text{ cal } (i, \theta) + a_s(i) \text{ sal } (i, \theta)],$$

where functions cal and sal are the analogs of cosine and sine. Cal and sal are defined by:

$$\left. \begin{array}{l} \text{wal } (2i, \theta) = \text{cal } (i, \theta) \\ \text{wal } (2i - 1, \theta) = \text{sal } (i, \theta) \end{array} \right\} \theta \in [-\tfrac{1}{2}, +\tfrac{1}{2}]$$

both being zero elsewhere.

The coefficients a_c and a_s are the analogs of the Fourier coefficients:

$$a(0) = \int_{-1/2}^{1/2} F(\theta) \, \text{wal} \, (0, \theta) \, d\theta = \int_{-1/2}^{1/2} F(\theta) \, d\theta$$

$$\begin{matrix} a_c(i) \\ s \end{matrix} = \int_{-1/2}^{1/2} F(\theta) \begin{vmatrix} \text{cal} \, (i, \theta) \\ \text{sal} \, (i, \theta) \end{vmatrix} d\theta.$$

Now if we wish to define a Walsh transform we are obliged to extend the definition for all i and θ. For any real μ we may write

$$\mu = \sum_{s=-\infty}^{+\infty} \mu_{-s} 2^{-s}$$

then

$$\text{cal} \, (\mu, \theta) = \prod_{s=-\infty}^{+\infty} \text{cal} \, (\mu 2^s, \theta) \quad \forall \theta$$

$$\text{sal} \, (\mu, \theta) = \begin{cases} -\text{cal} \, (\mu, \theta) & \text{for} \quad -\infty < \theta < 0 \\ \text{cal} \, (\mu, \theta) & \text{for } 0 < \theta < +\infty \end{cases}$$

where, g being an even number, $\mu = (g+1)/2^M$ is a dyadic rational.

Note that cal (μ, θ) and sal (μ, θ) are respectively symmetric and skew-symmetric when μ is considered as the variable. Then we may define the Walsh transform of $F(\theta)$ for all θ:

$$\begin{matrix} a_c(\mu) \\ s \end{matrix} = \int_{-\infty}^{\infty} F(\theta) \begin{vmatrix} \text{cal} \, (\mu, \theta) \\ \text{sal} \, (\mu, \theta) \end{vmatrix} d\theta$$

and the inverse Walsh transform:

$$F(\theta) = \int_0^{\infty} [a_c(\mu) \, \text{cal} \, (\mu, \theta) + a_s(\mu) \, \text{sal} \, (\mu, \theta)] \, d\mu.$$

Walsh functions are not periodic if μ is not binary rational. But, exactly as frequency is half the number of zero crossings per second, 'sequency' is half the *mean* number of zero crossings per second for a Walsh pulses waveform.

We also can define a power spectrum which is:

$$\int_{-\infty}^{+\infty} F(\theta)^2 \, d\theta = \int_0^{\infty} [a_c(\mu)^2 + a_s(\mu)^2] \, d\mu.$$

But we fail in trying to define an amplitude spectrum which is the square root of the power spectrum because for cal and sal there does not exist such a relation as:

$$A \cos x + B \sin x = (A^2 + B^2)^{1/2} \cos (x - \phi)$$
$$\text{with} \quad \phi = \tan^{-1} (B/A).$$

For the same reason the Wiener-Kintchine theorem does not apply. But it appears that we can find an analog theorem.

On the other hand the Walsh transform has a very interesting property not shared by Fourier transform. The WT of wal $(0, \theta)$, cal (i, θ), sal (i, θ) vanish outside $[-1, +1]$, $[-(i+1), i+1]$, $[-i, i]$. Hence functions that are sums of such Walsh functions, with finite i only are both time and sequency limited.

We also can elaborate a theory of filtering and modulation/demodulation by the Walsh functions. An interesting fact is that such a modulation yields only a single side band (instead of double for sine and cosine).

All the preceding properties are very well described by Harmuth [35, 36]. It is also possible to define a dyadic convolution [37] and to derive special convolution theorems for Walsh transforms analogous to those of Fourier transforms:

$$\text{WT} (f \otimes g) = \text{WT}(f) \cdot \text{WT}(g)$$

$$\text{WT}(f \cdot g) = \text{WT}(f) \otimes \text{WT}(g).$$

Let us now point out the relationship between Walsh functions and Hadamard matrices. Welch [38] has looked for a system of orthogonal functions operating on the set of integers $\{1, 2, \ldots, N\}$ with values restricted to $[-1, +1]$ and having a fast transform algorithm. Walsh functions satisfy all these properties. Now if we suppose they are arranged in a matrix H such that:

$$h_{i,k} = \text{wal} (i, k),$$

then the matrix H called a Hadamard matrix satisfies:

$$H \cdot H^t = nI.$$

Welch has also studied the Williamson matrices and how to arrange them to create high-order Hadamard matrices.

In the end, he defines the 'group characters' of Abelian groups, characters which must satisfy a certain number of properties that enable a functional decomposition on a set composed of N orthogonal functions. Then he shows that these sets of group characters all possess a FFT algorithm. And if N is factorized into primes:

$$N = p_1 \cdot p_2 \cdots p_k,$$

this algorithm needs only

$$N \sum_{i=1}^{k} (p_i - 1)$$

multiplications and additions.

A quite similar derivation of the analogy between FFT and FWT has been underlined by Shanks [39]. It is based only on the iterative equation defining the Walsh functions. The flowgraph is the same but you simply have to suppress the multiplications by sines and cosines.

Another interesting result [40] related to Fourier and Walsh transforms is the square wave Fourier transform. In certain circumstances (when, strictly speaking, Shannon's theorem does not apply) its results are more accurate than those of the DFT and in addition the square waves are easier to generate than sines, cosines, or Walsh functions.

Gebbie and Gibbs [41] notice that the WT of the 'logical' interferogram is the Walsh power spectrum. If the experiment provides one with the classical interferogram one has to change it by a simple linear transformation into the logical one before taking

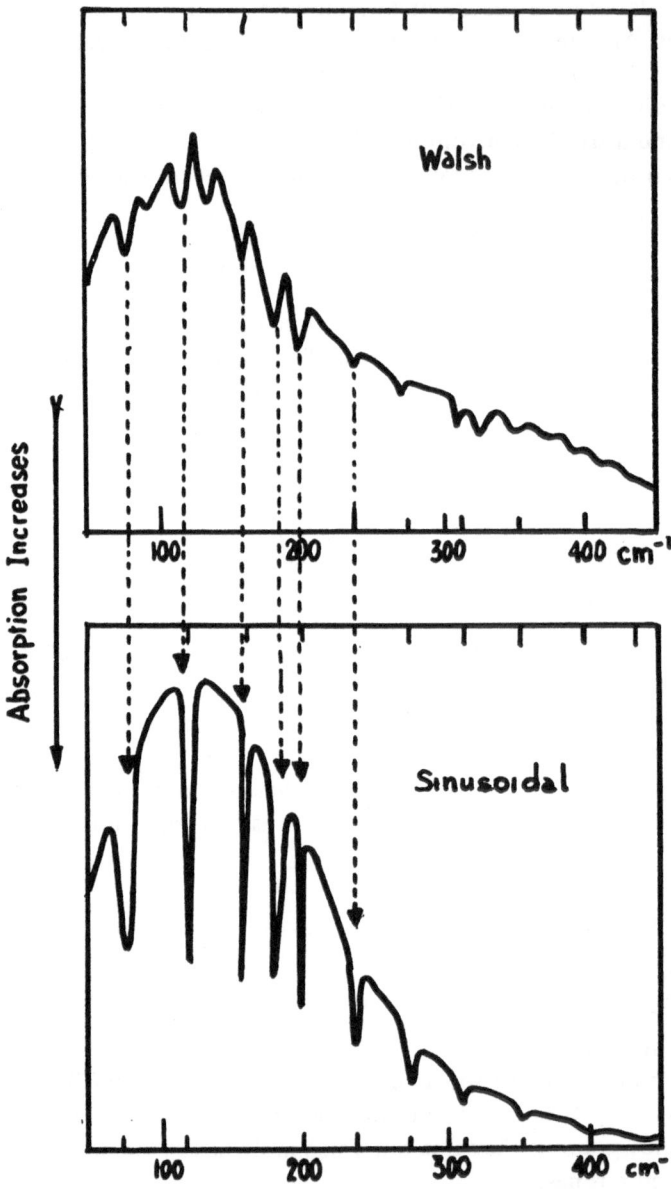

Fig. 10.   The Walsh transform of a 'logical' interferogram compared to the Fourier transform of the classical one.

the WT. The two authors compare the FT of the classical interferogram to the WT of the logical one. The features are the same in both, although the relative amplitudes are distorted. But if one has to solve yes/no decision problems, this method may be very interesting. (See Figure 10.)

6. Applications of Hadamard Matrices to Spectroscopy

The idea of multiplex-encoding the output, the input (or both) of a classical spectrograph is not new. Golay, Felgett, and Girard for example, have taken advantage of this method. If M is the number of spectral elements to be analyzed in the spectrum the signal-to-noise ratio is improved by a factor $(M)^{1/2}$ compared to a sequentially scanned spectrometer.

But the use of Hadamard matrices as masks is quite recent. Ibbet *et al.* [41] in 1968 have described an encoding scheme that appeared to correspond to the Walsh functions. At the same time Decker *et al.* [42] proposed a more flexible encoding based on the theory of simultaneous equations.

Let us first described such multislit-spectrometers. The incoming radiation passing through the entry slot is dispersed by a classical device (prism, grating, ...), then focused at the exit plane. The mask consisting of transparent and opaque slots parallel to the image of the entry slot is placed in this plane where it modulates the energy distribution in the spectrum. The exiting light transmitted by the mask is then focused on the detector. The different mask configurations are obtained by revolving the mask

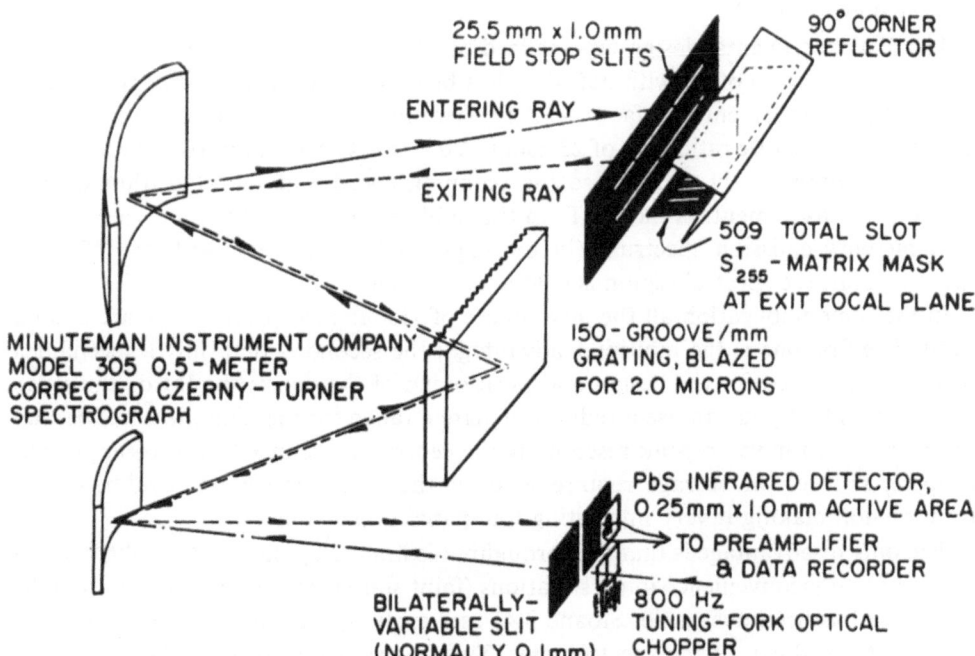

Fig. 11. Optical schematic of Comstock and Wescott/AFCRL HTS spectrometer.

(if it is circular) or translating it by a slot width, at each time, in order to get the M different masks. (See Figure 11.)

Let us examine the requirements needed for the mask design. It has to be formed of slots with transmittance 0 or 1. If we consider the matrix whose columns represent the different mask positions, then this matrix has to be cyclic. Each column is obtained from the preceding one by a shift of one position. This arrangement is due to the manner in which one drives the mask step by step. The other requirement is that since the detector is noisy, one has to find the best estimate of the spectrum in the mean square sense.

The matrices that fulfil these requirements are Hadamard matrices or are obtained from them [43]. If H is the normalized Hadamard matrix of order N, H is a first solution having only 1's in the first row and column. If one suppresses them a sub-matrix G results which is also a solution. The matrix S obtained from G by replacing +1's by 0's and −1's by +1's is another solution. And it is easy to show that the error ε made in the estimation is

$$\varepsilon = \sigma^2 \cdot \text{Trace} \left[(P^{-1})^t \cdot P^{-1} \right],$$

where

$$\sigma^2 = \langle n_i^2 \rangle$$

is the mean square of the noise

$$P = H \text{ or } G \text{ or } S.$$

The first rows of G are given by Sloane *et al.* [44] in order to get cyclic matrices of various dimensions.

The first experimental design was realized by Decker [43] and had a mask with only 19 slots. But now another with 255 slots has been constructed providing a signal-to-noise ratio improvement by a factor 8. A further modification of this spectrometer will have a signal-to-noise ratio gain of 23 using a 2047×2047 S type matrix. The mask will be driven continuously to increase the scanning speed and the algorithm used to invert the measurements is the FHT. In the publication the authors give the example of a mercury emission spectrum (in the region $1.6 \, \mu$) obtained with the 255 slot-spectrograph; the spectral region is $0.34 \, \mu$ wide. (Figure 12)

In the same publication all the advantages of Hadamard spectroscopy are emphasized. The first one is the multiplex advantage; the second is that only a mechanical precision is needed for the design and positioning of the mask; the third, and it is a great one, is that you can use a reduced dynamic range for the digitalization, because in this case you have no peak resembling the central fringe in FT spectroscopy. This property reduces the storage requirements or the data rate (or bandwidth) for the transmission making it very interesting for spatial applications.

The only disadvantage is that the throughput is limited by the entrance slit. Though this is not an inconvenience in all situations (faint star spectroscopy for example) this limitation has been overcome. Sloane [44] uses cyclic Hadamard matrix masks both at the entry and exit of the spectrometer. This allows one to recover the throughput advantage. Phillips *et al.* [45] have built such a laboratory experimental device. They

```
00000 00001 00000 00010 10000 00100 01000 01010 10100 10000
00011 01000 00111 00100 01101 11010 11101 01000 10100 00101
00010 01000 10101 10101 00001 10000 10011 11001 01110 01110
01011 11011 10010 01010 11101 10000 10101 11001 00001 01110
10010 01010 01101 10001 11101 11011 01101 01011 11000 00010
01100 00101 11110 01001 00011 10110 10110 10110 00110 00111
01111 01101 01001 01100 00110 01110 01111 11011 11000 01010
01100 10001 11111 01011 00011 10010 10110 11100 00110
10110 01110 00111 11011 01100 01011 01110 10011 01010 01111
00001 11001 10011 01111 11111 01000 00001 00100 00010 11010
00100 11001 01011 11110 00010 00011 00101 00111 11000 11100
01101 10110 11101 10110 10101 10110 00001 10111 00011 10101
10110 10001 10110 01011 10111 10010 10100 11100 00011 10110
00110 10111 01110 00101 01011 01000 00011 00100 00111 11010
01100 01001 11110 10111 00010 00101 10101 01001 10000 00111
11000 01100 01100 11110 11111 10010 10000 11100 01001 10110
10111 10110 00100 10111 01011 00101 00011 11000 10110 01101
00111 11100 11100 00111 10110 01100 10111 11111 00100 00001
11010 00011 01001 00111 10110 11101 11110 10101 00010 00000
10101 00001 00000 10010 10001 01100 01010 01110 10001 11010
01011 01001 10011 00111 11111 11100 00000 00110 00000 01111
00000 11001 10001 11111 11011 00000 00111 10000 10010 11001
01100 11110 01111 10011 11000 11110 01101 10011 11101 11110
00101 00011 01000 10111 00101 00101 11000 11001 01101 11110
01101 00011 11100 10110 00011 00111 01101 11101 01101 00100
01100 11010 11111 11000 10000 01101 01000 11100 00101 10110
01001 10111 10111 10100 10100 10011 01111 10111 10111 01000
10101 00101 00000 11000 10001 11101 01011 00100 00011 11010
00110 01001 01111 10110 01000 10111 10101 00100 10000 11011
01001 11011 00111 01011 11101 00010 00000 10101 01011 00000
00011 10000 00110 11000 01110 11100 11010 10111 11000 00100
01100 01010 11110 10000 10010 01001 01101 10110 01101 10111
11101 10100 00101 10010 01001 11101 00101 01011 10101 11001
10001 01111 11010 01000 01001 10100 10111 10011 00100 11111
11011 10000 01010 11000 10000 11101 01001 10100 00111 10010
01100 11101 11111 10101 00000 10101 01001 01010 00110
00001 01111 00010 01001 10101 10111 10001 10100 11011 10011
11010 11110 01000 10011 10101 01110 10000 01010 01000 10001
10101 01011 10000 00010 11000 11100 11001 10111 11000 00100
10110 01100 00111 11110 01100 00011 11110 00010 00011 01111
00111 01001 11101 00111 00100 11101 10101 01010 01010 10000
00000 01000 00000 10100 00001 00010 00010 10101 00100 00000
11010 00001 11001 00011 01110 10111 01010 00101 00001 01000
10010 00101 01101 01000 00101 00110 01110 01011 10011 10010
11110 11100 10010 10111 01100 00101 01110 01000 01011 10100
10010 10011 01100 01111 01110 11001 01010 11110 00000 10011
00001 01111 10010 01000 01011 10101 10011 10001 10001 11011
11011 01010 01011 00001 10011 10011 11110 11110 00010 10011
00100 01111 11010 11000 01000 11100 10101 10111 00001 10101
10011 10001 11110 11011 00010 11011 10100 11010 10011 11000
01110 01100 11011 11111 11010 00000 01001 00000 10110 10001
00110 01010 11111 10000 10000 11001 01001 11110 00111 00011
01101 10111 01001 10101 10000 10101 11000 11101 11101 01101
10100 01101 10010 11101 11100 10101 00111 00000 11101 10001
10101 11011 10001 01010 11010 00000 00001 11110 10011
00010 01111 10101 11000 10001 01101 01010 01100 00001 11110
00011 00011 00111 10111 11100 10100 00111 00010 01101 10101
11101 10001 00101 11010 11010 01001 00011 01101 10101
11111 00111 00001 11101 10011 00101 11111 11001 00000 01110
10000 11010 01001 11001 10111 01111 10101 01000 10000 00101
01000 01000 00100 10100 01011 10010 10100 01110 10010
11010 01100 11001 11111 11111 00000 00001 10000 00011 11000
00110 01100 01111 11110 11000 00010 11100 00100 10110 01011
00111 10011 11100 11110 10011 01111 11111 01111 11100
01000 11010 00101 11001 01001 01110 00110 01011 01111 10011
01000 11111 00101 10001 11001 11011 01111 01011 01001 00011
00110 10111 11110 00100 00011 01010 00111 00001 01101 10011
01101 11101 11101 00101 00100 11000 11011 11101 11010 00101
01001 01000 00110 00100 01111 01010 11010 00000 11110 10001
10010 01011 11101 10010 00010 11101 01001 00100 00110 11010
01110 11001 11010 11111 01000 10001 00101 01010 11000 00000
11100 00001 10110 00011 10111 00110 10101 11110 00001 00011
00010 10111 10100 00100 10010 01011 01101 10011 01101 11111
01101 00001 01100 10010 01111 01101 11001 01101 01110 01100
01011 11110 10010 00010 01101 00101 11100 11001 00111 11110
11100 00010 10110 00100 00111 01010 01101 00001 11100 10011
00111 01111 11101 01000 00100 00100 01010 01010 10001 10000
01011 11000 10010 01101 11110 00110 11100 11110 11001 11000
10111 10010 00100 11101 01011 10100 00010 10010 00100 01101
01010 11100 00000 10110 00001 00111 00010 11101 10100 10101
10011 00001 11111 10011 00000 11111 10001 10000 11011 11001
11010 01111 01001 11001 00111 01110 11101 01010 101
```

Fig. 12.   4093-slot encoding mask code used with 2047-slot HTS system

compare these doubly multiplexed spectrometers to Michelson's interferometers and conclude that they are quite similar in performance. But questions like reflection and transmission losses, sizes of gratings and beam splitters have to be considered in detail to provide a good comparison.

We also want to mention, although that is not the subject of this communication, that 2-dim. image processing using Walsh transformation for coding, data compression, transmission, pattern recognition ... is now frequently used. H. C. Andrews, W. K. Pratt, and K. L. Caspari for example, have published theoretical as well as practical studies related to these various applications.

7. Conclusion

We hope that these few examples have shown all the interest of signal processing. We also want to point out that we have noticed that the most striking progress has taken place only when a sufficient theoretical basis has been established. We think that plenty of interesting results have been achieved but that numerous problems need further examination.

References

[1] Connes, P. and Michel, G.: 1970 'Real-time Computer for Fourier Spectroscopy', Aspen International Conference on Fourier Spectroscopy.
[2] Connes, J.: 1970, 'Computing Problems in Fourier Spectroscopy', 1970 Aspen International Conference on Fourier Spectroscopy.
[3] Standish, C. J.: 1967, Two Remarks on the Reconstruction of Sampled Non-band Limited Functions, *IBM J.*, Nov.
[4] Morven Gentleman, W.: 1968, *Bell System Tech. J.* 47, No. 6.
[5] Hoffman, J. E.: 1968, AFCRL Instrumentation Paper No. 146.
[6] Bosomworth, D. R. and Gush, H. P.: 1965, *Can. J. Phys.* 43, 740.
[7] Delouis, H.: 1968, Thèse de 3ème Cycle, Paris.
[8] Biraud, Y.: 1969, 'Spectroscopie par transformation de Fourier en infrarouge lointain', IRS/10. Rapport interne.
[9] Biraud, Y.: 1969, 'Solution nouvelle au problème de déconvolution en présence de bruit', *Comptes-Rend. du Deuxième Colloque du Groupe d'Etude du Traitement du Signal. Nice, Mai.*
[10] Arsac, J.: 1961, *Transformation de Fourier et théorie des distributions*, Dunod, Paris.
[11] Jacquinot, P. and Roizen-Dossier, B.: 1964, *Progr. Optics* 3, 31.
[12] Barnes, Casper W.: 1966, *J. Opt. Soc. Am.* 56, No. 5.
[13] Scheidecker, J.-P.: 1968, Thèse de Doctorat de 3ème Cycle, Paris.
[14] Williams, R. A. and Chang, W. S.: 1966, *J. Opt. Soc. Am.* 56, No. 2.
[15] Frieden, B. R.: 1967, *J. Opt. Soc. Am.* 57, No. 58.
[16] Slepian, D. and Pollak, H. O.: Jan. 1961 and July 1962, 'Prolate Spheroidal Wave Functions, Fourier Analysis and Uncertainty', Three parts in *Bell System Tech. J.*
[17] MacPhie, R. H.: 1962, Tech. Report, No. 58, EERL, University of Illinois, Urbana, Ill.
[18] Ville, J. A.: 1956, *Cables et transmissions*, p. 1–44.
[19] Wolter, H.: 1964, *Progr. Optics* 1, 202.
[20] Harris, J. L.: 1964, *J. Opt. Soc. Am.* 54, No. 7.
[21] Robaux, O. and Roizen-Dossier, B.: 1970, Part 1: *Optica Acta*, 17, No. 10; Part 2: *Optica Acta* 17, No. 11.
[22] Biraud, Y.: 1969, *Astron. Astrophys.* 1, 124.
[23] Hay-Chee, Frank Wong: 'A Study and Evaluation of Biraud's Deconvolution Algorithm', M. of S. Thesis to be submitted.

[24] Hershel, R. S.: 1970, *Positive Restoration of Non-Coherent Object Scenes Degraded by Diffraction and Noise*, Optical Sciences Center, The University of Arizona, Tucson, Ariz.
[25] Frieden, B. R.: 1971, Tech. Rep. No. 67, Optical Sciences Center, The University of Arizona, Tucson, Ariz.
[26] Frieden, B. R.: 1971, *Maximum Entropy: a First Principle for Obtaining positive Restorations*, Optical Sciences Center, The University of Arizona, Tucson, Ariz.
[27] Hershel, R. S.: 1970, 'Numerical Restoration of Object Scenes', Optical Sciences Center, The University of Arizona, Tucson, Ariz.
[28] Jansson, P. A.: 1970, *J. Opt. Soc. Am.* **60**, No. 2.
[29] Jansson, P. A., Hunt, R. H., and Plyler, E. K.: 1970, *J. Opt. Soc. Am.* **60**, No. 5.
[30] Dolby, R. M.: 1958, *Proc. Phys. Soc.* **73**, 81.
[31] Walsh, J. L.: 1923, *Am. J. Math.* **45**, 5.
[32] Gebbie, H. A.: 1970, 'Walsh Functions and the Experimental Spectroscopist', *Applications of Walsh functions, Proc. Symp. NRL, Washington D.C.*
[33] Gibbs, J. E. and Gebbie, H. A.: 1969, *Nature* **224**.
[34] Koenig, M. and Zolesio, J.-P.: 1970, 'Les fonctions de Walsh', *Comptes-Rendus du Colloque sur le traitement du Signal*, Faculté de Nice.
[35] Harmuth, H. F.: 1970, *Transmission of Information by Orthogonal Functions*, Springer-Verlag.
[36] Harmuth, H. F.: 1968, *IEEE Trans. Information Theory IT* **14**, No. 3.
[37] Pichler, F.: 1970, 'Walsh functions and Optimal Linear Systems', *Applications of Walsh functions, Proc. Symp.*, NRL, Washington DC.
[38] Welch, R.: 1970, 'Walsh Functions and Hadamard Matrices', *Applications of Walsh functions, Proc. Symp. NRL, Washington D.C.*
[39] Shanks, J. L.: 1969, *IEEE Trans. Comput.*, C 18.
[40] Bates, R. H. T., Napier, P. J., and Chang, Y. P.: 1970, *Electron. Letters* **6**, No. 23.
[41] Ibbet, R. N., Aspinall, D., and Grainger, J. F.: 1968, *Appl. Opt.* **7**, 1089.
[42] Decker, Jr., J. A. and Harwit, M. O.: 1968, *Appl. Opt.* **7**, No. 11.
[43] Sloane, N. J. A., Fine, T., Phillips, P. G., and Harwit, M.: 1969, *Appl. Opt.* **8**, No. 10.
[44] Sloane, N. J. A.: 1970, *Aspen International Conference on Fourier Spectroscopy*.
[45] Phillips, P. G., Harwit, M., and Sloane, N. J. A.: 1970, 'A new multiplexing Spectrometer with Large Throughput', *Aspen International Conference on Fourier spectroscopy*.

REMARKS ON INFRARED TECHNOLOGY AND SPECTROSCOPY IN ASTROPHYSICS

F. KNEUBÜHL

Infrared Physics Group, Solid State Physics Laboratory,
Swiss Federal Institute of Technology (ETH), Zürich, Switzerland

1. Spectrometers and Interferometers

Today interferometers [1] widely replace the diffraction grating spectrometers [2] in space research. This trend is caused by the wide spectral grasp and the Fellgett advantage of Fourier spectroscopy. For these reasons interferometers are well suited for the detection and study of weak signals from space. Inherent problems of the commonly used Michelson interferometers are

– the determination of absolute intensities;
– the high-frequency cut-off filter to avoid computing errors by 'Umklapp' processes; and
– the beam splitter.

An additional difficulty arises if weak narrow-band absorptions ought to be detected in an intense broad-band emission, e.g. weak absorptions in the solar infrared emission. In this case a diffraction grating monochromator may be advantageous.

The absolute-intensity calibration of an interferometer represents a serious task [3] and requires a careful test of stray or false radiation. The lack of high-precision black-body radiation standards [3] will be discussed later. The need of a high-frequency cut-off for the interferometers and the elimination of higher orders of diffraction gratings explains the interest and the work of many infrared astrophysicists in filters of various types [4].

The uneasiness about beam splitters of Michelson interferometers can be avoided by the construction of lamellar-grating interferometers [3]. These instruments provide the Fourier spectrum by the reflection of the parallel beam on the plane metallic surfaces of the lamellar instead of the splitting of the parallel beam by the mylar films or metallic mesh of the Michelson interferometer. The metallic reflection is extremely broad-band and does not depend on the temperature. However the construction of a low-weight precise lamellar-grating is costly and time-consuming.

Rockets as platforms of infrared astrophysical instrumentation have to rely upon observation times of a few minutes [5]. Therefore rocket-spectrometry is based on broad-band radiometers or filter spectrometers. Grating spectrometers or interferometers can only be applied if they are light, compact and shock-proof as well as capable of rapid scanning [6].

2. Black-Body Radiation Standards

The construction of black-body radiation standards [3] for the far-infrared spectral region is impeded by the following facts:

Manno and Ring (eds.), Infrared Detection Techniques for Space Research, 332–334. All Rights Reserved

(a) Thermal homogeneity and space conditions require black-body dimensions of the order of magnitude one hundred far-infrared wavelengths. This implies that far-infrared black-body radiation standards can be compared neither with single-mode microwave noise standards nor with optical black bodies.

(b) Compact wall materials possess Reststrahlen reflections which are strong and narrow-band. Therefore broad-band radiation standards need sponge or felt-like walls.

The problem of the Planck radiation formula for finite cavity dimensions has been theoretically solved only recently [7, 3]. The thermal radiation in a finite cavity, shows deviations from Planck's formula of the second order for the intensity and of the first order for the homogeneity and polarization. These deviations must be taken into consideration for the study of the 'isotropic submillimetre background' of the space.

Far-infrared black-body radiation standards should not be fitted with inner obstacles or ring-shield as found in optical black bodies. These obstacles imply unwanted resonances and considerable changes of the mode density. The shapes to be recommended are long cones or cone-terminated cylinders with a circular aperture. As mentioned before the structure of the wall-material should be sponge- or felt-like.

3. Lasers and Tunable Sources in the Infrared

Heterodyne detection in the infrared is strongly desired by astrophysicists for the discovery and observation of faint stellar objects emitting monochromatic infrared. However, heterodyne detection requires stable narrow-band sources, which should be tunable.

Several of today's tunable infrared sources are based on the powerful well-known CO_2-laser. Special attention should be paid to the recently discovered transverse excitation [8] hitherto successfully applied to about 150 pulsed stimulated emissions of CO_2 and 12 other gases. With gas pressures up to 1 atmosphere radiation pulses between 0.2 and 1 μ s and with a power up to 10 MW were observed.

As today's tunable infrared sources we should mention

(a) difference frequency mixing with magneto-plasma effects for phase-matching [9];

(b) the tunable spin-flip Raman laser [9];

(c) the tunable $LiNbO_3$ source based on stimulated Raman and parametric effects [10].

All these sources are pulsed and require a Q-switched or transversely excited high-power laser.

Another kind of tunable source is represented by the pseudotunable $Pb_{1-x}Sn_x$ Te diode laser [11], which is tunable in steps, jumping from one cavity mode to the next, with a small amount of tuning possible about each cavity mode.

In the field of submillimetre wave lasers (SMASERS) the generation of new emissions pumped by the CO_2-laser has been reported [12]. The more simple H_2O, D_2O, HCN and ICN-submillimetre wave-lasers have been studied extensively [13]. The development of special hollow cathode [14] for stable continuous plasma discharges and the possibility of transverse pulse excitation provide a stimulus for further studies and future applications.

4. Infrared Spectra of Dust and Powders

It has been attempted [15] to assign the observed spectra of interstellar or intergalactic dust to materials like enstatite, silicates, olivine, on the basis of the known infrared spectra of bulk samples. Assuming a particle size of the dust to be of the order of magnitude of a few microns, we wish to draw attention to the fact that powder or dust spectra differ considerably from bulk spectra. Recent theoretical and experimental studies on small alkalihalide crystals [16] and other powdered materials have demonstrated, that the Reststrahlen structure of the spectra of the bulk material is completely lost for the spectra of small particles. Dominating features of the latter are due to the so-called surface and bulk modes. Exceptions are to be expected for the spectra of typically molecular crystals like naphthalene and anthracene. Therefore the future study of interstellar or intergalactic dust requires previous extensive laboratory measurements of the powder and dust spectra of the material under consideration.

References

[1] See this volume, Session on Interferometers, p. 265.
[2] Herse, M.: 1972, this volume, p. 306.
[3] Baltes, H. P. and Stettler, P.: 1972, this volume, p. 160.
[4] See this volume, Session on Filters, p. 197.
[5] Shivanandan, K.: 1972, this volume, p. 73.
[6] Smyth, M. J.: 1972, this volume, p. 284.
[7] Baltes, H. P. and Kneubühl, F.: 1970, *Proceedings XX PIB Symposium*, p. 667; *Opt. Commun.* **2** (1970), 14; **4** (1971), 9; *Helv. Phys. Acta*, 40 pp., in press.
[8] Wood, O R. *et al.*: 1971, *Appl. Phys. Letters* **18**, 261.
[9] Patel, C. K. N.: 1970, *Proceedings XX PIB Symposium New York*, p. 135.
[10] Sussman, S. S. *et al.*: 1970, *Proceedings XX PIB Symposium, New York*, p. 211.
[11] Hinkley, E. D. and Harman, T. C.: 1968, *Appl. Phys. Letters* **13**, 49; **16** (1970), 351.
[12] Chang, T. Y. and Bridges, T. J.: 1970, *Proceedings XX PIB Symposium, New York*, p. 93.
[13] Steffen, H. and Kneubühl, F.: 1968, *J. Quant. Electron.* **4**, 992.
[14] Schötzau, H. J., Grathwohl, Ch., and Kneubühl, F.: 1971, *J. Appl. Math. Phys.* (ZAMP) **22**, 778.
[15] Hoffmann, W. F.: 1972, this volume, p. 8.
[16] Kälin, R. Baltes, H. P., and Kneubühl, F.: 1970, *Solid State Commun.* **8**, 1495; *Helvetica Phys. Acta* **43** (1970), 487.

7. GENERAL DISCUSSION

GENERAL DISCUSSION

(*directed by Prof. F. Kneubühl*)

1. *Dr Hoffmann* on 'Future Perspectives of Infrared Astronomy'

I shall very briefly restate some of the issues in astrophysics to which space astronomy in the infrared will be able to contribute.

A. INFRARED EXCESSES IN STARS

Many stars exhibit an infrared excess radiation above the Planck curve at 10 μ. This has been attributed to circumstellar dust shells. Moderate resolution spectroscopy at 8 to 14 and 20 μ indicates a structure similar to silicate particles, for example Olivine. As Prof. Kneubühl has pointed out, there is a great deal of laboratory work to be done in spectroscopy of such materials, particularly as fine powders. Also in the astronomical observations one needs to have much more continuity in the spectral observations in order to fill in regions between 5 and 8 μ, 14 to 20 μ, and beyond 25 μ. This clearly will be one of the interesting projects from aircraft, balloons, or spacecraft.

B. DIFFUSE GALACTIC SOURCES

With our initial attempts at 100 μ we have observed 72 sources in the Milky Way. Many of these are identified with H II regions, where the material is surrounding a hot ionizing star. A few are identified with dark nebulae. In these cases we see cooler dust not associated with ionized regions. It appears that, if we can take another step with space or intermediate space technology to improve the sensitivity and resolution of these measurements it will be possible to map out the whole of the Galaxy in the far infrared obtaining interstellar dust distribution and temperature.

C. INFRARED EXCESSES FROM GALAXIES

The infrared excess from galaxies is so great, and in some cases so compact in extent that the emission may be non-thermal and might represent a new type of phenomena.

Since the flux density is rising rapidly through the atmospheric windows, space approaches will be relied upon to determine the characteristics of the dominant part of this emission. In some cases the infrared luminosity is 10^4 times greater than the visible emission from the galaxies.

D. COSMIC RADIATION

Another dramatic surprise from submillimetre or far infrared is what seems to be a great excess over the 2.8 K black body curve. There has been a good deal of effort to interpret this in cosmological terms. However from what we have seen from more conventional objects, it may very well be that there is a galactic and an extragalactic

Manno and Ring (eds.), Infrared Detection Techniques for Space Research, 337–344. *All Rights Reserved*
Copyright © 1972 *by D. Reidel Publishing Company, Dordrecht-Holland*

background created by numerous discrete sources. Since the details of the spectrum have not yet been sorted out, a good deal of work has to be done from rockets or balloons.

E. HIGH RESOLUTION SPECTROSCOPY OF
INTERSTELLAR LINES

In ionized regions, there are a large number of ions, which have high level quantum transitions that can radiate sharp lines in the far infrared region. When the sensitivity and spectral resolution become adequate, the observations will contribute to determination of the temperatures and compositions of these regions.

In infrared astronomy a good deal of work has still to be done from the ground through the atmospheric windows. Space techniques can extend and complement this ground work. Large sophisticated instruments, with good spectral resolution and cool optics are desirable in the stratosphere and above.

Prof. Smith: I have been particularly interested by the remark of Dr Hoffmann on the existence of narrow lines. The thing which appeals to me is the possibility of making narrow band and extremely sensitive heterodyne detectors.

I think it is possible that the instrument can work at 30 K, without a superconducting magnet, and wonder if it is impossible to consider flying such an object on a satellite for the purpose of looking at these lines.

Dr Hoffmann: I think it would be excellent to have an infrared heterodyne receiver in space. Such an instrument would be a good candidate for a manned space station.

Prof. Marsden: I tend to disagree with your suggestion that the excess submillimetre radiation could be extragalactic, because one is finding quite a large flux of 10^{19} eV cosmic ray particles. The point here is that if the high energy cosmic ray particles are predominantly protons—about which there is some doubt—but if they are protons, and also extragalactic in origin, then due to their interactions with an extragalactic component of the submillimetre radiation we should expect a change in the slope of the energy spectrum of the primary cosmic rays at $\sim 10^{19}$ eV. The most recent cosmic ray observations do not show this break.

This does not exclude completely the possibility of a universal excess of submillimetre radiation as one way out would be to suppose that the very high energy cosmic rays are being produced and confined locally in our galaxy. This then introduces problems about the isotropy of the particle radiation although this and the confinement problem would be eased if the highest energy particles were heavy nucleons.

Dr Hanel: Why did you exclude the planets from infrared astronomy?

Dr Hoffmann: Infrared studies of planets is a long story in itself. High resolution spectral measurements from the stratosphere will contribute much to our knowledge of the composition, temperature, and structure of planetary atmospheres.

2. *Prof. Ring* on 'Comparison of Aircraft and Balloon-Borne Telescopes for Infrared Astronomy'

Characteristic	Balloon			Aircraft		
Telescope diameter	12"–36"			12"–36"		
Accuracy of stellar autoguider	arcmin			arcsec		
Central facility available	no			yes		
Cost/h of observation	∼ $10²			∼ $10²–2 × 10³		
Residual flux		10μ	100μ		10μ	100μ
(Wcm⁻² sr⁻¹ μ⁻¹)	Atmos. $\varepsilon < 0.01$	10^{-6}	10^{-8}	$\varepsilon \sim 0.1$	10^{-5}	10^{-7}
	Optics $\varepsilon < 0.1$	10^{-5}	10^{-7}	$\varepsilon \sim 0.1$	10^{-5}	10^{-7}

Notes on the table

(1) The only balloon telescope larger than 18" is Stratoscope – not normally used in the IR.

(2) Cost of flight relates only to marginal costs – the aircraft is a much more expensive capital facility.

(3) The residual flux at a balloon telescope is dominated by emission from the mirrors.

CONCLUSIONS

There is no technological reason why larger balloon telescopes with cooled optic and more accurate guidance on stars should not be made available as a generally-accessible facility, and such a development is urgently needed. A good deal of useful astronomy can be done from both types of system reasonably cheaply. A balloon telescope, improved as described above, would be particularly useful at long wavelengths.

However, detectors are already available with NEP $< 10^{-13}$ W·Hz$^{-1/2}$ and improvements are likely, so that however much one improves the balloon or aircraft telescope, there will be a case for true space vehicles. This will of course make cooled optics a necessity.

Dr Jennings: As for the guidance which has normally been used on balloons, it would be an important improvement if one could have a triaxial system on balloons. I also have the feeling that the cost per hour of balloon flight is roughly a factor 10 too low. As regards cooling the optics it may not always be necessary at 100 μ.

Prof. Ring: I tend to agree but the lower limit of 100 dollars/hr is still feasible. On the triaxial guidance I have been told that there would be no difficulty to adapt rocket guidance technology to a balloon telescope. So I think all the technology for that is done.

Prof. Léna: CNES has in fact already built a gondola with inertial guidance, the total weight of the system is 300 kg, of which 40 kg is for the payload. It may be improved in the future, but so far the system is still under development. As far as the costs are concerned, I am not sure the balance is in favor of balloons. The amount of money involved in a balloon flight is of the order of 10 000 dollar, launch operations included. So if you take this figure and you compare it to the aircraft, then the balance is rather in favor of an aircraft.

Dr Bolle: It is to be made pretty clear what one wants to measure, whether point or extended sources.

In the first case one can compensate for the emission of the atmosphere by looking at the source and next to it. In the second case the problem is more severe, and one has to get rid of the atmospheric background. In this second case all measurements made from low flying balloons or from an aircraft, would increase the radiometric error. The transmission losses can anyhow only be estimated or corrected for, if very high resolution spectra are obtained.

As far as the pointing accuracy is concerned, there are means of overcoming the limitations due to the guidance subsystem. The telescope itself can be used in the visible range to compensate for errors, if the object is both a shortwave and an infrared emitter.

Coming back to the planetary atmospheres, one would need there very accurate radiometric measurements, in the region where the atmospheres have highest emission e.g. between 10 and 100 μ. This is also the region of strong absorption and emission of water vapour and other constituents such as carbon dioxide and oxygen in our atmosphere, with the exception of the 10–12 μ region, so that it is indeed rather difficult to make good radiometric measurements from an aircraft or a balloon, and one should aim to go higher in space.

I also think that high resolution interferometry is difficult to perform from an aircraft flying through the stratosphere because of the difficulties in evaluating the changes in water vapour emission and absorption during the rather long scanning time of the interferometer.

Dr Chanin: Some comments on the table. We have constructed an epoxy plastic mirror obtained by centrifuging. The size is 85 cm in diameter and its star image is between 1 and 2′. The total weight of the primary plus secondary mirror, is 27 kg. The telescope is compatible with the Astrolab pointer of the CNES in both direct and offset sighting.

Prof. Kneubühl at this stage enquired on the availability of Gondolas in Europe.

It appeared that these were available to the groups of Liège, University College London, Service d'Aeronomie du CNRS, University of Munich, Meudon Observatoire, Max Planck Institute of Heidelberg, Geneva Observatory, the Queen Mary College, Groningen University, University of Florence.

3. *Dr Shivanandan* on 'Rocket Astronomy'

I would like to emphasise that ground based observations in the wavelength range of 1 μ to 25 μ utilising existing 'infrared windows' and in the millimetre region could still be investigated using suitable telescopes in Europe and in the Southern Hemisphere.

In the wavelength range of 25 μ to 1 mm, rocket observations above the radiating atmosphere would be most useful for general background surveys. Dr Hoffmann has discussed in detail the purposes of infrared astronomy, and I shall briefly discuss what is necessary for an infrared rocket astronomy program.

Preliminary studies of cryogenic systems at liquid helium temperatures should be developed which should be lightweight, have long cryogenic endurance and with suitable optics for light collection in the infrared. Operation of detectors under background controlled conditions should be evaluated and modulation techniques and electronic systems to operate at liquid helium temperatures with low power dissipation should be developed. Proper choice of rockets should be made so as to give maximum observation times at high altitudes. This will require choosing proper launch sites both in terms of astrophysical needs as well as the recovery of the payload for subsequent flights. It should be understood that the techniques developed for rocket infrared astronomy should then be extended for future satellite experiments which involves another order of magnitude in complexity.

4. *Dr Bader* on 'Future perspectives of Infrared Astronomy from the Shuttle'

Dr Bader drew a parallel between the operations of a telescope on an aircraft and on the shuttle. He thought this could be very much the same. The advantages of the aircraft are (a) the capability of putting on board standard laboratory equipment, including standard electronics, and (b) the presence of man for scientific decisions, for equipment alignment and maintenance, and for retrieval of data. These advantages will probably apply to the shuttle and will contribute to lower the costs of the mission and improve scientific returns as compared to the fully automated systems. However, technological improvements in the detector and cryogenic fields had to be achieved before flying on the shuttle. Dr Bader thought that balloons and aircraft will continue to be useful tools both to do science and also to test new techniques.

He also felt that an intermediate step before the shuttle becomes operational could be a group of cryogenically cooled medium size telescopes on unmanned spacecraft, with lifetimes of 1 or 2 yr.

Dr Ortner remarked that ESRO is at present considering the possibilities for utilizing the Post Apollo facilities, following an invitation from NASA. A number of groups have been set up to study possible research and applications modules (RAMs), as they are outlined in the NASA 'Blue Book'. One of these groups is concerned with discussions on the possibilities of using these facilities for infrared astronomy.

This is in line with the scientific policy recommendations of the Launching Programmes Advisory Committee of ESRO which state that ESRO should foster the development of infrared techniques and should study how Europe could prepare itself to enter this field with a view to possible future participation in the Post Apollo programme.

It is unlikely, however, that ESRO could envisage the development of an automatic infrared satellite to be put into orbit before 1979 when the shuttle should become available. As interim steps to the shuttle, in addition to balloon and aircraft experiments already performed in national programmes, infrared experiments carried out by sounding rockets would be more realistic if they are technically feasible and scientifically useful.

Prof. Ring: commented that he would need to think very hard before advising new groups to start in rocket infrared astronomy. He thought that rocket astronomy will eventually yield to high altitude flying balloons and satellites or skylabs, except for some special missions.

Mr Peraldi: expressed his conviction that infrared satellites were feasible in Europe, the technological problems being already well at hand, except perhaps for the cryogenics.

5. *Prof. Léna* on 'Prospects from Different Platforms'

I will try to describe what are the limiting factors from the detectors and systems side, and try to find out if and for what reasons is it worthwhile to go into space.

A. BASIC SOURCES OF NOISE IN THE SYSTEMS

		Aircraft	Balloon	Space
Photon shot noise	Optics emission	$(\varepsilon \simeq 0.05)$ 220 K	$(\varepsilon \simeq 0.05)$ 220 K	$\begin{cases} <70 \text{ K passive cooling} \\ <30 \text{ K active cooling} \end{cases}$
	Sky D.C. emission	$(\varepsilon \simeq 0.1)$	$(\varepsilon \simeq 0.01)$	3 K
Non-thermodynamic source of noise	Optics noise	4×10^{-13}	4×10^{-13}	negligible
	Sky noise	8×10^{-13}	8×10^{-14}	negligible

Notes on the table
(a) ε = emissivity.
(b) Noise is in W Hz$^{-1/2}$, at $\lambda \simeq 50\ \mu$, for $\Delta\lambda/\lambda \simeq \frac{1}{2}$ and throughout $A\Omega = 1$ cm^2 sr. Noise varies as $\sqrt{(A\Omega)} \times \Delta\lambda/\lambda$.

B. NOISE SOURCES FOR DETECTORS

1. *Wide band detectors*

Assumption: System is diffraction limited $\Delta\lambda/\lambda = 1/10$
Photon noise due to background either from the optics or sky ($\epsilon = 0.1$)

Background temperature	Noise	
300 K	10^{-13}–10^{-14}	W Hz$^{-\frac{1}{2}}$
30 K	10^{-14}–10^{-15}	W Hz$^{-\frac{1}{2}}$
3 K	10^{-15}–10^{-16}	W Hz$^{-\frac{1}{2}}$

Cooling the optics produces figures compatible with the present state of bolometers.

Cooling the optics to 3 K allows us to reach the theoretical limit of background noise.

This argument shows the advantage of cooling the optics in aircraft and balloons; but whether this is feasible or not, is another question. Size of the system or frosting problems may make it very difficult.

If it was unfeasible, then the advantages of going into space would be even clearer, since otherwise there would be no hope of reaching the limits of our present detectors.

But for all known wide band detectors one cannot improve sensitivity beyond these figures because of background temperatures.

2. *Narrow band detectors*

The figures above become lower by a factor 100 or 1000 depending on the wavelength.

These detectors will be efficient to study lines reasonably above the thermal background.

C. AS A SUMMARY:

–cooling at 4.2 K in space, would probably mean short observation times, but it will eliminate the sky noise.

– cooling at 20 K (with detector also at 20 K), would mean higher sensitivity, higher background noise, but longer observation times.

– the problem is to decide how the level of background and the level of cooling balance each other.

Dr Hoffmann: There are other sources of noise besides the sky noise produced by microthermal fluctuations as seen from aircraft and balloons. This is the instrumental noise due to differential chopping or vibrations etc.

At the present time sky noise produced by microthermal fluctuations is not the limiting factor. The limit is due to imperfection of the instrumentation or by the basic sources of noise, in the detector.

6. *Prof. Kneubühl* summarized the session

Balloon and airborne systems for scientific missions should be the responsibility of National Programmes.

Sounding rockets, and even more spacecraft, should be the responsibility in Europe at least, of an International Organisation.

The Scientific Groups in Europe involved in this activity should cooperate among themselves and being coordinated in their efforts by the same International Organisation.

7. *Dr Trendelenburg* concluded with the following remarks:

(a) The infrared scientific community so far has not participated in ESRO affairs. The results of this Symposium show that a space activity at international level is required for the future of infrared astronomy. This community must therefore become involved in ESRO.

(b) Space science has lost the old glamour. There is a tendency to sacrifice space science in favour of space application.

This is a real danger, and this new scientific community should help in making clear that it is space science which has paved the way for space application and this will also be true for the future.

(c) What can ESRO do in the infrared field?

There seems to be agreement that the final objective will be the space shuttle or space station utilisation. However it became clear that some intermediate, more moderate steps have to be taken and these may consist of rocket or even satellite flights. It was not the aim of this Symposium to identify what had to be done in detail. This will have to be defined in a smaller circle to be organised by ESRO.

(d) Experience in the past has shown that a potential danger arises when groups start fighting each other in order to secure to themselves a part of hardware. This should be avoided by all means. There is a tendency now that ESRO should develop and finance under its own responsibility the so-called 'observatory type satellites'. The scientific community would therefore not be involved particularly in the development and construction of hardware, but would be powerful in influencing the mission and in the participation of data analysis and evaluation.

Dr Trendelenburg concluded by thanking the speakers, the chairmen and those attending, for the contribution they gave to the success of this Symposium.

ASTROPHYSICS AND SPACE SCIENCE LIBRARY

Edited by

J. E. Blamont, R. L. F. Boyd, L. Goldberg, C. de Jager, Z. Kopal, G. H. Ludwig, R. Lüst,
B. M. McCormac, H. E. Newell, L. I. Sedov, Z. Švestka, and W. de Graaff

p.t.o.

16. S. Fred Singer (ed.), *Manned Laboratories in Space. Second International Orbital Laboratory Symposium*. 1969, XIII + 133 pp.

17. B. M. McCormac (ed.), *Particles and Fields in the Magnetosphere. Symposium Organized by the Summer Advanced Study Institute, held at the University of California, Santa Barbara, Calif., August 4–15, 1969*. 1970, XI + 450 pp.

18. Jean-Claude Pecker, *Experimental Astronomy*. 1970, X + 105 pp.

19. V. Manno and D. E. Page (eds.), *Intercorrelated Satellite Observations related to Solar Events. Proceedings of the Third ESLAB/ESRIN Symposium held in Noordwijk, The Netherlands, September 16–19, 1969*. 1970, XVI + 627 pp.

20. L. Mansinha, D. E. Smylie and A. E. Beck, *Earthquake Displacement Fields and the Rotation of the Earth. A NATO Advanced Study Institute Conference Organized by the Department of Geophysics, University of Western Ontario, London, Canada, 22 June–28 June, 1969*. 1970, XI + 308 pp.

21. Jean-Claude Pecker, *Space Observatories*. 1970, XI + 120 pp.

22. L. N. Mavridis (ed.), *Structure and Evolution of the Galaxy. Proceedings of the NATO Advanced Study Institute, held in Athens, September 8–19, 1969*. 1971, VII + 312 pp.

23. A. Muller (ed.), *The Magellanic Clouds. A European Southern Observatory Presentation: Principal Prospects, Current Observations and Theoretical Approaches, and Prospects for Future Research. Based on the Symposium on the Magellanic Clouds held in Santiago de Chile, March 1969, on the Occasion of the Dedication of the European Southern Observatory*. 1971, XII + 189 pp.

24. B. M. McCormac (ed.), *The Radiating Atmosphere. Proceedings of a Symposium Organized by the Summer Advanced Study Institute, held at Queen's University, Kingston, Ontario, August 3–14, 1970*. 1971, XI + 455 pp.

25. G. Fiocco (ed.), *Mesospheric Models and Related Experiments. Proceedings of the 4th ESRIN-ESLAB Symposium, held at Frascati, Italy, July 6–10, 1970*. 1971, VIII + 298 pp.

26. I. Atanasijević, *Selected Exercises in Galactic Astronomy*. 1971, XII + 143 pp.

27. Constantin J. Macris (ed.), *Physics of the Solar Corona. Proceedings of the NATO Advanced Study Institute on Physics of the Solar Corona, held at Cavouri-Vouliagmeni, Athens, Greece, 6–17 September 1970*. 1971, XII + 345 pp.

28. Francis Delobeau, *The Environment of the Earth*. 1971, IX + 113 pp.

29. E. R. Dyer (ed.), *Solar-Terrestrial Physics/1970. Proceedings of the International Symposium on Solar-Terrestrial Physics, held in Leningrad, U.S.S.R., 12–19 May, 1970*. 1972, VIII + 942 pp.

31. Myron Lecar (ed.), *Gravitational N-Body Problem. Proceedings of IAU Colloquium No. 10, held in Cambridge, England, August 12–15, 1970*. 1972, XI + 441 pp.

In preparation:

32. B. M. McCormac (ed.), *Earth's Magnetospheric Processes*.

33. A. Rükl, *Maps of Lunar Hemispheres*. With a foreword by Z. Kopal.

SOLE DISTRIBUTORS FOR U.S.A. AND CANADA:

Vols. 2–6, and 8: Gordon and Breach Inc., 150 Fifth Ave., New York, N.Y. 10011
Vols. 7, and 9–29: Springer Verlag New York Inc., 175 Fifth Ave., New York, N.Y. 10011